DIAGNOSTIC PROCEDURES
IN VETERINARY BACTERIOLOGY
AND MYCOLOGY

Fourth Edition

DIAGNOSTIC PROCEDURES IN VETERINARY BACTERIOLOGY AND MYCOLOGY

By

G. R. CARTER, D.V.M., M.S., D.V.Sc.

Division of Pathobiology and Public Practice
Virginia-Maryland Regional College
of Veterinary Medicine
Virginia Tech
Blacksburg, Virginia

With

Fourteen Contributors

CHARLES C THOMAS • PUBLISHER
Springfield • Illinois • U.S.A.

Published and Distributed Throughout the World by
CHARLES C THOMAS • PUBLISHER
2600 South First Street
Springfield, Illinois 62717

© *1967, 1973, 1979, 1984 by CHARLES C THOMAS • PUBLISHER*
ISBN 0-398-04870-3
Library of Congress Catalog Card Number: 83-4805

With THOMAS BOOKS *careful attention is given to all details of manufacturing and
design. It is the Publisher's desire to present books that are satisfactory as to their physical
qualities and artistic possibilities and appropriate for their particular use.* THOMAS
BOOKS *will be true to those laws of quality that assure a good name and good will.*

First Edition, 1967
Second Edition, First Printing, 1973
Second Edition, Second Printing, 1975
Second Edition, Third Printing, 1978
Third Edition, 1979
Fourth Edition, 1984

Library of Congress Cataloging in Publication Data

Carter, G. R. (Gordon R.)
    Diagnostic procedures in veterinary bacteriology
and mycology.

    Includes bibliographical references and index.
    1. Veterinary microbiology — Technique.   2. Com-
munical disease in animals — Diagnosis.   I. Title.
SF780.2.C36 1984   636.089'601   83-4805
ISBN 0-398-04870-3

Printed in the United States of America
SC-R-3

CONTRIBUTORS

E. L. Biberstein
Department of Veterinary Microbiology
School of Veterinary Medicine
University of California
Davis, California 95616

Margery E. Carter
Division of Pathobiology and Public Practice
Virginia-Maryland Regional College of
Veterinary Medicine
Virginia Tech
Blacksburg, Virginia 24061

J. Manuel Chirino-Trejo
Departamento de Microbiología
Universidad Nacional Autónoma de México
Mexico, D.F.

John R. Cole, Jr.
Veterinary Diagnostic and Investigational
Laboratory
University of Georgia
Tifton, Georgia 31794

C. A. Kirkbride
Department of Veterinary Science
Diagnostic Laboratory
Box 2175
South Dakota State University
Brookings, South Dakota 57007

Harold A. McAllister
Department of Veterinary Pathology
College of Veterinary Medicine
Iowa State University
Ames, Iowa 50011

Margaret E. Meyer
Department of Epidemiology
and Preventive Medicine
School of Veterinary Medicine
University of California
Davis, California 95616

F. H. S. Newbould
*Department of Veterinary Microbiology
and Immunology
Ontario Veterinary College
University of Guelph
Guelph, Ontario, Canada N1G 2W1*

John Prescott
*Department of Veterinary Microbiology
and Immunology
Ontario Veterinary College
University of Guelph
Guelph, Ontario, Canada N1G 2W1*

Emmett B. Shotts
*Department of Medical Microbiology
College of Veterinary Medicine
University of Georgia
Athens, Georgia 30602*

Ole H. V. Stahlheim
*Veterinary Medical Officer
National Animal Disease Center
Ames, Iowa 50010*

Johannes Storz
*Department of Veterinary Microbiology
and Parasitology
School of Veterinary Medicine
Louisiana State University
Baton Rouge, Louisiana 70803*

Charles O. Thoen
*Department of Veterinary Microbiology
and Preventive Medicine
College of Veterinary Medicine
Iowa State University
Ames, Iowa 50011*

Harry W. Yoder, Jr.
*U.S. Department of Agriculture
Poultry Research Laboratory
Athens, Georgia 30605*

PREFACE

E ACH VETERINARY diagnostic laboratory establishes its own routine micro-biological procedures. The sources of these procedures are many, and modifications and innovations are made as experience and new developments indicate. The number of specimens and the availability of technical assistance, supplies, and equipment frequently determine the thoroughness with which procedures can be carried out.

In general, this manual describes methods that have been found effective in clinical microbiology laboratories serving clinical and pathology departments of a college of veterinary medicine as well as private veterinary practitioners. For the most part the emphasis has been on practical methods that can be applied in laboratories with average facilities. In some instances, workers may wish to amplify or vary the procedures described. With this in mind, sources of additional information have been provided in the lists of supplementary references.

I would like to acknowledge the invaluable assistance of Mr. Harold A. McAllister, who prepared many of the illustrations and as well wrote Chapters 29 and 35. In addition, he gave freely of his considerable practical experience on many aspects of the manual.

Because of the great increase in knowledge in some areas, it was thought advisable to increase the number of contributors. I am greatly indebted to them for the chapters they have contributed.

Thanks is also expressed to my associates, Dr. Margery Carter, Ms. Susan Stevens, and Ms. Karen George for their help and advice on many facets of the book.

G. R. Carter

CONTENTS

ix

DIAGNOSTIC PROCEDURES IN VETERINARY BACTERIOLOGY AND MYCOLOGY

Chapter 1

CLASSIFICATION AND NORMAL FLORA

CLASSIFICATION AND NOMENCLATURE

THE EIGHTH EDITION of *Bergey's Manual* (1) provides a comprehensive listing of many established species and their characteristics. Generally speaking, the names used in *Bergey's Manual* will be employed in this manual. For the most part, the names of bacteria used in the book will be those in the *Approved Lists of Bacterial Names* (1) published by the American Society for Microbiology. These are the names to be used in the new edition (9th) of *Bergey's Manual*. Only a few of the names will differ from those in the current 8th edition of the *Manual* (2). In some instances, in the absence of official names for some bacteria, names will be used that have received general acceptance.

An outline of the Bergey classification as it relates particularly to genera associated with diseases in animals and man is provided below. The various bacteria are discussed in the text, usually in the same order in which they appear in the outline. The classification of the fungi will be discussed in Chapter 28.

Kingdom Procaryotae

Division I. The Cyanobacteria
Division II. The Bacteria

Part 1. Phototrophic Bacteria
Part 2. Gliding Bacteria
Part 3. Sheathed Bacteria
Part 4. Budding and/or Appendaged Bacteria
Part 5. Spirochetes
    Order I. Spirochaetales
        Family I. Spirochaetaceae
            Genus I. *Spirochaeta*
            Genus III. *Treponema*
            Genus IV. *Borrelia*
            Genus V. *Leptospira*

Part 6. Spiral and Curved Bacteria
    Family I. Spirillaceae

Genus I. *Spirillum*
Genus II. *Campylobacter*

Part 7. Gram-negative, Aerobic Rods and Cocci
    Family I. Pseudomonadaceae
        Genus I. *Pseudomonas*
    (Four other families)
        Genera of Uncertain Affiliation (6 genera)
            Genus *Alcaligenes*
            Genus *Brucella*
            Genus *Bordetella*
            Genus *Francisella*

Part 8. Gram-negative, Facultatively Anaerobic Rods
    Family I. Enterobacteriaceae
        Genus I. *Escherichia*
        Genus II. *Edwardsiella*
        Genus III. *Citrobacter*
        Genus IV. *Salmonella*
        Genus V. *Shigella*
        Genus VI. *Klebsiella*
        Genus VII. *Enterobacter*
        Genus VIII. *Hafnia*
        Genus IX. *Serratia*
        Genus X. *Proteus*
        Genus XI. *Yersinia*
        Genus XII. *Erwinia*
    Family II. Vibrionaceae
        Genus I. *Vibrio*
        Genus II. *Aeromonas*
        (Three additional genera)
        Genera of Uncertain Affiliation (9 genera)
            Genus *Chromobacterium*
            Genus *Flavobacterium*
            Genus *Haemophilus*
            Genus *Pasteurella*
            Genus *Actinobacillus*
            Genus *Streptobacillus*
            (Three additional genera)

Part 9. Gram-negative, Anaerobic Bacteria
    Family I. Bacteroidaceae
        Genus I. *Bacteroides*

Family I. Bacillaceae
  Genus I. *Bacillus*
  Genus II. *Clostridium*

Part 16. Gram-positive, Asporogenous Rod-shaped Bacteria
  Family I. Lactobacillaceae
    Genus I. *Lactobacillus*
  Genera of Uncertain Affiliation
    Genus *Listeria*
    Genus *Erysipelothrix*
    Genus *Caryophanon*

Part 17. Actinomycetes and Related Organisms
  Coryneform Group of Bacteria
    Genus I. *Corynebacterium*
    Genus IV. *Kurthia*
  Family Propionibacteriaceae
    Genus I. *Proprionibacterium*
    Genus II. *Eubacterium*
  Order I. Actinomycetales
    Family I. Actinomycetaceae
      Genus I. *Actinomyces*
      Genus V. *Rothia*
    Family II. Mycobacteriaceae
      Genus I. *Mycobacterium*
    Family V. Dermatophilaceae
      Genus I. *Dermatophilus*
    Family VI. Nocardiaceae
      Genus I. *Nocardia*
    Family VII. Streptomycetaceae
      Genus I. *Streptomyces*

Part 18. The Rickettsias
  Order I. Rickettsiales
    Family I. Rickettsiaceae
      Tribe I. Rickettsieae
        Genus I. *Rickettsia*
        Genus II. *Rochalimaea*
        Genus III. *Coxiella*
      Tribe II. Ehrlichieae
        Genus IV. *Ehrlichia*
        Genus V. *Cowdria*
        Genus VI. *Neorickettsia*

Family II. Bartonellaceae
  Genus I. *Bartonella*
  Genus II. *Grahamella*
Family III. Anaplasmataceae
  Genus I. *Anaplasma*
  Genus IV. *Haemobartonella*
  Genus V. *Eperythrozoon*
Order II. Chlamydiales
  Family I. Chlamydiaceae
    Genus I. *Chlamydia*

Part 19. The Mycoplasmas
  Class I. Mollicutes
    Order I. Mycoplasmatales
      Family I. Mycoplasmataceae
        Genus I. *Mycoplasma*
      Family II. Acholeplasmataceae
        Genus I. *Acholeplasma*
      Genera of Uncertain Affiliation
        Genus *Thermoplasma*
        Genus *Spiroplasma*

The Normal Flora

It is important that the clinical microbiologist have some familiarity with the kinds of organisms encountered normally in and upon animals. Such knowledge is necessary in the interpretation of the results of microbiologic examinations.

The so-called normal flora consists of the wide variety of bacteria and fungi that live in or upon the normal animal without producing disease. Included in this flora are many potential pathogens and opportunistic organisms. The term *normal flora* is a convenient concept, but it should be kept in mind that the kinds and numbers of bacteria present vary greatly with different circumstances. The intestinal flora of the young animal differs markedly from that of the older animal. The flora is also influenced by geographic location, nutrition, and climate. The technical procedures employed to recover pathogenic organisms frequently give a distorted idea of the kinds and numbers of bacteria present. The older studies of the normal floras of the domestic animals have often neglected the obligate anaerobes, which in the intestine make up by far the largest number of bacteria (see Chapter 15).

The normal flora of the domestic animals have not been studied in as detailed a fashion as that of human beings. What little information that is available, and firsthand experience in the diagnostic laboratory, indicate a considerable similarity between the normal flora of humans (3) and the domestic animals.

Some of the kinds of bacteria that can be expected to occur normally in and upon domestic animals are tabulated below.

Mouth, Nasopharynx

Micrococci (aerobic and anaerobic, pigmented and nonpigmented); *Staphylococcus* spp.; hemolytic and nonhemolytic streptococci; *Bacillus* spp.; lactobacilli; fusiform bacilli; *Actinomyces; Veillonella* and other Gram-negative cocci; coliforms and *Proteus* spp.; spirochetes; mycoplasmas; *Pasteurella* spp.; diphtheroids; pneumococci; yeasts, including *Candida albicans; Haemophilus* spp.; *Simonsiella.*

Jejunum, Ileum

Only small numbers of bacteria are present in this portion of the intestinal tract of animals.

Large Intestine

Fecal streptococci; *E. coli; Klebsiella; Enterobacter; Pseudomonas* spp.; *Proteus* spp.; staphylococci; clostridia: *Cl. perfringens, Cl. septicum,* and other clostridia; Gram-negative anaerobes; spirochetes; lactobacilli.

Trachea, Bronchi, Lungs

Few, if any, bacteria and fungi reside in these structures.

Vulva

Diphtheroids; micrococci; coliforms and *Proteus* spp.; enterococci; yeasts; Gram-negative anaerobes. The same kinds of organisms and others can be recovered from the prepuce of the male.

Vagina

The numbers and kinds of bacteria vary with the reproductive cycle and age. The cervix and anterior vagina of the healthy mare possess few bacteria. Some of the organisms recovered from the vagina are hemolytic and non-hemolytic streptococci; coliforms and *Proteus* spp.; diphtheroids and lactobacilli; mycoplasmas; yeast and fungi.

Skin

Animals, by virtue of their habits and environment, frequently possess a large and varied bacterial and fungal flora on their hair and skin. *Staph. epidermidis* and *Staph. aureus* occur commonly, as do other micrococci. Of the many other organisms isolated, it is not known which make up the resident flora and which are "transients."

Milk

Micrococci, staphylococci, nonhemolytic streptococci, mycoplasmas, and diphtheroids including *Corynebacterium bovis* are frequently shed from the apparently normal mammary gland.

LABORATORY SAFETY

Many of the bacteria, fungi, and viruses encountered in the diagnostic veterinary microbiology laboratory have the potential for causing disease in humans. There have been numerous reports of laboratory-acquired infections. In recent years there has been much emphasis on laboratory safety, and an extensive literature is available on the subject. Some instructive references are listed at the end of this chapter under Supplementary References.

All employees working with potentially pathogenic agents should be instructed in proper safety procedures. Employees should be encouraged to read safety literature of the kind cited at the end of this chapter.

The Safety Program outlined below, although by no means definitive, has been found effective and workable in the author's laboratory.

Safety Program for the Prevention of Infections in the Clinical Microbiology Laboratory

All employees who are exposed to animals or animal tissues, clinical specimens, microbes, etc., in their work are urged to comply with the following recommendations:

1. A serum sample will be taken from each employee immediately prior to beginning work in the laboratory. This sample will be frozen for possible future reference.

2. A serum sample will be taken from each employee at the time of termination of employment. This will also be kept for reference purposes.

3. a. It is recommended that employees working in the microbiology laboratory be vaccinated against rabies and given the recommended booster vaccinations.

   b. A skin test for tuberculosis should be conducted annually.

4. Employees carrying out procedures and tests with known dangerous zoonotic agents, suspected or actual, must carry out such operations in the safety hood or hoods available. Among the agents that are particularly infectious for human beings are the following:

Bacteria: *Brucella, Leptospira,* mycobacteria, *Francisella tularensis, Salmonella, Bacillus anthracis, Clostridium botulinum, Chlamydia*

Pathogenic fungi: particularly *Coccidioides immitis*

Viruses: rabies virus, Eastern equine encephalomyelitis virus

There are many others that have been known to cause infrequent and usually not as severe infections.

5. All efforts should be made to prevent exposure to infectious organisms. Among measures that will reduce the risk of exposure are the following:
   a. Avoid procedures that produce infectious aerosols.
   b. Wear a laboratory coat or uniform.
   c. Avoid mouth pipetting of any infectious fluids.
   d. Do not eat or drink in the laboratory area.
   e. Wash benches with disinfectant at the end of the work day. Disinfect infectious spills or breakages.
   f. Thoroughly wash hands prior to eating and at the end of the work day.
   g. Take every precaution to prevent animal bites and wash thoroughly after handling laboratory or other animals.
   h. Report and record in an "accident book" any injuries, abrasions, eye contamination, etc., obtained in the laboratory.

6. Report any illness thought to be job-related to your immediate supervisor.

7. A record of compliance or lack of compliance with the above recommendations, along with details of vaccinations, tests, serum samples, reports of illness or exposure, etc., will be kept in the laboratory office.

8. The Clinical Microbiology Laboratory will arrange for tests for such diseases as brucellosis, leptospirosis, or mycobacteriosis if employees working with the causal agents of these diseases request such tests.

9. A file containing pertinent "biological safety" manuals and materials is available in the microbiology laboratory office. For their personal safety, faculty, staff, and students should familiarize themselves with this literature.

10. Each new employee should be briefed regarding biological risks and instructed in appropriate safety procedures.

REFERENCES

1. Skerman, V. B. D., McGowan, V., and Sneath, P. H. A. (Eds.): *Approved Lists of Bacterial Names.* Washington, D.C., American Society for Microbiology, 1980.
2. Buchanan, R. E., and Gibbons, N. E. (Eds.): *Bergey's Manual of Determinative Bacteriology,* 8th ed. Baltimore, Williams & Wilkins, 1974.
3. Rosebury, T.: *Microorganisms Indigenous to Man.* New York, McGraw, 1962.

SUPPLEMENTARY REFERENCES

Lab Safety. Atlanta, Georgia, U.S. Department of Health and Human Services, Public Health Service, Center for Disease Control, Office of Biosafety, 1974.

Phillips, G. B.: In Gerhardt, P. (Ed-in-chief): *Manual of Methods for General Bacteriology*. Washington, D.C., American Society for Microbiology, 1981, Chapter 24.

Richardson, J. N., and Huffaker, R. H.: In Lennette, E. H. (Ed-in-chief): *Manual of Clinical Microbiology*, 3rd ed. Washington, D.C., American Society for Microbiology, 1980, Chapter 96.

Shapton, D. A., and Board, R. G. (Eds.): *Safety in Microbiology*. New York, Academic Press, 1972.

Chapter 2

SELECTION AND SUBMISSION OF CLINICAL SPECIMENS

A RECURRING PROBLEM in clinical veterinary microbiology results from the submission of unsatisfactory specimens with little or no history or clinicians' comments. Improvements in the selection and submission of specimens can usually only be obtained by a constant educational effort. Veterinarians should be supplied with an adequate submission and history form with instructions on the selection and shipment of specimens. Instructions of the kind given below have been found to be of value.

Just prior to death, and shortly thereafter, a number of intestinal bacteria may invade the host's tissues. The significance of these organisms, some of which are potential pathogens, is difficult to assess when tissues have been taken even a short time after death. Live, sick animals presented for necropsy are usually the best source of specimens. In all instances, the importance of fresh tissues taken as soon as possible after death cannot be overemphasized.

SPECIMENS FOR BACTERIAL AND MYCOLOGICAL EXAMINATION

Preservation and Shipment

Tissues and Organs

Asepsis should be practiced as much as possible in collecting and handling materials for culture. Place tissues in individual plastic bags or leakproof jars. Portions of intestines should be packed separately. Specimens can be conveniently shipped in a Styrofoam® box or ice chest containing a generous amount of ice. Dry ice with plenty of insulation is preferred for longer preservation.

Brains sent for examination should be halved longitudinally. One half is refrigerated or frozen over dry ice, and the other is placed in 10% formalin for histopathological examination. Tissues in formalin should not be frozen.

Postmortem invasion of tissues by intestinal and other bacteria can be rapid, particularly in warm weather. The bone marrow is less accessible than other tissues to this invasion. An opened rib from a small animal, or a four to five inch aseptically cut piece of rib from a large animal, will often yield the causative bacterium in pure or nearly pure culture. The muscle or periosteal tissue should be removed from the rib before submission. If the rib is unopened the marrow can be exposed with a small bone saw.

Swabs

Swabs are of value in many instances for the transportation of infectious material to the laboratory. However, because many bacteria are susceptible to desiccation during shipment, it is advisable to place the swab in non-nutritional transport medium (see Appendix B). Swabs that utilize a transport medium are available commercially.* The survival rate of bacteria on the conventional cotton swabs is improved if they are boiled for five minutes in Sörensen's buffer pH 7.5 before autoclaving. Treatment with buffer is not required for calcium alginate swabs.

Equine Cervical Swabs

Special swabs are required for swabbing the cervices of mares and other large animals. These can be prepared by attaching absorbent cotton to the end of an eighteen to twenty-four inch length of wire with a rubber band. Then approximately a foot of the portion containing the cotton is enclosed with paper or a pipette paper cover and autoclaved. Sterile swabs with long handles, referred to as Disposable Guarded Culture Instruments, are available commercially.†

Specimens for Anaerobic Culture

See Chapter 15.

Diseases Requiring Special Consideration

Not all of the diseases requiring special consideration are listed below. For those not listed consult the appropriate chapter.

Clostridial Infections (Blackleg, malignant edema, etc.)

Fresh affected tissue is especially important in that the clostridia rapidly invade tissues after death. The muscle tissue involved may be difficult to locate.

Enterotoxemia (Clostridia)

Several ounces of fresh intestinal contents are required. This can be submitted in a jar or plastic bag, or a section of affected intestine may be tied off and submitted. This material should be refrigerated and dispatched to the laboratory as soon as possible.

---

*Cepti-seal Culturette®, Medi-Flex Division, Medical Supply Company, Rockford, Illinois.

†Kalayzian Industries, 6050 Appian Way, Long Beach, California.

Vibriosis, Campylobacteriosis (Cattle and Sheep)

To make possible the isolation of the causal agents, semen, preputial washings, fetal stomach content, or cervical mucus should reach the laboratory under refrigeration within five hours of collection. Procedures for the collection of specimens are described in Chapter 6. Failing recovery of live organisms, dead campylobacter can be recognized by an FA procedure.

Anthrax

Cotton swabs are soaked in exuded blood, or blood taken from a superficial ear vein, in acute or peracute anthrax. In swine, because the organisms may not be present in the blood, swabs should be taken from exudates and the cut surface of hemorrhagic lymph nodes.

Johne's Disease

The most suitable specimens are one to two feet of the terminal sections of the ileum with the ileocecal junction (ileocecal valve) and a similar length of the adjacent cecum, flushed free of intestinal content. Several mesenteric lymph nodes of the ileocecal region should also be included. For fecal culture, one-half ounce of feces should be submitted in a refrigerated, sealed container (see also Chapter 23).

Tuberculosis

See Chapter 23.

Swine Dysentery

Six to eight inches of spiral colon from an acutely affected pig should be submitted. The specimen should be fresh, and although it should reach the laboratory as soon as possible, it may be held at 4°C for two to three days.

Fungi (Ringworm)

Scrapings or epilations should be made at the edge of active lesions. Submit in a cotton-plugged test tube or paper envelope. Saprophytic fungi will frequently proliferate rapidly in a sealed tube because of the moisture.

Serum Samples

No anticoagulant should be used. Samples are allowed to clot and are shipped preferably in wet ice but not frozen.

Wet needles, syringes, and tubes will cause hemolysis and spoiling of blood for serological examination. Blood samples for serological examination that become overheated will also hemolyze. Care should be taken to prevent overheating from the time the samples are drawn. This is especially important with samples for complement fixation tests.

Swine blood is especially susceptible to hemolysis. It is advisable to pour the serum from clotted samples into clean, dry tubes for shipment.

Chapter 3

CULTURAL PROCEDURES EMPLOYED FOR CLINICAL SPECIMENS

T HE KINDS of specimens submitted to veterinary microbiology laboratories are various, and the procedure to be followed in processing each depends upon the disease or organism suspected. One of the problems is knowing—often in the absence of any clues provided by the pathologist or clinician—just what pathogen or disease to suspect. Because so little information is usually available on a given specimen and because the frequency of submissions does not always allow for special efforts, routine procedures are established by the laboratory for the bulk of specimens. Procedures for the processing of the more common kinds of specimens are summarized in Tables 3-I and 3-II. It is recommended that a CO_2 incubator be used to provide 5% CO_2 for the routine incubation of plates at 37°C.

DIRECT EXAMINATION OF MATERIALS

The materials most frequently submitted for examination are tissues, feces, swabs, milk, urine, pus, discharge, fetal stomach contents, cervical mucus, and skin scrapings. The microbiologist should carry out a direct examination for the agent that the veterinarian or veterinary pathologist may suspect. These examinations are dealt with in the manual under the appropriate pathogen or disease. The examination of stained smears and wet mounts should be routine with most materials. The findings may aid in the selection of appropriate media. When indicated, fluorescent antibody procedures are carried out on smears of tissues, fluids, or exudates.

TABLE 3-I
SUMMARY OF SOME ROUTINE CULTURAL PROCEDURES

Specimens	To Culture:	Media	Atmosphere	Incubation Temperature
Organs, tissues, pus, urine, swabs, etc.	Aerobes* (not enterobacteria)	Blood agar, Schaedler broth	Aerobic†	37°C
Feces, fecal swabs, intestine	Enterobacteria	Selenite, (18 hr.) to MacConkey's agar, brilliant green. Direct: MacConkey's brilliant green, and blood agar	Aerobic	37°C
Milk	Aerobes	Blood agar	Aerobic	37°C
Organs, tissues, pus, swabs, etc.	Microaerophiles (Brucella, Campylobacter)	Blood agar and selective media	10% CO₂; see special requirements for Brucella, Campylobacter	37°C
Organs, tissues, intestinal content, fecal swabs	Anaerobes (Clostridia)	Blood agar, cooked meat, Schaedler broth; FA procedures	Anaerobic Aerobic	37°C
Intestinal content: Suspect enterotoxemia		Mouse or rabbit inoculations: blood agar	Anaerobic	37°C
Organs, tissues, pus, swabs, etc.	Anaerobes (Gram-neg. and Actinomyces bovis)	Blood agar, cooked meat medium semisolid; Schaedler broth	Anaerobic and aerobic (blood agar)	37°C
Organs, tissues, pus, swabs, etc.	Fungi in general	Sabouraud agar, Sabouraud with inhibitors, blood agar, BHI semisolid	Aerobic	25°C or room temp. and 37°C
Skin scrapings, epilations, hair	Dermatophytes	Soubouraud agar (with inhibitors) Blood agar, BHI semisolid	Aerobic Aerobic	25°C or room temp. 37°C

*Including facultative anaerobes
†Air with 5% CO₂ preferred

TABLE 3-II
ADDITIONAL ROUTINE CULTURAL PROCEDURES

Ear and Cervical Swabs:

Plate on blood agar ....37°C

BHI semisolid or Schaedler broth ....37°

Sabouraud agar ....25°C

For *Listeria:*

Grind tissue in broth → store at 4°C and culture on blood
          ↓   agar as follows:

Blood agar — 37°   end of first week
                     end of third week
                     end of sixth week   } all at 37°C
                     end of twelfth week

For Mycoplasmas:

(See Chapters 26 and 27.)

For *Haemophilus* spp:

Streak beta-hemolytic *Staphylococcus aureus* over streak
lines of clinical material on blood agar. (See also Chapter 13.) } 37°C

Chapter 4

ISOLATION AND IDENTIFICATION
OF BACTERIA FROM CLINICAL SPECIMENS

S TEPS FOLLOWED in the isolation and identification of bacteria from clinical specimens are listed in Table 4-I. The selection of routinely used media for primary inoculation was referred to in Tables 3-I and 3-II.

PRIMARY INOCULATION OF MEDIA

Two procedures are frequently used to obtain material for inoculation from tissues and organs. One is to sear the surface of the specimen with a hot spatula, then incise with a sterile scalpel. From this incision, material is transferred to media with an inoculation loop or a Pasteur pipette. If the tissues are fresh, fluid media such as thioglycollate or semisolid brain-heart infusion are inoculated directly. This procedure is especially indicated if antimicrobial agents have been employed.

Another procedure, which is more convenient with small specimens, is to sterilize the external surfaces of the specimen by holding it with sterile forceps and passing it through a Bunsen flame several times. It is then sectioned with sterile scissors, and the exposed surface is impressed on the agar surface. The inoculum is then spread with an inoculating loop. Occasionally, media are inoculated from a pool of tissues previously ground in broth, especially if the tissues or organs are too small to carry out the other manipulations.

The two goals of primary inoculation are, first, to cultivate the organisms and, second, to obtain discrete colonies. From the latter, pure cultures are obtained. These aims are usually best accomplished by the inoculation of solid media in Petri dishes; however, there are instances in which "pour plates" and "agar shake cultures" are useful. A useful procedure after the inoculation of plate media is to place all swabs, except fecal swabs, in a tube of semisolid brain-heart infusion (BHI) broth or Schaedler broth. This medium supports the growth of many fastidious organisms, including aerobes, facultative anaerobes, microaerophiles, and anaerobic bacteria. It has the disadvantage that the pathogen, if present, may be overgrown by other bacteria. A smear is made from the semisolid broth culture and stained by Gram's method. If indicated, the culture is inoculated onto plate media, usually blood agar.

Special directions for the cultivation of the certain pathogens are provided in the chapters dealing with these particular organisms. Blood agar is the most useful and widely employed medium. Material is streaked out on the solid medium with an inoculating loop, a swab, or a glass spreader. Media for primary culture (not anaerobic) should be placed in an incubator with 5% carbon dioxide and adequate humidity.

Urine Culture

Most of the urine samples submitted for culture are from catheterized dogs. If not catheterized, midstream urine should be caught when feasible in a sterilized container. Centrifugation is carried out if a direct examination is required. The sediment is stained by Gram's method.

Generally speaking, bacteriological examinations of animal urine have not been quantitated. In order to carry out a semiquantitative procedure, known amounts of urine can be plated on blood agar and other media. A loop that delivers 0.01 ml may be used. The inoculum is spread thoroughly, and after incubation, the number of colonies is estimated.

A quantitative procedure that provides more precision involves the use of pour plates of trypticase soy agar or other media. Generally, the inoculum used for each of two pour plates is 1.0 ml of 10^{-3} dilution.

Results may be given as the approximate number of bacteria per milliliter of urine. In man, clinical bacteriuria is indicated by the presence of 100,000 organisms per milliliter. A count less than 1000/ml is not considered significant, while counts between 10,000 and 100,000 are suggestive of infection. It is now generally conceded that broth cultures are of little or no value.

For additional information on the laboratory diagnosis of bacteriuria, readers are referred to the *Manual of Clinical Microbiology* listed in the Supplementary References.

Blood Culture

Blood cultures are made whenever there is reason to suspect a clinically significant bacteremia. Because bacteremia may be intermittent, it is advisable to culture more than one blood sample. In human beings, as many as three to four blood cultures are recommended in the initial twenty-four-hour period (1). The author recommends the same regimen for animals.

Because only a small number of bacteria may be present in the blood of an animal with a bacteremia, 3–10 ml of blood, depending upon the size of the animal, is taken aseptically. If there is no anticoagulant in the medium, an anticoagulant should be added when the blood is taken. Suitable media may be prepared or purchased (see Appendix B).

The author has found the Vacutainer® Culture Tube* particularly convenient for small animals. It contains an anticoagulant and supports the growth of aerobes, facultative anaerobes, and anaerobes, and because the blood is inoculated directly from the animal, the chances for contamination are reduced. It is recommended that at least four culture tubes be used for a dog during a twenty-four-hour period.

EXAMINATION OF PLATE MEDIA

The kind and number of colonies or amount of growth is studied and recorded. Tissues taken from animals some time after death may yield a variety of colonies, usually indicating postmortem invasion from the alimentary tract. It is not always feasible to identify all of the different bacteria. Generally, only those colonies thought to represent the more significant organisms are identified. Smears are made from representative colonies and stained by Gram's method.

Before discarding plates, it is advisable to examine them for minute colonies with a stereoscopic microscope. By this procedure, colonies that cannot be discerned with the unaided eye, e.g. dwarf colonies, those of *Haemophilus* spp., and mycoplasmas, can be seen. All plates should be incubated, when feasible, for three to four days or longer if no growth is detected earlier. To prevent dehydration, plates should be placed in an airtight container, or individual plates should be sealed with tape or a large rubber band. It is advisable to hold plates at room temperature for one week before discarding. Some organisms such as *Yersinia enterocolitica* grow better at this temperature than at 37°C.

PURE CULTURES FOR IDENTIFICATION

A small tube of broth is inoculated from one colony of the culture to be identified. Because it may be difficult to initiate a broth culture from one colony, several colonies are sometimes used if it is evident that the plate culture is pure. The inoculated tube is placed in a container of water in the incubator at 37°C in order to accelerate growth. Tubes thus inoculated, if pure, can sometimes be inoculated into differential media the same day with a Pasteur or serologic pipette. An alternative procedure is to inoculate a slant of a suitable medium such as tryptose agar, a sector of a blood plate, or a TSI slant. Differential media are then inoculated from these with an inoculating needle. The usual sequence of procedures is summarized in Table 4-I.

---

*Becton-Dickinson, Rutherford, New Jersey.

TABLE 4-I

STEPS USUALLY FOLLOWED IN
THE ISOLATION AND IDENTIFICATION
OF BACTERIA FROM CLINICAL SPECIMENS

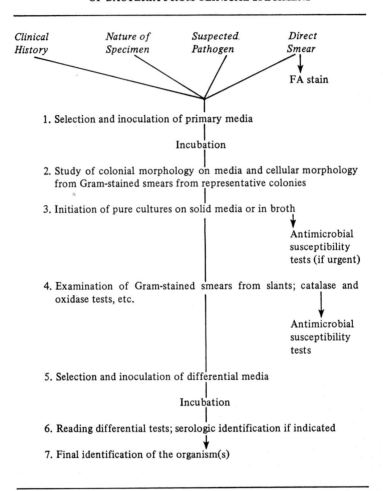

Clinical   *Nature of*   *Suspected.*   *Direct*
History   *Specimen*   *Pathogen*   *Smear*

FA stain

1. Selection and inoculation of primary media

Incubation

2. Study of colonial morphology on media and cellular morphology
   from Gram-stained smears from representative colonies

3. Initiation of pure cultures on solid media or in broth

Antimicrobial
susceptibility
tests (if urgent)

4. Examination of Gram-stained smears from slants; catalase and
   oxidase tests, etc.

Antimicrobial
susceptibility
tests

5. Selection and inoculation of differential media

Incubation

6. Reading differential tests; serologic identification if indicated

7. Final identification of the organism(s)

PROCEDURES FOLLOWED IN IDENTIFICATION

If bacteria have invaded tissues after death or if specimens have been insufficiently refrigerated, a considerable variety of bacteria will be recovered unless selective media are used. Contaminants of diverse origins are frequently encountered. Attempts to identify these organisms as to species are pointless, and the experienced veterinary bacteriologist will take into consideration the condition of the specimen, the number and variety of bacteria, the gross pathology, and the history in assessing their significance.

Experienced bacteriologists and technologists can generally tentatively recognize the colonies of the more commonly occurring bacteria, such as *Pasteurella multocida, Corynebacterium pseudotuberculosis, Actinobacillus equuli, Streptococcus* spp., *Bacillus* spp., etc. This ability, along with knowledge of the Gram reaction and morphology of a particular organism, will guide them in the selection of differential media and frequently make for a saving in the time required for final identification.

In order to determine the genus involved, some basic characteristics must be determined. These are dealt with below.

GRAM-POSITIVE AEROBIC BACTERIA

The genus of the aerobes is determined by the features listed in Table 4-II and by means of King's (2) key listed at the end of this chapter.

The mycobacteria, although Gram-positive, are not included in Table 4-II, as they are identified as to genus by the acid-fast stain. *Nocardia* spp. are partially acid-fast and display characteristic branching and beading. *Erysipelothrix rusiopathiae* and *Listeria* can be distinguished from *Actinomyces* on the basis of microscopic morphology, colony characteristics, and oxygen requirements, and from *Lactobacillus* by the fact that the latter have a characteristic morphology and different growth requirements. The *Clostridium* spp. are anaerobic and catalase negative.

When the genus is determined, reference should be made to the appropriate generic tables where the essential reactions are indicated.

GRAM-NEGATIVE AEROBIC BACTERIA

The most frequently recovered aerobic Gram-negative bacteria are members of the family *Enterobacteriaceae*. They are rod-shaped organisms producing colonies of moderate size. They are fermenters, and it is customary to inoculate triple sugar iron (TSI) agar slants from colonies of suspected "enterics." As well as fermenting one or more sugars of the TSI slants, they grow on MacConkey's agar and are oxidase negative. The reactions of various Gram-negative organisms on TSI are given in Table 10-IV. Identification of enteric bacteria as to species or group is accomplished by the criteria listed in Table 10-V. The remaining genera of Gram-negative aerobes can be recognized by the characteristics listed in Table 4-III and by King's key (2) at the end of this chapter. Not included in Table 4-III are the genera *Campylobacter, Vibrio,* and *Francisella*. Members of the genera *Vibrio* and *Campylobacter* are recognized by their characteristic microscopic morphology and cultural characteristics. The animal source, history, and lesions usually indicate the likelihood of a culture being *Francisella tularensis*. The requirement of the latter for cystine is highly supportive.

As one gains experience, the different genera become more easily recog-

TABLE 4-II

DETERMINATION OF THE GENERA OF
GRAM-POSITIVE AEROBES AND FACULTATIVE ANAEROBES

	Spores	Motility	Catalase	Oxidase	Glucose (acid)	O - F Test (Glucose)	Growth In		Remarks
							Air	Anaerobic	
Cocci									
Streptococcus	-	-	-	-	+	NR	+	+	
Micrococcus	-	-	+	-	(+)	O/-	+	-	
Staphylococcus	-	-	+	-	+	-/F	+	+	
Rods									
Corynebacterium	-	-	(+)	-	+*	NR	+	+	*C. suis,* anaerobe
Listeria	-	+	+	-	+	NR	+	+	
Erysipelothrix	-	-	-	-	+	NR	+	+	
Bacillus	+†	(+)	+	(+)	(+)	NR	+	V	
Lactobacillus	-	-	-	-	+	NR	+	+	
Kurthia	-	+	+	-	-	NR	+	-	
Nocardia	-	-	+	-	+	O/-	+	-	*N. asteroides* usually partially acid-fast

+ = positive reaction; - = negative reaction; (+) = majority of strains positive but some exceptions;

(-) = majority negative but some exceptions; NR = not required.

O = glucose oxidized; F = glucose fermented; V = variable.

* except *C. equi*

† asporogenous strains occur

nizable. A number of features aid in the tentative recognition of Gram-negative genera. For example: (1) *Pseudomonas aeruginosa* produces a characteristic colony with hemolysis and a fruity odor on blood agar; (2) *Acinetobacter* and *Neisseria* are predominantly coccal in morphology; (3) *Aeromonas* spp. are usually hemolytic and resemble species of the *Enterobacteriaceae*; (4) *Chromobacterium* may produce a characteristic pigment; (5) *Flavobacterium* spp. usually produce colonies containing yellow pigment; (6) *Pasteurella multocida* has a characteristic odor, and strains from the lungs of cattle and swine frequently produce mucoid colonies; (7) *Pasteurella haemolytica* usually produces beta-hemolysis of bovine blood agar; (8) *Actinobacillus* spp. are most frequently recovered from the horse and from lesions of actinobacillosis; colonies of *A. equuli* frequently have a "sticky" or tenacious character; (9) The common pathogenic *Campylobacter* are slow-growing, fastidious organisms requiring reduced oxygen tension for growth; (10) *Bordetella bronchiseptica* is most commonly recovered from respiratory infections of swine, rabbits, and guinea pigs; (11) *Moraxella bovis* is recovered almost exclusively from the bovine eye; (12) *Haemophilus* spp. produce minute colonies and may require V factor in addition to the X factor; (13) *Brucella* spp. grow slowly and are usually recovered from fetal and genital tissues.

Gram-negative anaerobes are often encountered in pus and necrotic tissue, which may have an offensive odor. Cultures of *Fusobacterium necrophorum* have a foul odor, and colonies of *Bacteroides melaninogenicus* have a characteristic dark pigment.

GENERAL COMMENTS ON IDENTIFICATION

The differential characteristics of the organisms referred to in the manual are listed under the particular organism and in the corresponding differential tables.

A perusal of various textbooks, scientific papers, and *Bergey's Manual* discloses many inconsistencies. Many of these may be attributed to strain variation, the use of different methods, and the occurrence of different biotypes. For example, *Pasteurella multocida* does not usually produce acid from lactose, but aberrant strains are encountered that do. Variation in fermentation or oxidation of sugars may depend upon the time of incubation. Other species may require incubation for a week or two before sufficient fermentation or oxidation takes place. Organisms requiring supplemented media may give negative results if the test medium is not supplemented. The considerable variation among strains of bacteria should be kept in mind when using tables of differential characteristics.

TABLE 4-III

DETERMINATION OF THE GENERA OF GRAM-NEGATIVE BACTERIA

	O—F Test	MacConkey's	Oxidase	Catalase	Motility	Species and Remarks
Predominantly Rods:						
Enterobacteriaceae	F	+	−	+	(+)	⎫
	F	+	−	+	−	See Table 10-V
	F	+	−	−	−	⎭
Yersinia	F	+	−	+	−	Y. pestis
	F	+	−	+	+	Y. pseudotuberculosis and Y. enterocolitica
Pseudomonas	O	+	+	+	+	Ps. aeruginosa
	O	+	(+)	(V)	+	Ps. pseudomallei
	O	+	−	+	−	Ps. mallei
Aeromonas	F	+	+	(+)	+	A. hydrophila
	F	+	+	(+)	−	A. salmonicida
Chromobacterium	F	−	−	+	+	Purple or violet pigment
Flavobacterium	O	−	+	+	−	Yellowish pigment
Pasteurella	F	−	+	+	−	P. multocida, P. pneumotropica
	F	+	+	+	−	P. haemolytica
	−	−	+	+	−	P. anatipestifer
Actinobacillus	F	+	+	+	−	
Bordetella	−	+	+	+	+	B. bronchiseptica
Alcaligenes	−	+	+	+	+	A. faecalis
Brucella	−	−	+	+	−	Oxidase may be weak
Moraxella	−	−	+	+	−	M. bovis, see King's Table and Table 8-I for other varieties
Haemophilus	NT	−	(+)	−	−	Fastidious; require X and/or V factors
Predominantly Cocci:						
Acinetobacter calcoaceticus subsp. anitratus	O	+	−	+	−	
A. calcoaceticus subsp. lwoffi	−	(+)	−	+	−	
Neisseria	O	−	+	+	−	Usually not pathogenic to animals

For explanation of symbols see Table 4-II; NT = not testable

IDENTIFICATION BY KEYS

In practical diagnostic bacteriology, the great majority of bacteria encountered from fresh specimens are those listed in the manual. The methodical keying-out of an organism is only occasionally required.

The key provided in *Bergey's Manual* (3) may be helpful in the identification of an unknown organism. In addition to its value as a key, it is a useful reference work for the diagnostic laboratory. Skerman (4) has prepared a key for the identification of genera with a complete list of the procedures used. The book by Cowan (5) contains much valuable information on the identification of bacteria of medical and veterinary significance.

King's (2) tables dealing with the identification of pathogenic Gram-negative aerobes contain a wealth of differential data on bacteria recovered from man and animals. In her tables are listed a number of infrequently recovered, as yet unnamed, Gram-negative aerobic bacteria. Some of these will be officially named in the forthcoming ninth edition of *Bergey's Manual*. Her useful key for the identification of aerobic bacteria is listed at the end of this chapter.

Commercial Identification Systems

There are a number of convenient and relatively rapid miniaturized systems available commercially for the identification of bacteria of veterinary and medical significance. Several have been specially designed for the identification of enterobacteria, anaerobic bacteria, and yeasts. Cost precludes their use in many laboratories, but they are particularly useful for the small laboratory that does not have the capability to prepare conventional media.

Some workers have reported variation in the reliability of these systems, and it is advisable to try and to compare several before making a selection.

A number of the most widely used systems are listed below along with their sources.

Detailed information on their use including charts, tables, and computerized coding and characterization profiles can be obtained from the manufacturer. The list below is not complete and additional, equally effective systems are available.

1. *The API 20E System.* It is used for enteric and nonfermenting bacteria. It consists of a plastic strip of twenty microtubes containing various dehydrated media, with which twenty-three biochemical tests can be performed. The API 20A system is used for anaerobes, and the API 20C system is available for the identification of yeasts and yeastlike organisms. The API 20S system for streptococci and the STAPH–IDENT system are additional microidentification systems offered by Analytab Products. Source: Analytab Products, 200 Express Street, Plainview, NY 11803.

2. *The Entero-Set 20 System for Enterobacteriaceae.* Similar to the API System. Source: Diagnostic Division, Fisher Scientific Co., 526 Route 303, Orangeburg, NY 10962.

3. *The Minitek Systems.* These are available for enteric bacteria, nonfermenters, yeasts, and anaerobes. They consist of disposable plastic plates with wells. A dispensing apparatus deposits into the wells various discs that have been impregnated with substrates. A suspension of the test organism is then inoculated into the wells.
   Source: BBL, P. O. Box 243, Cockeysville, MD 21030.

4. *The PathoTec Rapid I-D System.* This is used for enteric and other bacteria. It consists of ten test strips impregnated with reagents. The strips are dipped into tubes containing a suspension of the test organism.
   Source: General Diagnostics, Morris Plains, NJ 07950.

5. *The Enterotube.* This is used for enteric bacteria. It consists of a tube containing eight compartments, each with a different agar-based medium. The compartments are traversed by a thin metal rod. The protruding end of the rod is inserted into a colony of the test organism, and the rod is then drawn through each compartment, thus inoculating each medium.
   Source: Roche Diagnostics, Nutley, NJ 07110.

6. *The Oxi/Ferm System.* This is used for nonfermentative Gram-negative bacteria and is similar to the Enterotube.
   Source: Roche Diagnostics, Nutley, NJ 07110.

7. *The Corning r/b Enteric Differential System* for Enterobacteriaceae and the *Corning N/F System for Oxidative Gram-negative Bacteria.* These systems use several tubes of conventional size, which in the former system allow for fifteen tests and in the latter seventeen tests.
   Source: Corning Medical Microbiology Products, Roslyn, NY 11576.

8. *The Micro-ID System.* This is used for enteric bacteria. It consists of a card with fifteen test chambers. A suspension of the test organism is used to inoculate the chambers.
   Source: General Diagnostics, Morris Plains, NJ 07950.

9. *Sensititre ID Plate.* This system is designed for enteric bacteria and closely related taxa. It provides for twenty-four biochemical reactions.
   Source: Gibco Laboratories, 421 Merrimack Street, Lawrence, MA 01843.

L-FORMS OF BACTERIA

L-type colonies indistinguishable from colonies of mycoplasmas are occasionally seen on culture media, especially from clinical materials. Although these colonies appear spontaneously with considerable frequency from certain species, e.g. *Fusobacterium necrophorum* and *Streptobacillus moniliformis*, they are also produced as a result of phage activity, penicillin, antibody, and various antimicrobial substances. Most L-forms revert to the parent bacterium on subculture, but occasionally they do not.

In diagnostic work, one occasionally notices bizarre and highly pleomorphic forms in smears from solid media or broth cultures. Some of these are the forms that under certain circumstances give rise to L-type colonies. They may consist of long filaments that show beading. Some filaments break up and produce large bodies and coccal forms. Sometimes single bacilli will give rise to large round or pear-shaped structures. Further subcultures will usually produce a preponderance of the bacillary form of the organism. Workers are referred to the Supplementary References for further information on L-forms.

Generally speaking, L-forms do not pose problems for the practical microbiologist, but the possibility of their occurrence, particularly with certain bacterial species, should be kept in mind.

OCCURRENCE OF PATHOGENS
AND POTENTIAL PATHOGENS
IN ANIMAL SPECIES

In many clinical microbiology laboratories, the greater part of the routine work is carried out by technicians with little knowledge of animal diseases. The experienced workers will have gained considerable knowledge as to the occurrence of different organisms in the tissues of various animal species, but novices may be at a loss as to the probable organisms involved. In order to help cope with this deficiency, some information is provided below on the kinds of organisms most frequently associated with infections in various organs and systems of the more important animal species.

Those bacteria such as coliforms, *Pseudomonas aeruginosa, Salmonella*, streptococci (except those with a species predilection), and staphylococci are not always referred to because of their wide distribution. Not all fungi, mycoplasmas, and chlamydial agents are included. Gram-negative anaerobes that are frequently present in purulent materials have not been included.

THE RESPIRATORY SYSTEM

Bovine

Pasteurella multocida, P. haemolytica, Corynebacterium pyogenes, Bordetella bronchiseptica, Haemophilus somnus, Actinobacillus actinoides (rare), *Mycoplasma mycoides*, other mycoplasmas.

Ovine

P. haemolytica, P. multocida, C. pyogenes, mycoplasmas, chlamydia.

Porcine

P. multocida, P. haemolytica, C. pyogenes, Haemophilus suis, Mycoplasma hyorhinis, M. hyopneumoniae, Actinobacillus suis.

Equine

Streptococcus equi, Str. equisimilis, Corynebacterium equi (foals), *Actinobacillus equuli* (foals), *P. multocida, Pseudomonas mallei, Klebsiella, Bord. bronchiseptica, Cryptococcus neoformans, Aspergillus.*

Canine

Bordetella bronchiseptica, P. multocida, Klebsiella, Str. canis, Nocardia asteroides, mycoplasmas, *Cryptococcus neoformans, Blastomyces dermatitidis, Actinomyces viscosus.*

Feline

P. multocida, Nocardia asteroides, Bord. bronchiseptica, chlamydia, *Cryptococcus neoformans.*

Chickens and Turkeys

Haemophilus gallinarum, P. multocida, P. gallinarum, various mycoplasmas, *Aspergillus fumigatus.*

ORGANISMS ASSOCIATED WITH CANINE SKIN INFECTIONS

Staphylococcus aureus, pyogenic streptococci, dermatophytes, *Pseudomonas aeruginosa, Candida albicans* (usually an extension from mucous membrane infection). Various bacteria are probably opportunists: fecal streptococci, coliforms, *Proteus* spp., diphtheroids.

BACTERIA ASSOCIATED WITH INFECTIONS
OF THE GASTROINTESTINAL TRACT

Bovine

Salmonella, Clostridium perfringens, types B and C, *Mycobacterium paratuberculosis.*

Ovine

Cl. perfringens type B (lamb dysentery), type C (struck), type D (pulpy kidney), type E; *Mycobacterium paratuberculosis; Salmonella.*

Porcine

Salmonella (especially *S. choleraesuis*), *Cl. perfringens* type C.

Equine

Salmonella, Corynebacterium equi (foals), *Actinobacillus equuli* (foals).

Canine

Salmonella, possibly other enteric bacteria, *Staph. aureus* (puppies), *Borrelia canis, Spirillum* spp., *Campylobacter.*

Feline

Salmonella, Candida albicans (kittens).

ORGANISMS ASSOCIATED WITH ABSCESSES AND ULCERS OF THE SKIN AND SUBCUTIS

Streptococci and staphylococci are the most common causes of abscesses involving the skin and subcutis of most animal species. *Actinomyces bovis* and *Actinobacillus lignièresi* occur rarely as causes of abscesses in species other than the bovine. Likewise *Nocardia asteroides* is an infrequent cause of abscesses in domestic animals other than the dog and cat. *Pseudomonas aeruginosa* may be associated with abscesses in all of the domestic animals.

Bovine

Corynebacterium pyogenes, Actinomyces bovis, Actinobacillus lignièresi, Sporothrix schenckii.

Ovine

Corynebacterium pseudotuberculosis.

Porcine

Group E streptococci (jowl abscesses), *Corynebacterium pyogenes.*

Equine

Corynebacterium pseudotuberculosis (chest abscesses), *Histoplasma farciminosum, Sporothrix schenckii.*

Canine

Nocardia asteroides, Blastomyces dermatitidis, Sporothrix schenckii, Actinomyces viscosus.

Feline

Pasteurella multocida, Nocardia asteroides.

ORGANISMS ASSOCIATED WITH GENITAL INFECTIONS

Streptococci, staphylococci, enteric bacteria, and *Pseudomonas aeruginosa* are commonly associated with genital infections in all species.

Bovine

Campylobacter fetus subsp. *fetus* and *intestinalis, Brucella abortus,* chlamydia, *Mycoplasma bovigenitalium, Listeria monocytogenes, Corynebacterium pyogenes.*

Ovine

Brucella ovis, Br. melitensis, Campylobacter fetus subsp. *intestinalis, Listeria monocytogenes, Actinobacillus seminis, Corynebacterium pseudotuberculosis, Chlamydia psittaci.*

Porcine

Brucella suis, mycobacteria, *Pseudomonas aeruginosa, C. pyogenes, Pasteurella multocida,* pyogenic streptococci.

Equine

Actinobacillus equuli, Klebsiella, Corynebacterium equi, Salmonella abortus-equi, Candida albicans, pyogenic streptococci.

Canine

Brucella canis, other brucella species (rare), *Klebsiella, Enterobacter, Proteus* spp., *Candida albicans, Pseudomonas aeruginosa,* mycoplasmas.

ORGANISMS ASSOCIATED WITH MASTITIS

Bovine

See Chapter 32.

Ovine

Staphylococcus aureus, Pasteurella haemolytica, Pasteurella multocida, Streptococcus agalactiae, Str. uberis, Str. dysgalactiae, Corynebacterium pyogenes, C. pseudotuberculosis, mycobacteria, mycoplasmas.

Porcine

Streptococci, *Staph. aureus, Fusobacterium necrophorum, Actinomyces bovis, Actinobacillus lignièresi, C. pyogenes,* mycobacteria, coliforms.

Equine

Streptococci, *Staph. aureus,* mycobacteria.

Canine and Feline

Streptococci, *Staph. aureus.*

ORGANISMS RECOVERED FROM THE CENTRAL NERVOUS SYSTEM

Bovine

Listeria monocytogenes, Haemophilus somnus, streptococci, *P. multocida, Chlamydia psittaci, Staphylococcus aureus.*

Ovine

Listeria monocytogenes, Staph. aureus.

Porcine

L. monocytogenes, P. multocida, streptococci.

Equine

L. monocytogenes, Str. equi, Staph. aureus.

Canine and Feline

Bacteria rarely involved.

ORGANISMS RECOVERED FROM URINARY INFECTIONS

Urine from dogs is most commonly submitted. Some of the organisms implicated in the dog are also recovered from the urine of the other species on the infrequent occasions that such specimens are submitted.

Canine

Proteus (usually *mirabilis*), *Pseudomonas aeruginosa, Staphylococcus aureus,* enterococci, *E. coli, Enterobacter,* pyogenic and fecal streptococci, *C. renale* (rare).

Feline

Urinary infections uncommon.

Bovine

Corynebacterium renale.

Ovine

C. renale (rare).

Porcine

Corynebacterium suis, C. renale (infrequent).

Equine

C. renale (rare).

ORGANISMS RECOVERED FROM JOINTS

Bovine

E. coli, pyogenic streptococci, *Salmonella, Corynebacterium pyogenes, Staphylococcus aureus,* chlamydia, mycoplasmas, *Haemophilus somnus.*

Ovine

E. coli, pyogenic and fecal streptococci, *Erysipelothrix rhusiopathiae, Haemophilus agni,* chlamydia, *Streptococcus dysgalactiae, Mycoplasma agalactiae, Corynebacterium pyogenes.*

Porcine

Erysipelothrix rhusiopathiae, Corynebacterium pyogenes, various pyogenic streptococci, *Mycoplasma hyorhinis, M. hyosynoviae, Staph. aureus, Haemophilus suis, Brucella suis, E. coli, Actinobacillus suis.*

Equine

Actinobacillus equuli, Staph. aureus, pyogenic and fecal streptococci, *E. coli, Corynebacterium equi, Klebsiella, Salmonella.*

CANINE OTITIS EXTERNA

Staphylococcus aureus, Pseudomonas aeruginosa, various streptococci, *Candida albicans, Malassezia pachydermatis, Proteus* spp., *Clostridium perfringens.*

BACTERIA FROM THE MARE'S CERVIX

Staphylococcus aureus, various streptococci, *Klebsiella, Pseudomonas aeruginosa, Salmonella abortus-equi,* various fungi, *Corynebacterium equi, Candida albicans, Enterobacter, E. coli, Actinobacillus equuli, Haemophilus genitalis.*

BACTERIA RECOVERED FROM EYES

Some of the bacteria have not been included.

Bovine

Moraxella bovis, Neisseria ovis (or closely related species), chlamydia.

Ovine

Neisseria ovis, Moraxella spp., chlamydia.

Equine

Streptococcus equi, Str. equisimilis, Staphylococcus aureus.

Canine and Feline

Staph. aureus, Staph. epidermidis, Pseudomonas aeruginosa, Clostridium perfringens, Candida albicans, chlamydia, mycoplasmas, *Moraxella* spp., and *Neisseria flavus* have been reported from the cat.

BACTERIA ASSOCIATED WITH INFECTIONS IN LABORATORY ANIMALS

Rats and Mice

Salmonella, pyogenic streptococci, *Bacillus piliformis, Pasteurella pneumotropica, Pasteurella multocida, Corynebacterium kutscheri, Bordetella bronchiseptica, Streptobacillus moniliformis, Streptococcus pneumoniae, Mycoplasma pulmonis, M. arthritidis, M. neurolyticum, Yersinia pseudotuberculosis.*

Guinea Pigs

Salmonella, Bordetella bronchiseptica, Streptococcus pneumoniae, pyogenic streptococci, *Klebsiella pneumoniae, Yersinia pseudotuberculosis, Streptobacillus moniliformis.*

Rabbits

Salmonella, Pasteurella multocida, Bordetella bronchiseptica, Yersinia pseudotuberculosis, pyogenic streptococci, *Yersinia enterocolitica, Haemophilus, Bacillus piliformis, Fusobacterium necrophorum, Treponema cuniculi.*

KING'S KEY* FOR IDENTIFICATION OF AEROBES

Numbers that follow the bacteria refer to pages in the original publication (2). Please note that some of the names, letters, or numbers used to identify some organisms will not correspond to those in the *Approved Lists of Bacterial Names*.

GRAM–NEGATIVE FERMENTERS
    MacConkey positive
        Oxidase negative

---

*Revised by Weaver, R. E., Tatum, H. W., and Hollis, D. G.; reproduced with permission.

Enterobacteriaceae (1), *Yersinia pestis* (2), *Y. pseudotuberculosis* (2), *Y. enterocolitica* (2), *Chromobacterium violaceum* (2), HB-5 (4).

Oxidase positive

Polar flagellated

Aeromonas hydrophila (2), *A. (Plesiomonas) shigelloides* (2), *Vibrio cholerae* (2), Noncholera *Vibrio* (2), *V. parahaemolyticus* (2), *V. alginolyticus* (2).

Polar and lateral flagella

Chromobacterium violaceum (2)

Peritrichous

Providence sp. (1) (oxidase weak when positive)

Nonmotile

Pasteurella haemolytica (3), EF-4 (3), *Actinobacillus lignieresi* (3), *A. equuli* (3), *A. suis* (3), HB-5 (4).

MacConkey negative

Oxidase negative

Haemophilus aphrophilus (4), *H. vaginalis* (4), *Actinobacillus actinomycetemcomitans* (4), HB-5 (4), Bacillus sp. (12) (may stain weakly Gram-positive or Gram-negative)

Oxidase positive

Pasteurella multocida (3), *P. pneumotropica* (3), *P. ureae* (3), *P. gallinarum* (3), *P.* "gas" (3), *Cardiobacterium hominis* (4), EF-4 (3), HB-5 (4), *Neisseria sicca* (5), *N. mucosa* (5), *N. subflava* (5), *N. flava* (5), *N. lactamica* (5), *Vibrio cholerae* (2), *Bacillus sp.* (12)

GRAM–NEGATIVE QUESTIONABLE FERMENTERS (Early reaction that of an oxidizer)

MacConkey positive

Oxidase positive

IIk, type 2 (10), rare strain of IIk, type 1 (10), *Flavobacterium meningosepticum* (7), *Fl.* sp., IIb (7).

MacConkey negative

Oxidase positive

IIk, type 1 (10), *Flavobacterium meningosepticum* (7), Fl. sp., 11b, (7), 11c (7), IIe (7), IIh (7), IIi (7), *Neisseria perflava* (5), *N. meningitidis* (5), *N. gonorrhoeae* (5), *N. lactamica* (5), *N. subflava* (5), *N. flava* (5), *N. sicca* (5), *N. mucosa* (5).

Oxidase negative

IIk, type 1 (10).

GRAM–NEGATIVE GLUCOSE OXIDIZERS

MacConkey positive

Oxidase negative

Herellea (Acinetobacter) (5), *Pseudomonas cepacia* (9), *P. (Actino-*

bacillus) *mallei* (9), Ve-1 (10), Ve-2 (10).

Oxidase positive

Polar flagellated

Pseudomonas aeruginosa (9), *P. fluorescens* (9), *P. putida* (9), *P. pseudomallei* (9), *P. stutzeri* (9), Vb-2 (9), Vb-3 (9), *P. cepacia* (9), *P. vesiculare* (10), Va (9).

Peritrichous

Vd (8).

MacConkey negative

Oxidase negative

Pseudomonas mallei (9).

Oxidase positive

Moraxella kingii (5), *Vibrio extorquens* (4), *Brucella* (4).

GRAM–NEGATIVE GLUCOSE NONOXIDIZERS

MacConkey positive

Oxidase negative

Mima (Acinetobacter) polymorpha (6), *Pseudomonas maltophilia* (10), *Bordetella parapertussis* (6 & 8).

Oxidase positive

Aerobic

Moraxella osloensis (6), *M. phenylpyruvica* (6), *M. nonliquefaciens* (6), *Mima polymorpha* var. *oxidans* (6), *Bordetella bronchiseptica* (8), *Comamonas terrigena* (Pseudomonas acidovorans, P. testosteroni) (10), *Alcaligenes faecalis* (8), *A. odorans* (8), *A. denitrificans* (8), *Pseudomonas alcaligenes* (10), *P. pseudoalcaligenes* (10), *P. denitrificans* (10), *P. diminuta* (10), *P. putrefaciens* (10), IIIa (8), IIIb (8), IVc-2 (8), IVe (8).

Microaerophilic

Vibrio fetus (4), "Related" *Vibrio* (4).

MacConkey negative

Oxidase negative

Mima (Acinetobacter) polymorpha (6), *Bordetella parapertussis* (6 & 8), *Bacillus sp.* (12).

Oxidase positive

Moraxella nonliquefaciens (6), *M. lacunata* (6), *M. bovis* (6), *M. osloensis* (6), *M. phenylpyruvica* (16), *Neisseria catarrhalis* (6), *N. flavescens* (6), IIf (7), IIj (7), HB-1 (*Bacteroides — Eikenella corrodens*) (4), *Brucella* sp. (4), *Vibrio extorquens* (4), *Bacillus sp.* (12), *Pasteurella anatipestifer* (3).

GRAM–POSITIVE ORGANISMS*

---

*From original King key.

MacConkey negative
   Catalase negative rods
      Lactobacilli; form chains, grow on tomato juice agar, H_2S negative in butt of TSI slant
      Erysipelothrix; short chains, no growth on tomato juice agar, form H_2S in butt of TSI agar
      Corynebacterium pyogenes; beta-hemolytic, gelatin positive
      Corynebacterium hemolyticum; beta-hemolytic, gelatin negative
      Clostridium tertium; grows aerobically, sporulates only anaerobically
   Catalase negative cocci
      Streptococci; chains of cocci, not bile soluble
      Pneumococci; chains of cocci, bile soluble
   Catalase positive rods
      Bacillus spp.; forms spores
      Corynebacterium spp.; sometimes solid staining, sometimes barred or clubbed, never chaining, palisading
      Listeria; usually very short rods, may form chains in broth or palisade
      Mycobacteria; acid-fast
   Catalase positive cocci
      Staphylococcus; fermenters
      Micrococcus; oxidizers
MacConkey positive
   Catalase negative
      Streptococcus; usually group D
      A very few other Gram-positive organisms will grow on MacConkey

REFERENCES

1. Bartlett, R. C., Ellner, P. D., and Washington, J. A.: *Cumitech I.* Ed. by J. C. Sherris. Washington, D.C., American Society for Microbiology, 1974.
2. Weaver, R. E., Tatum, H. W., and Hollis, D. G.: *The Identification of Unusual Pathogenic Gram-Negative Bacteria* (Elizabeth O. King), Preliminary Revision, 1972. Published as training materials by the Center for Disease Control, Public Health Service, Department of Health Education and Welfare, Center for Disease Control, Atlanta, Georgia.
3. Buchanan, R. E., and Gibbons, N. E. (Eds.): *Bergey's Manual of Determinative Bacteriology*, 8th ed. Baltimore, Williams & Wilkins, 1974.
4. Skerman, V. B. D.: *A Guide to the Identification of the Genera of Bacteria*, 2nd ed. Baltimore, Williams & Wilkins, 1967.
5. Cowan, S. T.: *Manual for the Identification of Medical Bacteria*, 2nd ed. Cambridge, Cambridge University Press, 1974.

SUPPLEMENTARY REFERENCES

Lennette, E. H. (Ed-in-chief): *Manual of Clinical Microbiology*, 3rd ed. Washington, D.C., American Society for Microbiology, 1980.

Madoff, S.: In Stan, M. P. et al. (Eds.): *The Prokaryotes*, Vol. 11. New York, Springer-Verlag, 1981, Chapter 166, p. 2225.

Chapter 5

SPIROCHETES

JOHN R. COLE, JR.

T HE SPIROCHETES are classified as bacteria in the order Spirochaetales and the family Spirochaetaceae. Three genera in this family, *Leptospira*, *Treponema*, and *Borrelia*, contain pathogenic species. The other two genera, *Spirochaeta* and *Cristispira*, are considered free-living and commensal, respectively.

Spirochetes are slender, spiral in shape, round on cross section, and multiply by transverse fission. Movement is active and accomplished by spinning and flexing about the long axis.

They are found in water, soil, decaying organic matter, plants, animals, and humans. These microorganisms are relatively inactive biochemically, and identification is based on staining reactions, growth requirements, morphology, pathogenicity, and serology.

Distinguishing characteristics of the three pathogenic genera, *Leptospira*, *Treponema*, and *Borrelia*, are shown in Table 5-I. They are Gram-negative but are observed best by darkfield or phase microscopy or by staining with silver impregnation or Giemsa stain. Only *Borrelia* stains with aniline dyes.

LEPTOSPIRA

The genus *Leptospira* is divided into saprophytic and pathogenic groups. The pathogens are included in the species *L. interrogans* and separated into approximately 160 serovars (synonym: serotype) on the basis of cross-agglutination and agglutinin adsorption reactions. Some important serovars and their hosts are summarized in Table 5-II. Those most commonly associated with diseases of domesticated animals are listed in Table 5-III.

Pathogenicity

Leptospirosis is generally characterized by having two distinct phases: (1) leptospiremia and fever for about seven days, followed by (2) leptospiruria, which may persist two to three months. Organisms may be recovered from the blood during the first phase and the kidney or urine during the second phase.

CANINE. The four clinical syndromes recognized are the acute hemorrhagic, icteric, subacute or uremic, and inapparent forms. The first two forms are

TABLE 5-I

DISTINGUISHING CHARACTERISTICS OF *LEPTOSPIRA, TREPONEMA,*
AND *BORRELIA**

Characteristic	Leptospira	Treponema	Borrelia
Morphology			
Length	6μm-20μm	5μm - 20μm	3μm - 20μm
Width	0.1μm-0.2μm	0.09μm - 0.5μm	0.2μm - 0.5μm
Ends	A semicircular hook on one or both ends	Pointed, may have terminal filaments	Taper terminally to fine filaments
Spirals			
Number	Many, fine, tight	6–14, regular, angular	4–8, loose
Amplitude	0.4μm - 0.5μm	1μm	3μm
Motility	Spinning, undulating	Rotating undulating, stiffly flexible	Lashing, cork-screwlike
Growth Conditions	Aerobic	Anaerobic	Anaerobic

*Modified from G.R. Carter, *Essentials of Veterinary Bacteriology and Mycology,* 1976. Courtesy of Michigan State University Press, East Lansing, Michigan.

caused primarily by *icterohaemorrhagiae*, while the latter two are caused by *canicola*. In the initial stages of the disease, the first three forms of leptospirosis are usually clinically indistinguishable; all are characterized by depression, anorexia, vomiting, and diarrhea or constipation. Signs of the specific clinical syndrome appear in later stages of the disease.

BOVINE. Clinical signs are fever, diarrhea, depression, anorexia, infertility, and sometimes abortion. Hemoglobinuria, icterus, and decreased milk production may also develop.

PORCINE. Infections are usually subclinical or asymptomatic. Abortions late in pregnancy are sometimes the only sign of infection. Occasionally, metritis, icterus, anemia, fever, and meningoencephalitis are observed.

EQUINE. The disease is characterized by fever, depression, anorexia, and icterus. Periodic ophthalmia or abortion may occur after the fever subsides.

LABORATORY EXAMINATION

Direct Examination

Leptospires can be demonstrated in tissue and body fluids by darkfield, phase, or fluorescence microscopy (Fig. 5-1). Leptospires observed by direct examination should be confirmed by isolation procedures.

Blood

Five milliliters of blood drawn during the febrile phase is mixed with 0.5 ml of a 1.0% solution of sodium oxalate or 0.1 ml of a 1.0% solution of heparin (sodium citrate may be inhibitory).

TABLE 5-II

IMPORTANT LEPTOSPIRAL SEROVARS AND THEIR HOSTS*

Serovar	Known Host	Occurrence in†			
		Man	Dogs	Cattle	Swine
icterohaemor-rhagiae	Rat, mouse, raccoon, opossum	Common	Occasional	Reported	Reported
canicola	Dog, cattle, swine, skunk	Common	Common	Rare	Occasional
pomona	Cattle, swine, skunk, raccoon, wildcat, deer, opossum, horse	Occasional	Rare	Common	Common
autumnalis	Opossum, raccoon, mouse	Rare	?	?	?
ballum	Mouse, gray fox, rat opossum, raccoon, wild-cat, skunk, gray squir-rel, rabbit	?	?	?	?
grippotyphosa	Raccoon, mouse, fox squirrel, rabbit, bobcat	Rare	Reported	Sporadic	Sporadic
bataviae	Rat, field mouse	Rare	?	?	?
hardjo	Cattle	Rare	?	Common	?
sejroe	Opossum, raccoon, mouse	?	?	Sporadic	?
hebdomadis	Opossum, raccoon	?	?	?	?
australis	Opossum, raccoon, fox	?	?	?	?
szwajizak	Cattle	?	?	Reported	?
balcanica	Cattle	?	?	Reported	?

*Data principally applicable to the United States.
†Based in some instances on serological evidence.
Modified from Carter, G.R. *Essential of Veterinary Bacteriology and Mycology,* 1976. Courtesy of Michigan State University Press, East Lansing, Michigan.

1. Centrifuge at 1500 rpm for fifteen minutes.
2. Transfer supernatant fluid to another tube and centrifuge at 3,000 rpm for thirty minutes.
3. Discard the supernatant fluid and prepare a wet mount from the sediment.
4. Examine by darkfield or phase microscopy.
   Caution: Protoplasmic extrusions from blood cells and other artifacts may be confused with leptospires.
5. Smears can be prepared from the sediment for fluorescent antibody staining. The procedures included with the fluorescent antibody conjugate should be followed.

Urine

Fresh urine is neutralized with N/10 HCl or N/10 NaOH.
1. Centrifuge at 3,000 rpm for ten minutes.

TABLE 5-III

PRINCIPAL LEPTOSPIRAL SEROVARS ASSOCIATED WITH DISEASES OF
DOMESTICATED ANIMALS

Animal Species Affected	Serovars
Cattle	*pomona*
	hardjo
	grippotyphosa
	icterohaemorrhagiae
	canicola
Sheep	*pomona*
Swine	*pomona*
	grippotyphosa
	canicola
	icterohaemorrhagiae
Horse	*pomona*
Dog	*canicola*
	icterohaemorrhagiae

2. Discard supernatant fluid and prepare a wet mount from the sediment.
3. Examine as described for blood.

Tissues

A portion of kidney or liver must be taken aseptically shortly after death.
1. Grind a weighed portion of tissue in a stomacher, mortar and pestle, or Ten Broeck tissue grinder.
2. Prepare a 10% tissue suspension in Stuart's (Difco Laboratories, Detroit, Michigan) or EMJH (Difco) liquid medium or 1% bovine serum albumin (BSA, Fraction V® powder; Miles Laboratories, Inc., Elkhart, Indiana) and mix well.
3. Follow steps 1 through 5 for blood.

Isolation Procedures

Leptospires can be isolated from blood, urine, and tissue suspensions by inoculation into any of the following media: (1) EMJH medium; (2) Fletcher semisolid medium (Difco); (3) Tween 80®-albumin medium (OAC) (1). The semisolid forms of EMJH and OAC media are recommended for the isolation of leptospires. They are prepared by adding 1.5 g agar per liter of medium.

Samples that are contaminated with other microorganisms should be filtered through a 0.45 μm bacteriological filter and inoculated into medium containing 5-fluorouracil in a concentration of 200 μg/ml, or neomycin sulfate, 300 μg/ml. A microsyringe filter holder (Millipore Corporation,

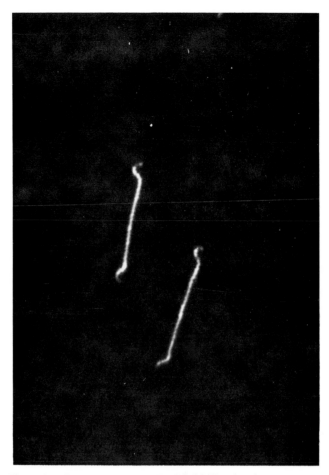

Figure 5-1. *Leptospira pomona.* Darkfield illumination, × ca. 845.

Bedford, Massachusetts) attached to a syringe can be used.

Animal inoculation can also be used when samples are contaminated or when leptospires are present in small numbers.

Blood

One to two milliliters of blood is drawn aseptically during the febrile phase.

1. Inoculate 1 or 2 drops of oxalated or heparinized blood (see Direct Examination—Blood) into each of three tubes of EMJH, Fletcher, or OAC semisolid medium.
2. Incubate at 28–30°C or at room temperature in the dark.
3. At seven-day intervals, examine a drop of medium from each tube by darkfield or phase microscopy. Tubes are discarded if growth does not appear within six weeks.

Urine

Aseptically collect a minimum of 1.0 ml of urine. The sample is inoculated both undiluted and diluted. However, if urine cannot be inoculated into the medium immediately, dilute the urine 1:10 with 1% BSA to maintain viable leptospires (2). The diluted material may be held at room temperature.

1. With a 2 ml syringe and 20 gauge needle, inoculate 1 or 2 drops of fresh undiluted urine into each of three tubes of EMJH, Fletcher, or OAC semisolid medium.
2. Discard all but 0.1 ml of urine from the syringe.
3. Draw 0.9 ml of EMJH liquid medium or 1% BSA into the syringe (1:10 dilution).
4. Discard a few drops (2 or 3) from the syringe.
5. Inoculate 1 or 2 drops of the 1:10 dilution into each of three tubes of semisolid medium.
6. Prepare and inoculate four additional tenfold dilutions (10^{-2}, 10^{-3}, 10^{-4}, 10^{-5}) as in steps 2 through 5.
7. Incubate and examine as described for blood.

Tissues

The kidney, liver, and brain are the organs preferred for isolation attempts.

1. Grind a weighed portion of tissue in a stomacher, mortar and pestle, or Ten Broeck tissue grinder.
2. Prepare a 10% tissue suspension in EMJH liquid medium or 1% BSA and mix well.
3. Follow steps 1 through 6 as described for urine.
4. Incubate and examine as described for blood.

Isolation attempts should be made as soon as possible after death because leptospires survive only a short time in autolytic tissues. After four hours, recovery attempts become impractical.

Laboratory Animals

Weanling hamsters, gerbils, or guinea pigs are preferred for the isolation of leptospires associated with infection in domesticated animals, especially *hardjo*.

Two or three animals are used for each isolation attempt.

1. Inoculate laboratory animals intraperitoneally with 0.5–1.0 ml of unclotted blood, neutralized urine, or a 10% tissue suspension in EMJH medium or 1% BSA.
2. Take cardiac blood aseptically on postinoculation days 5, 8, 10, and 14, or when an increase in temperature is detected.

3. Immediately inoculate at least two tubes of Fletcher or EMJH semisolid medium with 2 or 3 drops of blood.
4. Prepare and examine wet mounts from unclotted blood as described in steps 1 through 5 of Direct Examination — Blood.
5. Collect blood samples for serology and kill animals surviving twenty-one days postinoculation.
6. Attempt isolation from tissues and urine as previously described.

Identification

Identification of isolated leptospires is based on serologic reactions with specific antiserum and is usually performed by the WHO/FAO Collaborating Laboratory for the Epidemiology of Leptospirosis, Center for Disease Control, Atlanta, Georgia.

SEROLOGIC PROCEDURES

Macroscopic and microscopic agglutination tests are the most commonly used procedures for detection of serum leptospiral antibodies. The enzyme-linked immunosorbent assay (ELISA) technique has recently been reported to be of value in the detection of specific immunoglobulins (3,4,5). Its use as a routine procedure for diagnosis of leptospirosis is limited at this time.

The macroscopic agglutination test is used for screening either single or pooled serum samples and employs commercially available killed antigens. The macroscopic test is used because of availability of antigens, ease of performance, and safety. A major disadvantage of the test is its lack of specificity. Sera may react with multiple serovars, especially if the samples are obtained during the acute phase of the disease.

The microscopic agglutination test, which utilizes live leptospires as antigen, is highly sensitive and serovar-specific. The time and attention required to maintain viable, pure cultures of several serovars is the major disadvantage of this test.

Sera that are positive on screening with the macroscopic test should be confirmed by the microscopic test. These positive sera may be negative by the microscopic test because of low serum antibody levels and nonspecific reactions.

Macroscopic Agglutination Test

1. Antigens
   A. The Galton antigens (Difco Laboratories, Detroit, Michigan) are available as single serovars or as pools designated I, II, III, IV, V, and VI. The pools most commonly used are I through IV, which contain the following serovars:

Pool I	Pool II	Pool III	Pool IV
ballum	bataviae	autumnalis	australis
canicola	grippotyphosa	pomona	hyos (currently *tarassovi*)
icterohaemorrhagiae	pyrogenes	wolffi	mini georgia (currently *georgia*)

    B. The Stoenner antigens (Fort Dodge Laboratories, Fort Dodge, Iowa) are available as individual serovars of *pomona, hardjo, grippotyphosa, canicola,* and *icterohaemorrhagiae.*

2. Plate screening test
    A. On a glass plate, place serum (0.01 ml Galton or 0.005 ml Stoenner) to be tested on a separate square for each antigen or pool.
    B. Using supplied dropper, add one drop of each single or pooled antigen to each drop of serum.
    C. Using a clean portion of applicator stick, mix antigen and serum.
    D. Rotate the plate by hand five to ten times.
    E. For the Galton method, place on an electric rotator for four minutes at 125 rpm. Using the Stoenner procedure, incubate for six minutes at room temperature in a moist chamber. Remove and slowly rotate the plate ten to fifteen times by hand.
    F. Observe reaction over indirect light.
    G. Record reaction as follows:
        Positive: Agglutination (peripheral*)
        Negative: Even suspension
    H. Use known positive and negative sera as controls.
    I. Confirm positive samples with the microscopic agglutination test.

3. Determination of titer by macroscopic methods
    The plate dilution (Galton) and rapid plate (Stoenner) tests may be used for the determination of titers. Initial serum-saline dilutions of 1:5 and subsequent twofold (Galton) or fourfold (Stoenner) dilutions are prepared. Specific instructions for conducting these tests are included with the antigens.

4. Evaluation of macroscopic methods for serum titrations
    The macroscopic tests may be helpful in establishing a presumptive diagnosis if a significant titer increase is detected with acute and convalescent serum samples or if a high titer is obtained with a single sample.
    Using the Stoenner antigen, a titer of 40 or greater is considered

---

*Nonspecific clumping due to bacteria and cell debris usually occurs in the center of the drop. Specific agglutination occurs at the edge of the drop. Confidence in reading this test is obtained with experience and use of adequate control antiserum.

positive with serovars *pomona, grippotyphosa, icterohaemorrhagiae,* and *canicola.* A titer of 10 or greater is considered positive with *hardjo.* For the Galton method, a dilution of 1:32.5 is considered positive.

The microscopic agglutination test is preferred for serum titrations because of its greater specificity.

Microscopic Agglutination Test

1. Antigens

    The antigens are five-day-old cultures grown in EMJH, Stuart, or OAC medium. The serovars used will depend upon those suspected of being prevalent in the animal population in the particular location. Suggested serovars are *pomona, hardjo* or *wolffi, grippotyphosa, icterohaemorrhagiae, canicola,* and *autumnalis.*

2. Antigen preparations

    A. Examine cultures microscopically for purity, homogeneity, and density.

    B. Transfer sufficient antigen for the test into tubes (13 × 100 mm).

    C. Centrifuge at 1500 rpm for fifteen minutes to remove debris.

    D. Transfer supernatant to another tube and adjust to an antigen concentration of 100–200 organisms per high-power field (× 450). This concentration is equivalent to a McFarland number 0.5, a light transmission of 60–75% on a Spectronic 20 or equivalent spectrophotometer set at 400 nm, or a Nephelometer set to 25 with either dry well or wet well.

3. Plastic tray

    This procedure utilizing the plastic tray (No. 96U–CV, Linbro Scientific Company, New Haven, Connecticut) was developed at the Center for Disease Control (6) and is a modification of the conventional tube test.

    A. Serum dilution. Prepare fourfold serum dilution as follows (see Table 5-IV):

    (1) Place four tubes in a rack behind each serum sample.

TABLE 5-IV
MICROSCOPIC AGGLUTINATION: SERUM DILUTION PROCEDURE

Tube No.	Phosphate buffered saline (PBS)	Volume of serum dilution transferred	Initial serum dilution	Final serum dilution*
4	1.5 ml	–	1:3200	1:6400
3	1.5 ml	0.5 ml	1:800	1:1600
2	1.5 ml	0.5 ml	1:200	1:400
1	4.9 ml	0.5 ml	1:50	1:100
Serum sample		0.1 ml		

*Final serum dilution after addition of antigen.

(2) Using a 5.0 ml Cornwall automatic syringe adjusted to deliver 4.9 ml, add 4.9 ml PBS (pH 7.2) to tube 1 for each serum sample.

(3) Using a 2.0 ml Cornwall automatic syringe adjusted to deliver 1.5 ml, add 1.5 ml PBS to tubes 2, 3, and 4 for each serum sample.

(4) Using an appropriate pipette, add 0.1 ml serum to tube 1. The same pipette may be used throughout for transferring serum by discarding the serum remaining in the pipette and drawing and expelling distilled water into the pipette several times to rinse.

(5) Using a 1.0 ml Cornwall automatic syringe with "filling outfit" removed and a four-inch cannula attached, mix serum-PBS solution in tube 1 by filling and expelling the solution from the syringe five to six times.

(6) Transfer 0.5 ml of dilution in tube 1 to tube 2 and mix.

(7) Repeat for tubes 3 and 4.

(8) After mixing tube 4, rinse syringe five to six times in distilled water.

(9) Repeat steps 5, 6, 7, and 8 for each serum sample.

(10) Include known positive and negative sera for controls.

B. Serum addition

(1) Using a 0.5 ml disposable plastic pipette (Mohr-type, Falcon Plastics, Oxnard, California) to which a rubber bulb has been attached, withdraw 0.3–0.5 ml of the serum dilution from tube 4.

(2) Holding pipette *vertically*, dispense 1 drop (0.05 ml) to one well (labeled 4 in Fig. 5-2) for each antigen tested. Discharge the remaining serum in the pipette back into tube 4.

(3) Using the same pipette, dispense serum dilution from tubes 3 through 1 (wells 3 through 1).

(4) The same pipette may be used for all serum samples if it is rinsed in distilled water between samples (after each tube 1).

(5) When the addition of serum to each tray is completed, the antigens are added.

C. Antigen addition

(1) Using a 0.5 ml disposable plastic pipette with a rubber bulb, withdraw 0.5 ml of antigen A.

(2) Holding the pipette *vertically*, add 1 drop (0.05 ml) to column A, wells 4 through 1, for each serum sample (see Fig. 5-2). Return pipette to the tube containing the antigen used.

(3) Repeat steps 1 and 2 for the remaining antigens in their corresponding columns.

(4) Shake the trays gently to mix contents, cover to exclude debris, and incubate at room temperature for two hours.

Serum Additions ⟶

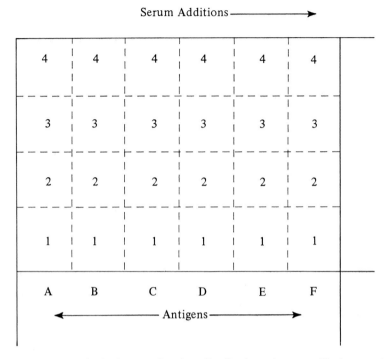

Figure 5-2. Quadrant of plastic tray showing distribution of serum dilutions and antigens in wells.

D. Reading of test
   (1) Place a loopful of the mixture from well 1 for each antigen on a microscope slide previously cleaned with alcohol. Between antigens, the loop must be flamed and cooled by touching to a paper towel moistened with water. The wire must be bent so that the loop will rest flat on the bottom of the well.
   (2) Examine slides by darkfield microscopy ($\times 100$).
   (3) If agglutination or lysis (clearing of the field) is observed at the 1:100 dilution (well 1), examine wells 2, 3, and 4 for that antigen.
E. Interpretation of test results
      The end point is the highest dilution in which at least 50 percent of the organisms are agglutinated. End points of 1:100 are suspicious, and those greater are positive. Lysis may occur at low dilutions with some serovars.
      The vaccination history of the herd must be considered in evaluating results. End points of 1:1000 or greater can be detected in animals that have been recently vaccinated (7,8).
4. Microtiter plate
      This technique (9) is a modification of a procedure initially reported by workers at the Center for Disease Control (10). A disposable microtiter

pipette equipped with a 0.025 ml dropper (Linbro Scientific Company, New Haven, Connecticut) is used for dispensing the serum and antigen into microtiter plates with flat-bottom wells (Linbro or Microtest II®, Falcon Plastics, Oxnard, California). A multimicrodiluter handle equipped with 0.025 ml microdiluters (Cooke Engineering Company, Alexandria, Virginia) is used to dilute the serum.

A. Serum dilution (twofold)
   (1) In a test tube (13 × 100 mm), prepare a 1:25 dilution using 2.4 ml PBS and 0.1 ml serum.
   (2) Add 1 drop (0.025 ml) of PBS to each well in the plate except for the wells in the first row (row H*).
   (3) Add 2 drops (0.05 ml) of the 1:25 serum dilution to wells in row H. One well is used for each antigen tested.
   (4) Using the 0.025 ml microdiluter, mix the dilutions in row H by twirling the diluters ten to fifteen times.
   (5) Transfer diluters to row G and mix.
   (6) Repeat step 5 for the desired number of dilutions.
   (7) After mixing the last dilution, rinse diluters by twirling in distilled water and blot dry.
   (8) Include known positive and negative sera for controls.

B. Addition of antigen (see section 2 for preparation of antigen)
   (1) Using a 0.025 ml dropper, add 1 drop of antigen to each dilution in the first column of each serum sample.
   (2) Repeat for the remaining antigens in their corresponding columns.
   (3) Gently shake the plates to mix contents, cover with plastic lid to exclude debris, and incubate at room temperature for two hours.

C. Reading of test
   The plate is placed on the stage of a darkfield microscope equipped with a long-working-distance 10-power objective (No. 519-438 or 559-003, E. Leitz, Inc., Rockleigh, New Jersey; or No. AO 1019 or AO 1076, American Optical Corp., Buffalo, New York) and 10-power eyepieces, and the wells are examined for agglutination (Fig. 5-3). A 3.5-power objective with 15-power eyepieces may be used; however, at this magnification, only agglutination will be observed.

D. Interpretation of test results
   Interpretation of the test is the same as described for the plastic trays (section 3E).

E. Screening test
   If a large number of sera are to be tested, it may be desirable to screen them first at the 1:50 dilution to eliminate the negatives. When

---

*Standard microtiter plates are labeled top to bottom, rows A through H.

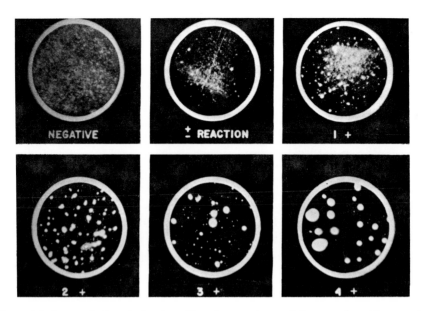

Figure 5-3. Leptospiral agglutination: Negative reaction and degrees of positive reaction.

screening, 0.025 ml of the 1:25 dilution is placed in each well of a column, e.g. serum 1 in column 1 and so on. The antigens are added to the rows, e.g. *pomona* in row A, *hardjo* in row B, and so on. This procedure allows twelve sera to be screened against eight antigens on one plate. Sera that are positive against one or more serovars are then titered only against the serovars to which they are positive.

F. Transfer plate system

A modification of the microtechnique using the transfer plate system (Cooke Laboratory Products, Alexandria, Virginia) for direct reading on microscope slides using a 10-power objective, 10-power eyepieces, and dry, darkfield condenser has been described (11).

5. Maintenance of cultures

Leptospires are sensitive to hydrogen peroxide; therefore, rabbit serum containing catalase from hemolyzed erythrocytes must be used in the culture medium.

Stock cultures in EMJH, Fletcher, or OAC semisolid medium should be maintained, since leptospiral transfers occasionally do not grow.

TREPONEMA

Treponema hyodysenteriae is the name that has been proposed for the large spirochete responsible for swine dysentery. This disease has been recognized since 1921 and was initially thought to be caused by *Vibrio coli*.

However, in the early 1970s, workers in the United States and Great Britain demonstrated that *T. hyodysenteriae* was the primary etiologic agent.

Pathogenicity

Swine dysentery is usually observed in 15–70 kg pigs but may affect suckling as well as adult swine. There is marked catarrhal hemorrhagic enteritis, which is confined to the large intestine. Death may occur, but high morbidity leading to poor weight gain is the usual finding.

LABORATORY EXAMINATIONS

Direct Examination

Treponema hyodysenteriae can be observed in the mucosal lesions of the large intestine by darkfield or phase-contrast microscopy using the following procedure (12).

1. Rinse or lightly scrape a portion of the affected mucosa to remove debris.
2. Suspend a portion of a deep scraping from the mucosa in a drop of saline or water on a microscope slide.
3. Examine by darkfield or phase-contrast microscopy at a magnification of 400 to 1000 power.
4. Observe three to five spirochetes per high-power field.

It is important to differentiate *T. hyodysenteriae*, which is 7–8 µm long, loosely coiled, motile by flexing movements, and tapered at the ends, from the smaller, tightly coiled spirochetes normally found in swine. Mucosal or fecal smears may be stained with crystal violet, carbol fuchsin, or Victoria blue 4-R stains, although wet mount preparations to observe motility are preferred. These spirochetes may be observed in histologic sections of the colonic mucosa stained with the Warthin-Starry, Goodpasture's, or Victoria blue 4-R stains (Fig. 5-4).

Rectal swabs or feces may be utilized for examination from a live animal, but numbers of organisms in the feces may be small and the spirochetes not detectable.

Isolation Procedures

Treponema hyodysenteriae can be isolated from the intestinal mucosa using the following procedure (12, 13).

1. Take six to eight inches of spiral colon from an acutely affected animal. The sample may be held for two to three days at 4°C before isolation attempts. Do not freeze.
2. Open the colon longitudinally and remove the mucosa with a sterile microscope slide.

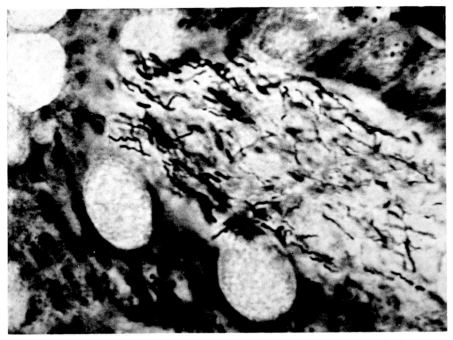

Figure 5-4. *Treponema hyodysenteriae* in a section of swine intestine. Warthin-Starry stain, × ca. 1000.

3. Prepare a 1:10 suspension of the colonic mucosa in saline.
4. Centrifuge slowly for ten minutes to remove the large particulate matter.
5. Pass the supernatant through a series of cellulose acetate filters: prefilter, 8.0 μm, 5.0 μm, 3.0 μm, 1.2 μm, 0.8 μm, 0.65 μm, and 0.45 μm.
6. Streak material from the filtrate that passed the 0.8 μm, 0.65 μm, and 0.45 μm filters onto freshly prepared or prereduced trypticase soy agar containing 5% defibrinated bovine or equine blood. The addition of 400 μg/ml of spectinomycin (The Upjohn Company, Kalamazoo, Michigan) to the medium suppresses most of the contaminating flora and does not adversely affect the isolation of *T. hyodysenteriae*.
7. Incubate the plates at 42°C in an anaerobic container. A vented Gas-Pak® jar with cold palladium catalyst (BBL, Cockeysville, Maryland) may be used to obtain a hydrogen and carbon dioxide atmosphere of 80:20 H_2-CO_2 by evacuation and refilling. A H_2-CO_2 generator envelope (GasPak—BBL) is also acceptable.

Cultural Characteristics and Identification

Growth of *T. hyodysenteriae* on blood agar is evidenced by a zone of clear (beta) hemolysis, which may contain small, white, translucent colonies. This is in contrast to *Treponema innocens*, which is weakly beta-hemolytic and considered nonpathogenic (14). *T. hyodysenteriae* is Gram-negative but is more readily observed using the stains listed under Direct Examination. *T. hyodysenteriae* is 6–8.5 µm long, 0.32–0.38 µm in diameter, loosely coiled, motile, cytochrome-oxidase negative, catalase negative, stimulated by hydrogen, and anaerobic.

Treponema paraluis—cuniculi is pathogenic for animals and the cause of rabbit syphilis. Diagnosis is based on lesions around the genitalia and demonstration of the organisms in these lesions by staining or darkfield microscopy.

Treponema suis has been observed in washings of ulcerated preputial diverticula in pigs (15).

Treponema succinifaciens, a small anaerobic spirochete, has been isolated from the colon of a pig and is considered nonpathogenic (14).

Serologic Procedures

Tests that have been adapted for use in the diagnosis of swine dysentery or for detection of carrier animals are the tray agglutination test (16), microtitration agglutination test (17), and enzyme-linked immunosorbent assay (ELISA) (18). The ELISA procedure appears to be the most sensitive of these tests, and it may be useful for detecting individually infected animals by diagnostic laboratories that have personnel and equipment necessary to perform this test.

BORRELIA

Borrelia anserina causes fowl spirochetosis, a highly fatal disease in chickens, turkeys, geese, and other fowl. The disease is characterized by acute septicemia with concomitant fever, diarrhea, listlessness, and emaciation. Anemia is commonly present. The spleen is usually enlarged and mottled.

The spirochete is transmitted primarily by fowl ticks. Other arthropods and feces of infected birds may also transmit *B. anserina*.

The disease is diagnosed by demonstration of the spirochetes, which are 3–20 µm in length, in blood, spleen, and liver smears stained by Giemsa's slow method (19). *B. anserina* can be isolated from the blood and grown in six– to twelve-day-old embryonated chicken or turkey eggs (20, 21) (Fig. 5-5).

Borrelia hyos has been observed in the blood of swine but has not been identified with a specific disease. A *Borrelia* species, possibly *B. hyos*, has been demonstrated in the large intestine of pigs with swine dysentery (22).

A spirochete, which has been tentatively classified as *B. suilla*, causes ulcerative granuloma in wounds of pigs and can be observed in sections by silver impregnation techniques (23).

The association of spirochetes with intestinal disorders of dogs has been observed by several workers. The organisms, which probably represent

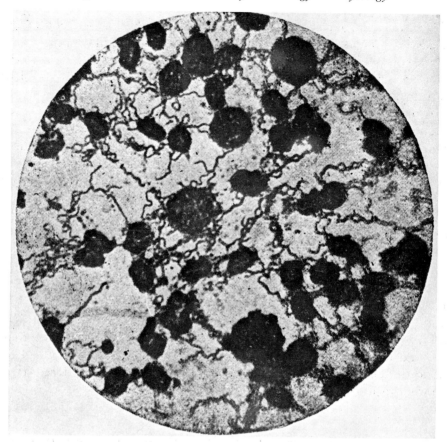

Figure 5-5. *Borrelia anserina:* Smear of chicken embryo blood, ×1300. From S. B. V. Rao, B. M. Thrakal, and M. R. Dhanda, *Indian Vet J, 31*:1, 1954.

different species although they have not been well characterized, have been given the tentative names *Borrelia canis, Spirillum eurygyrata,* and *Spirillum minutum.* Their occurrence and significance have been discussed by Craige (24).

Borrelia vincentii can be found in the mouth of normal dogs. In combination with *Fusobacterium fusiforme, B. vincentii* is thought to cause ulceromembranous stomatitis in debilitated dogs (25).

REFERENCES

1. Ellinghausen, H. C., Jr., and McCullough, W. G.: *Am J Vet Res, 26*:45, 1965.
2. Nervig, R. M., and Ellinghausen, H. C., Jr.: *Proc Am Assoc Vet Lab Diag, 19*:57, 1976.
3. Waltman, W. D., Dawe, D. C., and Shotts, E. B.: *Proc Internatl Pig Vet Soc*:200, 1982.
4. Adler, B., Murphy, A. M., Locarnini, S. A., and Faine, S.: *J Clin Microbiol, 11*:452, 1980.
5. Adler, B., Faine, S., and Gordon, L. M.: *Austral Vet J, 57*:414, 1981.

6. Sulzer, C. R., and Jones, W. C.: *Health Lab Sci, 10*:13, 1973.

7. Cole, J. R., Jr., Ellinghausen, H. C., and Rubin, H. L.: *Proc US Animal Health Assn, 83*:189, 1979

8. Tripathy, D. N., Hanson, L. E., and Mansfield, M. E.: *Proc US Animal Health Assn, 83*:180, 1979.

9. Cole, J. R., Jr., Sulzer, C. R., and Pursell, A. R.: *Appl Microbiol, 25*:976, 1973.

10. Galton, M. M., Sulzer, C. R., Santa Rosa, C. A., and Fields, M. J.: *Appl Microbiol, 13*:81, 1965.

11. Carter, P. L., and Ryan, T. J.: *J Clin Microbiol, 2*:474, 1975.

12. Harris, D. L., and Glock, R. D.: Swine Dysentery. In Dunne, H. W., and Leman, A. D. (Eds.): *Diseases of Swine,* 4th ed. Ames, Iowa St U Pr, 1975, Chapter 28.

13. Songer, J. G., Kinyon, J. M., and Harris, D. L.: *J Clin Microbiol, 4*:57, 1976.

14. Harris, D. L., and Glock, R. D.: Swine dysentery. In Leman, A. D., Glock, R. D., Mengeling, W. L., Penny, R. H. C., Scholl, E., and Straw, B. (Eds.): *Diseases of Swine,* 5th ed. Ames, Iowa St U Pr, 1981, Chapter 41.

15. Kujumgiev, I., and Spassova, N.: *Zentralbl Veterinaermed* (B), *13*:357, 1967.

16. Hunter, D., and Saunders, C. N.: *Vet Rec, 93*:107, 1973.

17. Joens, L. A., Harris, D. L., Kinyon, J. M., and Kaeberle, M. L.: *J Clin Microbiol, 8*:293, 1978.

18. Joens, L. A., Nord, N. A., Kinyon, J. M., and Egan, I. T.: *J Clin Microbiol, 15*:249, 1982.

19. Kelly, R. T.: Borrelia. In Lennette, E. H., Spaulding, E. H., and Truant, J. P. (Eds.): *Manual of Clinical Microbiology,* 2nd ed. Washington, D.C., American Society for Microbiology, 1974. Chapter 35.

20. Knowles, R., Das Gupta, B. M., and Basu, B. C.: *Indian J Med Res, 22*:1, 1932.

21. McKercher, D. G.: *J Bacteriol, 59*:446, 1950.

22. Todd, J. N., Hunter, D., and Clark, A.: *Vet Rec, 86*:228, 1970.

23. Jubb, K. V. F., and Kennedy, P. C.: *Pathology of Domestic Animals,* 2nd ed. New York, Academic Press, 1970, vol. 2, p. 617.

24. Craige, J. E.: In Hoskins, H. P., Lacroix, J. V., Mayer, K., Bone, J. R., and Golick, P. F. (Eds.): *Canine Medicine,* 2nd ed. Santa Barbara, California, Am Vet Publications, 1959, Chapter 4.

25. Chester, D. K.: Spirochetal diseases. In Catcott, E. J. (Ed.): *Canine Medicine,* 4th ed. Santa Barbara, California, Am Vet Publications, 1979, Chapter 2.

SUPPLEMENTARY REFERENCES

Alexander, A. D.: Leptospira. In Lennette, E. H. (Ed-in-chief): *Manual of Clinical Microbiology,* 3rd ed. Washington, American Society for Microbiology, 1980, Chapter 32.

Alexander, A. D.: Serological diagnosis of leptospirosis. In Rose, N. R., and Friedman, H. (Eds.): *Manual of Clinical Immunology,* 2nd ed. Washington, American Society for Microbiology, 1980, Chapter 75.

Alston, J. M., and Broom, J. C.: *Leptospirosis in Man and Animals.* Edinburgh, E. and S. Livingstone, 1958.

Burgdorfer, W.: Borrelia. In Lennette, E. H. (Ed-in-chief): *Manual of Clinical Microbiology,* 3rd ed., Washington, American Society for Microbiology, 1980, Chapter 33.

Ellinghausen, H. C., Jr., and Top, F. H., Sr.: Leptospirosis. In Top, F. H., Sr., and Wehrle, P. F. (Eds.): *Communicable and Infectious Diseases,* 8th ed. St. Louis, Mosby, 1976, Chapter 40.

Ellinghausen, H. C., Jr., Thiermann, A. B., and Sulzer, C. R.: Leptospirosis. In Balows, A., and Hausler, W. J., Jr. (Eds.): *Diagnostic Procedures for Bacterial, Mycotic and Parasitic Infections,* 6th ed. Washington, American Public Health Association, 1981, Chapter 30.

Galton, M. M., Menges, R. W., Shotts, E. B., Nahmias, A. J., and Heath, C. W.: *Leptospirosis: Epidemiology, Clinical Manifestations in Man and Animals, and Methods in Laboratory Diagnosis*. Washington, D.C., Public Health Service Publ. No. 951, 1962.

Johnson, R. C. (Ed.): *The Biology of Parasitic Spirochetes*. New York, Academic Press, 1976.

Johnson, R. C.: Aerobic Spirochetes: The genus *Leptospira*. In Starr, M. P., Stolp, H., Truper, H. G., Balows, A., and Schlegel, H. G. (Eds.): *The Prokaryotes: A Handbook on Habitats, Isolation, and Identification of Bacteria*. New York, Springer-Verlag, 1981, Chapter 51.

Roth, E., Adams, W. V., Greer, B., Sanford, G. E., Newman, K., and Moore, M.: Comments on the laboratory diagnosis of leptospirosis in domestic animals with an outline of some procedures. *Proc US Livest Sanit Assoc*, 67:520, 1961.

Solorzano, R. F.: A comparison of the rapid slide agglutination test for leptospirosis. *Cornell Vet*, 57:239, 1967.

Sulzer, C. R., and Jones, W. L.: *Leptospirosis—Methods in Laboratory Diagnosis*. Washington, D.C., U.S. Department of Health, Education and Welfare No. (CDC) 76-8275, 1962.

White, F. H., Sulzer, K. R., and Engel, R. W.: Isolations of *Leptospira interrogans* serovars *hardjo, balcanica*, and *pomona* from cattle at slaughter. *Am J Vet Res*, 43:1172, 1982.

World Health Organization: Current problems in leptospirosis. *Report of WHO Expert Group Tech Rep Ser No. 380*, 1967.

Chapter 6

CAMPYLOBACTER

J. F. PRESCOTT

Organisms in this genus are thin (0.2–0.8 μm × 0.5–5 μm), Gram-negative, motile, curved rods. The cells are often S-shaped or seagull shaped, but are occasionally long (8 μm) spiral rods. They are vigorously motile by a single polar flagellum. Microaerophilic (3–15% O_2) conditions are generally required for growth, and optimal conditions are 6% O_2. Some species, however, (e.g. *C. sputorum* subsp. *mucosalis*) are anaerobic, while others (e.g. aerotolerant *Campylobacter*) are aerobic. All species are cytochrome oxidase positive.

The genus contains species causing important genital and intestinal infections of animals, as well as saprophytic species. The recent interest in *C. jejuni* as a zoonosis in humans has resulted in improved classification of the genus, for long generally neglected in veterinary bacteriology. At one time *Campylobacter* were classified with *Vibrio*, but the former are nonoxidizers and the latter fermenters, and for this and other reasons the two genera are recognized as distinct.

CAMPYLOBACTER FETUS

Classification and Pathogenicity

The classification of *C. fetus* is distinguished by confusion, summarized in Table 6-I. The taxonomy recommended in the *Approved List of Bacterial Names* (1) and used here is that *C. fetus* is divided into two subspecies, *C. fetus* subsp. *venerealis* and *C. fetus* subsp. *fetus*. The former organism is found in the prepuce of carrier bulls and the genital tract of infected cows and is an important cause of infectious infertility and sporadically of abortion. A biotype of *C. fetus* subsp. *venerealis*, biotype *intermedius*, is recognized (2). *Campylobacter fetus* subsp. *fetus* is found in the intestine of cattle, sheep, and humans. In cattle and sheep, it is isolated sporadically from cases of abortion and in humans from cases of septicaemia in immunosuppressed people.

The distinction between the subspecies on the basis of pathogenic potential may be somewhat arbitrary since there are a few reports of the ability of some *C. fetus* subsp. *fetus* strains to cause localized genital tract infection in cows (3). The differentiation between the subspecies on biochemical grounds

Table 6-I

Comparison of Some Classification Schemes for the Catalase-Positive Campylobacter Species

ICSB* (Modern)	Smibert†	Florent‡	Veron and Chatelaine§	King//
C. fetus ss. *venerealis*	*C. fetus* ss. *fetus*	*Vibrio fetus* var. *venerealis*	*C. fetus* ss. *venerealis*	*Vibrio fetus*
C. fetus ss. *fetus*	*C. fetus* ss. *intestinalis*	*Vibrio fetus* var. *intestinalis*	*C. fetus* ss. *fetus*	*Vibrio fetus*
C. jejuni	*C. fetus* ss. *jejuni*	—	*C. jejuni*	'Related Vibrios'
C. coli	*C. fetus* ss. *jejuni*	—	*C. coli*	'Related Vibrios'

*International Journal of Systematic Bacteriology, 30:270, 1980.
†Bergey's Manual, 8th ed., 1974.
‡Cited in Annual Review of Microbiology, 32:673, 1978.
§International Journal of Systematic Bacteriology, 23:122, 1973.
//Journal of Infectious Disease, 101:119, 1957.

is not precise, and there may be variants of *C. fetus* subsp. *venerealis* that are glycine resistant (4).

DIAGNOSIS OF *CAMPYLOBACTER FETUS* SUBSP. *VENEREALIS* INFECTION

A definitive diagnosis of genital campylobacteriosis can be difficult. In the past, diagnosis has been made on the basis of cervical mucus agglutination, on fluorescent antibody staining, and on culture. The introduction of a selective transport medium has improved diagnosis in carrier bulls.

Direct Examination

Campylobacter can be demonstrated in Gram-stained smears of fetal stomach contents.

Fluorescent antibody procedures have been widely employed for the identification of *C. fetus* in preputial washings, cervical mucus, and fetal stomach contents. Smears are prepared from the sediment of centrifuged preputial washings and from other materials in the conventional manner. Filtration of cervicovaginal mucus through a cellulose acetate disc (5 μm pore size) improves the quality of the smears and contributes to greater accuracy in the FA test (5). The conjugate does not distinguish between subsp. *venerealis* and subsp. *fetus*. The FA reagent is not available commercially but is available through the courtesy of some laboratories. The FA procedure has been found to be highly effective, probably because dead as well as living campylobacter are stained, and distinguishes *C. fetus* from the nonpathogenic *C. sputorum* subsp. *bubulus* sometimes present in preputial washings of bulls.

Collection of Cervical Mucus for Mucus Agglutination Test

The test is convenient and accurate if animals are tested between two and seven months after infection and are not in estrus, and if no blood is present. On a herd basis the test is useful if ten animals in the above category are tested, or twenty if no selection is possible.

A number of methods have been described. The following simple procedure is given by Laing and associates (6). The mucus is collected by means of a glass tube about 50 cm in length and 1 cm in diameter. This pipette has a slight bend about 10 cm from one end, which is lightly plugged with cotton wool at one end; the other end is lightly plugged with cotton wool to act as a stop for the mucus. Before use, the pipettes are wrapped in greaseproof paper and sterilized by autoclaving.

A piece of pressure rubber tubing about 50 cm long is attached to the straight end of the pipette, the plug at the bent end is removed, and this end is passed into the vagina as far as the cervix. Then with a rubber bulb at the free end of the rubber tubing and by moving the pipette backward and forward in the vagina, a portion of mucus is loosened and drawn into the pipette. The suction is maintained as the pipette is withdrawn from the vagina; a small, sterile rubber stopper is inserted into the pipette and a label with the identity of the animal is attached. In the laboratory, the mucus is forced from the pipette into a test tube by applying pressure on the cotton wool plug at the straight end of the pipette with a length of flexible wire.

Hoerlein and Kramer (7) describe the collection of cervical mucus using sterile artificial insemination pipettes (large bore 1.5 ml) introduced through a twelve inch long (8 mm diameter) Pyrex® speculum. The mucus is drawn into the pipette by means of a small syringe attached by a short length of rubber tubing. If mucus cannot be cultured within four hours after collection, it is recommended that it be frozen on dry ice while in transit to the laboratory.

Collection of Preputial Washings

Bartlett and others (8) suggest the following method. A plastic or Perspex® glass pipette of approximately 8 mm outside diameter and 1.5 mm thick wall is used. It should be about twenty-one inches long with a 15 degree bend in the tube three inches from the end, to where a two ounce rubber bulb is attached. The other end is shaped so that a blunt chisel edge is formed, the bevel being on the same side as the reflex angle located near the opposite end of the tube (3).

Sterile pipettes should be used for each bull. Samples are obtained by inserting the pipette into the prepuce and surrounding preputial membrane with the chisel edge, and the rubber bulb is used to suck into the pipette the approximately 1 ml of smegma that can be obtained. The pipette is flushed gently with 4 ml of sterile saline on withdrawal from the prepuce.

PREPARATION OF TRANSPORT MEDIUM. The transport medium of Clark and Dufty (9) allows excellent recovery of *C. fetus* and is convenient because it can be transported in air at room temperature for two to three days without loss of organisms. The medium consists of freshly prepared bovine serum with 300 µg 5-fluorouracil, 100 units polymixin B sulfate, 50 µg brilliant green, 3 µg of nalidixic acid, and 100 µg of cycloheximide per ml. It is dispensed in 10 ml quantities in widemouth vaccine bottles (30 ml capacity), and the rubber stoppers are inserted. The bottles are placed for two minutes in a boiling water bath so that the serum solidifies. After cooling, the medium is dispersed by stirring with a sterile glass rod. An 18-gauge needle is inserted through the firmly fixed rubber stoppers, and the vials are placed in an anaerobe jar, the air evacuated and replaced with a mixture of 2.5% O_2, 10% CO_2, and the balance N_2. On opening the jar the needles are immediately removed and the vials then stored at 4°C for up to three months. The vials should be stored for one week at 4°C before use; during this time the medium becomes dark green in color.

INOCULATION OF PREPUTIAL FLUID INTO TRANSPORT MEDIUM. Preputial fluids are washed into 4 ml of physiological saline. The sample is allowed to stand for fifteen to twenty minutes to allow epithelial cells to settle, and then 1 ml of supernatant liquid is withdrawn by sterile syringe and injected through the rubber stopper of the vial. Duplicate vials are recommended. The sample is mixed into the medium by shaking. The vial is then transported to the laboratory at temperatures of 18–37°C (*not* cooled). At the laboratory, the container is incubated at 37°C for four days. Then 2–3 ml of physiological saline is mixed thoroughly with the sample and all fluid removed by pipette to a sterile tube. The fluid is then examined for *C. fetus* by culture and immunofluorescence. For culture, the fluid is filtered through a 0.65 µm Millipore® filter as described below.

Isolation Procedures

Filtration is recommended to reduce contaminating bacteria. After prefiltration to remove debris, preputial fluid, cervical mucus, or enrichment broth is filtered through a Millipore filter, pore size 0.65 µm. A 0.1 ml volume is distributed over the surface of solid media using a swab.

Isolation media should be rich. Examples are brucella, cystine heart, serum dextrose, Columbia or brain-heart infusion agar, with 5–10% blood. The basal media are available as dry powders from suppliers such as BBL, Difco, or Oxoid (except serum dextrose agar). A plate of both nonselective and selective medium should be inoculated with each sample. Media is made selective by the addition of polymixin B sulfate (2 units per ml), novobiocin (2 μg per ml), and cycloheximide (20 μg per ml).

Plates are incubated at 37°C for four to six days under a microaerophilic atmosphere of 6% O_2, 10% Co_2, and 84% N_2. This atmosphere can be obtained ready-mixed from gas suppliers. An alternative is the use of CampyPak II (BBL) or a GasPak (BBL) H_2-CO_2 generator envelope in an anaerobic jar *without* the catalyst. Least preferable is the use of a candle jar. 10% CO_2 in air is not satisfactory; the O_2 content must be reduced. Other methods of producing a microaerophilic environment are described under *C. jejuni*.

Identification

Campylobacter fetus produces fine, nonhemolytic, round, 1 mm, slightly raised, smooth, translucent colonies. A few colonies may be rough or granular. Colonies contain Gram-negative, motile, curved rods (Fig. 6-1). Some cultures may show long filaments. *Campylobacter fetus* can be readily distinguished from *C. sputorum* subsp. *bubulus*, a saprophyte, by FA staining of smears from colonies or by biochemical tests (Table 6-II). Tests for growth in glycine and for growth temperature effect should be done in thioglycollate medium and for H_2S production by the lead acetate strip method after heavily inoculating semisolid (0.16% agar) brain-heart infusion or brucella broth each supplemented with 0.02% cysteine, and incubated for four to six days. The *intermedius* biotype of subsp. *venerealis* gives a positive test for H_2S in this medium.

Karmali and others (10) have reported distinguishing the two subspecies of *C. fetus* (and also *C. jejuni*) by measurement of cell size and the wavelength and amplitude of the spirals. A trained observer can distinguish the three organisms by size.

Mucus Agglutination Test

For the vaginal mucus agglutination test, 0.2–0.3 g of mucus are extracted overnight with seven times its volume of phenolized (0.37%) physiological saline. A four-tube final dilution series (1/32–1/256) is recommended. The antigen and the test are prepared according to the method described by Moynihan and Stovell (11). A serum agglutination test for the diagnosis of genital campylobacteriosis has been described but is unreliable.

CAMPYLOBACTER JEJUNI

Campylobacter jejuni is a common and important cause of acute diarrhea in people in developed countries, and an important zoonosis. Cases of diarrhea in humans have been reported to be acquired, directly or indirectly, from all

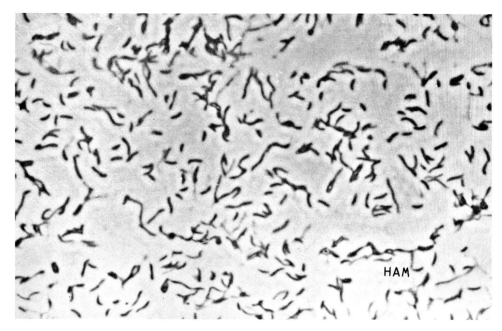

Figure 6-1. *Campylobacter fetus* subsp. *fetus* from a blood agar plate culture. Dark-phase illumination, × 2440 (H. A. McAllister).

common domestic animals. In animals the role of *C. jejuni* as a cause of diarrhea is not as well established as in humans, but it is clear that the organism will cause enteritis in cattle and dogs, particularly in young animals, and it has been suggested as a cause of diarrhea in kittens, foals, lambs, and pigs. Apart from enteric disease, the organism is a cause of outbreaks of abortion in sheep and also of "vibrionic hepatitis" of laying chickens, a disease now rarely diagnosed. The organism can be isolated from feces or intestinal tracts of many different species, particularly birds. Its presence in apparently healthy animals makes a diagnosis of *C. jejuni* enteritis on cultural grounds alone difficult. The situation is thus similar to the carriage of *Salmonella* by healthy animals.

Direct Examination

Diluted fecal smears may be examined by darkfield or phase-contrast microscopy for large numbers of characteristically darting, motile, corkscrewed organisms.

In aborted sheep fetuses, the organism can be shown in Gram-stained smears of stomach contents as a slender curved rod, usually present in large numbers. In chickens with "vibrionic hepatitis" the organism can be demon-

Table 6-II

Differentiation of Campylobacter species

Species	Principal Host	Catalase	H₂S, lead acetate strip*	H₂S in TSI	Nalidixic acid sens.†	Cephalothin sens.†	Growth in: 1% glycine	Growth in: 3.5% NaCl	Growth in: 1% bile	Hippurate hydrolysis‡	Growth at: 25°C	Growth at: 42°C	Nitr. Red'n
C. fetus subsp. *venerealis*	Cattle	+	–	–	R		–	–	+	–	+	–	–
C. fetus subsp. *fetus* Z060-000	Cattle, sheep, man	+	+	–	R	S	+	–	+	–	+	–	–
C. jejuni	Universal	+	+	–	S//	R	+	–	+	+	–	+	–
C. coli	Pigs	+	+	–	S	R	+	–	+	–	–	+	–
C. fecalis	Cattle, sheep	+	+	+			+	V	V		–	+	
C. hyointestinalis	Pigs	+	+	+	R	S	+	–		–	+	–	–
C. sputorum subsp. *sputorum*	Man	–	+	+			+	–	+		+	–	+
C. sputorum subsp. *bubulus*	Cattle, sheep	–	+	+			+	+	–		+	V	+
C. sputorum subsp. *mucosalis*	Pigs	–	+				–	–	–				+
Aerotolerant *Campylobacter*	? Universal	+	–	–	S		V	–			+	–	

*Albimi brucella broth (Pfizer) with 0.02% cysteine
†30 µg disc
‡See text
§Biotype *intermedius* positive, see text
//Some exceptions

strated in the bile by Gram staining, or by phase-contrast microscopy after diluting bile 1:1 with saline.

Isolation Procedures

Rectal swabs or feces are suitable for culture and should be stored at 4°C before plating. If feces are not cultured soon, then storage of swabs in Cary Blair medium at 4°C is recommended (12).

The medium is made by adding 1.5 g sodium thioglycollate, 1.1 g Na_2HPO_4, 5 g NaCl, and 1.6 g agar to 991 ml of water. Heat until the agar is dissolved, cool to 50°C, add 9 ml of a fresh solution of 1% $CaCl_2$. Adjust to pH 8.4 and steam heat for fifteen minutes. Store at 4°C.

Fecal specimens should be inoculated onto selective plate media as soon as possible; selective media is not, however, necessary for isolation of *C. jejuni* from ovine fetuses or placenta, or for "vibrionic hepatitis."

Several selective media have been described (12), but Blaser's Campy-BAP medium is preferred. It is available commercially. The medium contains brucella agar base (BBL), 5% sheep blood, and the following antibiotics: vancomycin (10 µg per ml), polymixin B sulfate (2–5 units per ml), trimethoprim lactate (5 µg per ml), cephalothin (15 µg per ml), amphotericin B (2 µg per ml).

Plates should be incubated in evacuatable anaerobe jars under an atmosphere of approximately 6% O_2, 10% CO_2, and 84% N_2. Such gas mixtures are available commercially. The jars should be evacuated and then filled with the gas mixture on three occasions. Alternative means of producing suitable microaerophilic conditions are the use of a GasPak or similar type of anaerobe jar with the specifically designed CampyPak II (BBL) envelopes. GasPak envelopes or Oxoid gas-generating envelopes can be used in anaerobe jars (*without* the palladium catalyst), but there is a danger of buildup of potentially explosive hydrogen in the jars. Candle jars are least preferable, but if they are to be used they must be incubated at 42°C.

Suitable conditions can be gained by incubating the selective plate with a second blood agar plate streaked with a pure culture of a facultative anaerobe, such as *Proteus*. If the two plates are incubated in an airtight plastic bag, the facultative anaerobe will lower oxygen concentrations and produce a suitable environment (13). *Campylobacter jejuni* grows best at 42°C and this temperature should be used; 37°C is suboptimal. Plates should be incubated for forty-eight hours and then examined. They need not be incubated further.

Colonies are flat, nonhemolytic, watery, gray (sometimes with a pink tinge), spreading, and often large. At times they appear like drops of water, spreading along the streak marks. If plates are not moist, colonies do not spread, but rather appear discrete (2 mm), convex, raised, round, and mucoid.

Identification Procedures

Campylobacter jejuni appear as slender, curved-to-spiral Gram-negative rods, 0.2–0.5 × 1.5–5 µm. Colonies older than forty-eight hours or exposed to air for hours show coccoid transformation, although some typical corkscrew *Campylobacter* will still be present. *Campylobacter jejuni* is cytochrome oxidase and catalase positive and shows motility if saline hanging drops are made of plate cultures and examined by darkfield or phase-contrast microscopy. Criteria necessary to confirm identity as *C. jejuni* (Table 6-II) are growth at 42°C but not 25°C, sensitivity to nalidixic acid but resistance to cephalothin, and positive hippurate hydrolysis test. Cultures grow better at 42°C than at 37°C.

The rapid hippurate hydrolysis test distinguishes *C. jejuni* from *C. coli* (14). For this test a 1% aqueous solution of sodium hippurate is prepared and distributed into glass tubes in 0.4 ml amounts, capped and frozen at −15°C or lower until use. After thawing, a large loopful of organisms from a twenty-four– to forty-eight– hour culture is emulsified in the substrate and incubated for two hours at 37°C. Then, 0.2 ml of a ninhydrin solution is added and tubes are observed after ten minutes at 37°C for development of a deep purple color. Ninhydrin solution is prepared by adding 3.5 g of ninhydrin in 100 ml of a 1:1 mixture of acetone and butanol.

Biotypes of *C. jejuni* have been described, but their usefulness has not yet been determined. For epidemiological purposes, *C. jejuni* and *C. coli* can be typed antigenically. The methods of Penner with heat-stable antigens (15) and of Lior with heat-labile antigens (16), when available on a large scale, offer great promise.

In view of the presence of *C. jejuni* in the intestines of many healthy animals, their recovery from diarrheic animals may not be convincing evidence of their role in disease. If required, demonstration of a rising agglutinating antibody titre (fourfold increase) would confirm their involvement (17).

CAMPYLOBACTER SPUTORUM SUBSP. MUCOSALIS

This organism has been isolated from the lesions of porcine intestinal adenomatosis and its manifestations. The other manifestations of the infection are proliferative hemorrhagic enteropathy, necrotic ileitis, and regional ileitis. In these infections, large numbers of very slender *Campylobacter* can be seen on silver stains within the apical cytoplasm of the proliferating epithelial cells. Failure to reproduce the disease with pure cultures of the organism and the subsequent recovery of another *Campylobacter* species from the lesions have led some to doubt its etiological importance. The characteristics of *Campylobacter hyointestinalis*, which is thought to be the agent of proliferative adenomatosis (18), are given in Table 6-II.

In the current uncertainty as to the etiological role of *C. sputorum* subsp. *mucosalis*, diagnosis of intestinal adenomatosis would most certainly be made by histopathological examination and use of Warthin-Starry silver staining, or by direct examination.

Direct Examination: Modified Acid-fast Stain

Heat-fixed smears are made from mucosal scrapings of affected areas of intestine. Smears are stained with 5% freshly diluted carbol fuchsin for five minutes, decolourized with 0.5% acetic acid for thirty seconds, and counter-stained with 1% methylene blue for ten seconds. Examination shows large numbers of bright pink-staining, *very* slender curved rods, predominantly intracellularly (19). Some *Campylobacter* will be seen unassociated with cells, but the diagnosis should only be made on the presence of intracellular tangled clumps of organisms.

Isolation Procedures

Mucosal surfaces of affected intestine are washed repeatedly in sterile saline to remove extracellular organisms. The mucosa is then scraped with a sterile scalpel, and cells in the scrapings are washed five times in twenty volumes of nutrient broth, centrifuging at 1000 G for ten minutes between washings. After the final washing, the cells are suspended in ten volumes of nutrient broth and homogenized in a glass Ten Broeck homogenizer. Homogenates are plated onto Columbia blood agar, with and without 1:60,000 brilliant green as a selective agent. Plates are incubated at 37°C for two to five days anaerobically, which gives better isolation results than microaerophilic incubation. Once the organism has been recovered, it will grow readily microaerophilically.

Identification Procedures

Colonies are about 1.5 mm at forty-eight hours and are circular and raised, with a flat surface and shiny gray appearance. Scrapings on the bacteriological loop appear dirty yellow; *C. coli* shows a pink tan pigment. They are cytochrome-oxidase positive, although the test may take sixty seconds to be positive. In young cultures, strains are Gram-negative, short, irregularly curved rods, 0.25–0.30 × 0.95–2.8 µm. Older cultures showed more coccoid and filamentous forms. Differentiation of the catalase-negative *Campylobacter* is shown in Table 6-II. Slide agglutination using antisera prepared in rabbits against one of the three serological types will allow rapid diagnosis; type A is the most frequently encountered serological type (20).

CAMPYLOBACTER SPUTORUM SUBSP. BUBULUS

A nonpathogenic saprophyte of the bovine genital tract must be distinguished from *C. fetus*.

CAMPYLOBACTER COLI

Campylobacter coli is occasionally a cause of *Campylobacter* enteritis in people. It is found commonly in the intestines of pigs, but its role in disease is not clear. At one time it was thought to be the cause of swine dysentery, now known to be caused by *Treponema hyodysenteriae*.

Isolation and identification are as described for *C. jejuni*, with the exception that *C. coli* is hippurate negative (Table 6-II).

CAMPYLOBACTER FECALIS

Campylobacter fecalis has been isolated from ovine feces and is regarded as a nonpathogen, but it may be found as a contaminant in bovine semen and vaginas. It has been reported to cause enteritis in cattle (17). They differ from *C. jejuni* or *C. coli* in production of large quantities of H_2S in TSI medium and variable resistance to 3.5% NaCl, as well as their ability to reduce nitrate to nitrite (21).

AEROTOLERANT CAMPYLOBACTER

Ellis, Neill, and O'Brien and others have isolated *Campylobacter* other than *C. fetus* subspecies from aborted bovine and porcine fetuses (22), and more recently from bovine mastitis. It seems likely that these organisms are a cause of abortion in cattle and pigs, and possibly sheep and horses (23). They are isolated microaerophilically but grow readily in air on subculture. In the current absence of a species name they will be called "aerotolerant Campylobacter."

Isolation procedures

The method described is that of Neill (23). Leptospira EMJH medium with 100 µg/ml of 5-fluorouracil with added 1% rabbit serum and 0.15% agar is used as an initial enrichment medium; 0.25 ml fluids (e.g. fetal fluids, preputial or vaginal washings) are inoculated directly into 2 ml of this medium. Fetal tissues or placentas can be homogenized (10% w/v) in quarter strength Ringer's solution and 0.25 ml of suspension inoculated. The inoculum is placed in the bottom part of the enrichment broth. This is done gently to avoid undue aeration of the medium.

The medium is incubated in air at 30°C, and growth of microaerophilic organisms is detected as a distinct zone beneath the surface. A wet mount is prepared after forty-eight hours and examined by darkfield or phase-contrast microscopy for the characteristically shaped and motile *Campylobacter*. Growth generally occurs within two weeks, but cultures should not be discarded for five weeks. Growth may sometimes be detected microscopically rather than macroscopically.

The second stage in isolation requires that the enrichment broth be subcultured to an enriched agar medium (e.g. Oxoid No. 2, brucella broth, brain-heart infusion) with 7% lysed horse blood and carbenicillin (125 µg per ml). Plates are inoculated with material drawn from the microaerophilic zone beneath the broth surface by Pasteur pipette and streaked in the usual manner. Plates are incubated at 30°C in a microaerophilic environment (see *C. jejuni*) for forty-eight to seventy-two hours.

Identification

Colonies are 1 mm, nonpigmented, and convex, but become more irregular on subculture. Once the organism has been recovered on solid medium at 30°C microaerophilically, it can then be subcultured aerobically at 37°C, although 30°C gives optimal growth. Only a small proportion of isolations will be made directly on solid media without going through the enrichment procedure. Organisms have typical *Campylobacter* morphology and are catalase and cytochrome-oxidase positive (Table 6-II). They do not react with antisera prepared against *C. fetus*. They differ from *C. fetus* (Table 6-II) in their ability to grow in air without added CO_2 and in their sensitivity to nalidixic acid, among other characteristics (22).

REFERENCES

1. Skerman, V. D. B., McGowan, V., and Sneath, P. H. A.: *Int J Syst Bact, 30*:225, 1980.
2. Bryner, J. H., Frank, A. H., and O'Berry, P. A.: *Am J Vet Res, 23*:32, 1962.
3. Schurig, G. C. D., Hall, C. E., Burda, K., et al.: *Am J Vet Res, 34*:1399, 1973.
4. Ogg, J. E., and Chang, W. J.: *Am J Vet Res, 33*:1023, 1973.
5. Shires, G. M. H., and Kramer, T. T.: *J Am Vet Med Assoc, 164*:398, 1974.
6. Laing, J. A. (Ed.): *Vibrio Fetus Infection of Cattle.* Rome, F. A. O., United Nations, 1960.
7. Hoerlein, A. B., and Kramer, T. T.: *J Am Vet Med Assoc, 143*:868, 1963.
8. Bartlett, D. E., Hasson, E. V., and Teeter, K. G.: *J Am Vet Med Assoc, 110*:114, 1947.
9. Clark, B. L., and Dufty, J. H. *Aust Vet J, 54*:262, 1978.
10. Karmali, M. A., Allen, K. A., and Fleming, P. C.: *Int J Syst Bact, 31*:64, 198.
11. Moynihan, I. W., and Stovell, P. L.: *Can J Comp Med, 19*:223, 1955.
12. Luechtefeld, N. W., Wang, W–L. L., Blaser, M. J., et al: *Lab Med, 12*:481, 1981.
13. Karmali, M. A., and Fleming, P. C.: *J Clin Microbiol, 10*:245, 1979.
14. Skirrow, M. B., and Benjamin, J.: *J Clin Path, 33*:1122, 1980.
15. Penner, J. L., and Hennessey, J. N.: *J Clin Micro, 12*:732, 1980.
16. Lior, H., Woodward, D. L., Edgar, J. A. et al.: *Lancet ii*:1103, 1981.
17. Al-Mashat, R. R., and Taylor, D. J.: *Vet Rec, 109*:97, 1981.
18. Gebhart, C. J., Ward, G. E., Chang, K., and Kurtz, H. J.: *Am J Vet Res,* (In press 1983).
19. Love, R. J., Love, D. N., and Edwards, M. J.: *Vet Rec, 100*:65, 1977.
20. Lawson, G. H. K., Leaver, J. L., Pettigrew, G. W. et al: *Int J Syst Bact, 31*:385, 1981.

21. Firehammer, B. D.: *Cornell Vet, 55*:482, 1965.
22. Neill, S. D., Ellis, W. A., and O'Brien, J. J.: *Res Vet Sci, 25*:368, 1978.
23. Neill, S. D., O'Brien, J. J., and Ellis, W. A.: *Vet Rec, 106*:152, 1980.

Chapter 7

PSEUDOMONAS, AEROMONAS, AND PLESIOMONAS

Organisms of the genus *Pseudomonas* are aerobic, non-sporeforming, glucose nonfermenting, Gram-negative rods. They are catalase positive and split sugars by oxidation; all have polar flagella except *Pseudomonas mallei*, which is nonmotile. Only the three species listed immediately below are recognized as important pathogens or potential pathogens of animals. Several other species, listed in Table 7-I, occur occasionally in clinical specimens but rarely cause disease.

Pseudomonas aeruginosa. This important potential pathogen is widespread, occurring on mucous membranes and widely in nature.

Ps. pseudomallei. This potential pathogen is found in soil and water in Southeast Asia, Australia, and central Africa.

Ps. mallei. This species differs from the others in that it is an obligate pathogen that does not occur in nature. In previous years, it has been included in different genera, including *Actinobacillus*.

PSEUDOMONAS AERUGINOSA

PATHOGENICITY. *Ps. aeruginosa* is a frequent opportunistic pathogen. Because of its relative resistance to drugs, it may persist in infectious processes from which other more susceptible organisms have been eliminated by treatment. This organism has been reported responsible for wound infections in various animal species, recovered from pigs with atrophic rhinitis, listed as an occasional cause of bovine mastitis and abscesses, and incriminated in abortions in the cow and mare, otorrhea in dogs and cats, septicemia in chickens, and a variety of other infections in animals and humans. Because of its widespread occurrence in the environment, it is a frequent contaminant, and consequently its recovery from clinical specimens is frequently not significant.

PSEUDOMONAS PSEUDOMALLEI

PATHOGENICITY. The disease in humans is referred to as melioidosis or pseudoglanders. It may assume a benign, chronic, or septicemic form in humans and animals. Infections have been reported in primates, cattle, sheep, goats, pigs, horses, dogs, cats, and rodents. In the more common chronic form, nodules and abscesses occur in the lungs, liver, spleen, lymph nodes, and subcutis.

PSEUDOMONAS MALLEI

PATHOGENICITY. *Ps. mallei* is the cause of glanders, a disease principally of the *Equidae*. It has been eradicated from North America, central and western Europe, but it still occurs in eastern Europe and Asia. The disease in horses, mules, and asses has three forms: pulmonary glanders, nasal glanders, and cutaneous glanders or farcy. Infections are characterized by the formation of encapsulated nodules that contain yellow caseous pus. Carnivores and humans become infected as a result of contact with infectious materials. Swine, cattle, rats, and birds are considered to be resistant.

OTHER *PSEUDOMONAS* SPECIES

There are numerous free-living pseudomonads. They occur occasionally as contaminants in clinical materials. Several species other than *Ps. aeruginosa* have been incriminated in infrequent infections, mostly in humans. The most common of these is *Ps. maltophilia. Ps. stutzeri, Ps. fluorescens, Ps. acidovorans,* and *Ps. putida* are usually contaminants in clinical specimens. Readers should consult the Supplementary References for information on other species.

ISOLATION PROCEDURES

Pseudomonas spp. grow well on blood, tryptose, or trypticase agar, and less complex media. They may be recovered on various enteric media. Glycerine stimulates the growth of *Ps. mallei*, and for this reason, a glycerol agar is sometimes used. Several of the less frequently occurring pseudomonads grow poorly at 37°C.

If glanders is suspected, pus or caseous material is removed from a nodule aseptically and inoculated onto blood agar. Place the plates in a sealed container to prevent dehydration. Incubate up to nine days and examine daily.

SELECTIVE MEDIA. A medium for the recovery of *Ps. mallei* from contaminated materials has been described by Miller et al. (1).

ANIMAL INOCULATION. Hamsters and guinea pigs are highly susceptible to *Ps. mallei*. The guinea pig is favored because of the well-known "Straus' phenomenon." Cultures are inoculated into male guinea pigs intraperitoneally, and pus is inoculated subcutaneously. The glanders organisms produce a septic orchitis, "Straus' phenomenon," in three to four days. The organism is readily recovered from the testicle. Lesions are also found in the spleen, liver, and other visceral organs.

CULTURAL CHARACTERISTICS

Pseudomonas aeruginosa. Large grayish colonies with irregular spreading margins. Some cultures are markedly mucoid. On blood agar, colonies are frequently beta-hemolytic. Cultures have a fruitlike odor. Greenish and/or yellowish green pigments may diffuse throughout clear media. On MacConkey's and SS agar, colonies are colorless, while on brilliant green agar, they are red.

Ps. aeruginosa produces two pigments: pyocyanin—green or blue green, water and chloroform soluble; fluorescein—water soluble, chloroform insoluble, fluoresces under ultraviolet light. A similar pigment is produced by *Ps. fluorescens*, a nonpathogenic organism that does not usually grow at 37°C. Some strains of *Ps. aeruginosa* produce the pigments pyorubin (red) and pyomelanin.

Ps. maltophilia. Colonies are round, smooth, nonpigmented, and glistening with regular margins. They are not hemolytic but may display a greenish discoloration of blood agar.

Ps. pseudomallei. Colonies are evident after twenty-four hours incubation on blood, tryptose, or trypticase agar. After forty-eight hours incubation, colonies obtain a size of 1–2 mm in diameter and are smooth, umbonate, and cream colored. Colonies enlarge markedly when left at room temperature for about a week. Cultures have an earthy or ammoniacal odor.

Ps. mallei. Colonies are shallow, round convex, opaque becoming yellowish green or brown on ageing. The colonies have a tendency to be slimy and tenacious in consistency.

IDENTIFICATION

The species referred to above, with the exception of *Ps. mallei*, are motile, and stained smears disclose Gram-negative, straight or slightly curved rods.

Ps. aeruginosa. Characteristic colonies, sometimes with a metallic sheen, are usually beta-hemolytic. A blue green or yellowish green pigment is often produced in clear media. Other characteristics include alkaline reactions on TSI with no gas or H_2S produced, positive oxidase reaction, and glucose used oxidatively in the OF test.

OTHER REACTIONS:
- Grows well on a medium (Cetrimide) containing cetyltrimethyl ammonium bromide; other pseudomonads are generally inhibited.
- Nitrates are reduced to nitrites.
- Indole is not produced.
- Gelatin is liquefied.
- Litmus milk first becomes alkaline, followed by coagulation and peptonization.
- Alkaline reaction is seen in TSI agar.

Seller's medium may be used to differentiate *Ps. aeruginosa* from *Alcaligenes faecalis* and *Acinetobacter* spp. The differential characteristics are listed in Table 7-I.

TABLE-7-I

DIFFERENTIAL FEATURES OF *PSEUDOMONAS AERUGINOSA,*
*ALCALIGENES FAECALIS.*AND OTHER SPECIES ON
SELLER'S MEDIUM (2)

Species	Slant Color	Butt Color	Band Color*	Under UV Light	Nitrogen Gas
Pseudomonas aeruginosa	Green	Blue or no change	Sometimes blue	Yellow green	+†
Alcaligenes faecalis	Blue	Blue or no change	Absent	Absent	+
Acinetobacter calcoaceticus subsp. anitratus	Blue	No change	Yellow	Absent	—
Acinetobacter calcoaceticus subsp. lwoffii	Blue	No change	Absent	Absent	—

*Seen at base of slant.
†Gas bubbles in butt.

To demonstrate pyocyanin production, infusion broth or a tryptose agar slant is inoculated and incubated overnight. If the organism is *Ps. aeruginosa,* a bluish green pigment, pyocyanin, may diffuse into the media. Identity of the pigment as pyocyanin is confirmed by the chloroform solubility test. Add 1–2 ml of chloroform to the broth culture or the slant culture and shake or agitate. If pyocyanin is present, it will be evident in the chloroform layer. Not all strains of *Ps. aeruginosa* produce pyocyanin. This test is not usually carried out as routine procedure.

Characteristics that separate *Ps. aeruginosa* from three other species are summarized in Table 7-II.

Ps. maltophilia. For identification, see Table 7-II.

Ps. mallei. See Table 7-II. Identification is also based on the features listed below. The production of the characteristic infection in the guinea pig is confirmatory.

- Nonmotile.
- Growth is enhanced by glycerin.
- Growth on glycerol potato agar is characteristic: a yellow viscid growth appears, which progresses to a yellowish brown and ultimately a dark brown.
- Carbohydrates for the most part are not broken down, although several may be split by oxidation.
- Slight acidity in litmus milk.
- Indole is not produced; oxidase test is negative.

TABLE 7-II
DIFFERENTIATION OF SOME *PSEUDOMONAS* SPECIES

	Fluorescein	Pyocyanin CHCl₃ Soluble	Gelatin Hydrolysis	Oxidase	Nitrate Reduction	Gas From Nitrate	Growth On SS	Maltose Oxidation	Other
Ps. aeruginosa	(+)*	(+)	+	+	+	+†	+	−	
Ps. fluorescens	(+)*	−	+	+	−	(−)	+	v	Brown, water-soluble pigment
Ps. maltophilia	−	−	+	(−)	(−)	−	−	+	
Ps. putida	+	−	−	+	(−)	−	+	v	
Ps. stutzeri	−	−	−	+	+	+	(+)	+	
Ps. cepacia	−	−	(−)	− or w‡	v	−	−	+	Some strains produce indigotin, an insoluble blue pigment
Ps. acidovorans	−	−	−	+	−	−	(+)	−	
Ps. pseudomallei	−	−	+	+	+	+	−	+	
Ps. mallei	−	−	+	−	+	−	v	v	+ Straus' reaction

*Can be detected with Wood's lamp
†Seller's medium
‡Weak

- Small amounts of H$_2$S, catalase, and ammonia are produced.
- Nitrate is reduced.
- Gelatin is not liquefied.

PSEUDOMONAS AERUGINOSA: SEROLOGIC AND PYOCIN TYPES

Serotyping and pyocin typing have been used mainly for epidemiological studies and investigations involving humans. Thirty-three O antigens distributed between thirteen serogroups have been proposed (3). The bacteriocins of *Ps. aeruginosa* are called pyocins, and pyocin typing has also been useful in epidemiological studies (4).

AEROMONAS AND *PLESIOMONAS*

These facultatively anaerobic, Gram-negative rods with polar flagella are recovered occasionally from clinical specimens and feces of animals and humans. They occur widely in water, sewage, and soil. *Bergey's Manual* includes them in the Enterobacteriaciae. The name *Plesiomonas shigelloides* has replaced *Aeromonas shigelloides.* The former is the only species in the genus *Pleisomonas.*

Three species are of occasional clinical significance: *A. hydrophila, A. salmonicida,* and *P. shigelloides.*

PATHOGENICITY

A. hydrophila or *P. shigelloides* have been isolated from several of the following human specimens: normal stools, bile, blood, throat swabs, osteomyelitic pus, feces from cases of dysentery and gastroenteritis, and cerebrospinal fluid (5, 6). *A. hydrophila* is a pathogen of reptiles and amphibians, while *A. salmonicida* causes disease in fish.

Reports of the isolation of these organisms from animals are scanty. It would seem that, as in humans, *Aeromonas* can, on occasions, account for infections in animals. The following are among the sources of *Aeromonas* that have been identified in the author's laboratory:

A. hydrophila	*A. salmonicida*	*P. shigelloides*
Various avian species, cattle, swine, dog, horse; wild, zoo, and laboratory animals	None	Penguin, cattle, swine, dog

They were isolated from a variety of specimens, and less than 5 percent of the isolations were obtained in pure culture.

ISOLATION PROCEDURES

The organisms grow well on the common laboratory media, including media used for the enteric bacteria. Although they grow best at room temperature or lower, they can be cultivated satisfactorily at 37°C.

CULTURAL CHARACTERISTICS

Aeromonas strains initially produce grayish white, translucent, moist, and stippled colonies. The appearance of colonies on enteric media vary with the capacity and rapidity with which strains ferment lactose. Some strains of *A. hydrophila* produce beta-hemolysis on blood agar. Many strains of *A. salmonicida* produce a soluble brown melaninlike pigment in two to three days on infusion agar.

IDENTIFICATION

As mentioned above, *Aeromonas* and *Plesiomonas* colonies resemble those of enterobacteria. They can be distinguished from the latter by their capacity to produce oxidase. Reactions on TSI are given in Table 10-IV. Many strains of *A. hydrophila* are beta-hemolytic. The three species are identified by the characteristics listed in Table 7-III.

REFERENCES

1. Miller, W. R., Pannell, L., Cravitz, L., Tanner, W. A., and Ingalls, M. S.: *J Bacteriol,* 55:115, 1948.
2. *Difco Supplementary Literature.* Detroit, Mich., Difco Laboratories, 1968.
3. Lanyi, B., and Bergan, T.: Serological Characterization of *Pseudomonas aeruginosa.* In Bergan, T., and Norris, J. R. (Eds.): *Methods in Microbiology,* Vol. 10, London, Academic Press, 1978, pp. 93–168.
4. Govan, J. R. W.: Pyocin Typing of *Pseudomonas aeruginosa.* In Bergan, T., and Norris, J. R. (Eds.): *Methods in Microbiology,* Vol. 10, London, Academic Press, 1978, pp. 61–91.
5. Ewing, W. H., Hugh, R., and Johnson, J. G.: *Studies on the Aeromonas Group.* Atlanta, Center for Disease Control, U.S. Public Health Service.
6. Ewing, W. H., and Hugh, R.: In Lennette, E. H., Spaulding, E. H., and Truant, J. P. (Eds.): *Manual of Clinical Microbiology,* 2nd ed. Washington, D.C., American Society for Microbiology, 1974, Chapter 20.

SUPPLEMENTARY REFERENCES

Pseudomonas

Franklin, M., and Franklin, M. A.: *A Profile of Pseudomonas.* Clifton, N.J., Beecham Pharmaceuticals (Division of Beecham Inc.), 1971.
Gilardi, G. L.: In Lorian, L. (Ed.): *Significance of Medical Microbiology in the Care of Patients,* 2nd ed. Baltimore, Williams & Wilkins, 1982, Chapter 7.

TABLE 7-III

COMPARISON OF BIOCHEMICAL REACTIONS OF
SPECIES OF *AEROMONAS* AND *PLESIOMONAS*

Substrate of Test	A. hydrophila (67)		A. salmonicida (21)		P. shigelloides (50)	
		% +		% +		% +
Glucose	+	100	+	100	+	100
Gas from glucose	d	49	+	100	—	0
" " other carbohydrates	d	76	+	100	—	0
Lactose	d	31	—	0	+	92
Sucrose	+	91	—	0	—	6
Maltose	+	98.5	+	100	+ or —	54
Mannitol	+	98.5	+	100	—	0
Dulcitol	—	0	—	0	—	0
Rhamnose	—	1.5	—	0	—	0
Arabinose	d	45	(+)	100	—	0
Inositol	—	0	—	0	+	100
Xylose	—	0	—	0	—	0
Raffinose	—	3	— or +w	28.5	—	0
Sorbitol	d	10.5	—	0	—	0
Salicin	d	62.5	(+)	95	— or (+)	34
Adonitol	—	0	—	0	—	0
Trehalose	+	100	+ or (+)	100	+	98
Cellobiose	— or +	34	—	0	—	0
Glycerol	d	85	d	85.5	d	82
Indole	d	86.5	—	0	+	100
Methyl Red 37°C	+	94			+	100
22°C	— or +	45	+	100	+	100
Voges-Proskauer 37°C	— or +	22.5			—	0
22°C	+ or —	72	—	0	—	0
Citrate, Simmons'	d	74.5	—	0	—	0
" , Christensen's	d	76	—	0	—	0
Hydrogen sulfide (TSI)	—	0	—	0	—	0
Urease	—	0	—	0	—	0
Nitrite	+	100	+	100	+	100
Ammon, salts glucose agar	d	78	—	0	+	90
Gelatin 22°C	+	98.5	+	100	—	0
Phenylalanine deaminase	— or +W	21	— or +w	12	+w or —	78.5
Malonate	—	0	—	0	—	0
KCN broth	— or +	48	—	0	—	0
Phenylproprionic acid reaction	—	0	see text		—	0
Lysine decarboxylase	—	7.5w	— or (+) w	19	+	100
Arginine dihydrolase	d	86.5	—	5	+	98
Ornithine decarboxylase	—	0	—	0	— or +	45
Mucate	—	0	—	0	—	0
Cetrimide agar	—	0	—	0	—	0
Catalase	+ or —	80	+ w or —	62	+	100
Cytochrome oxidase	+	100	+	100	+	100
Motility	+	100	—	0	+	86
ONPG	+	100	—	0	+	100

d = different biochemical types; (+) = delayed positive, 3 or more days; w = weak reaction.
From Ewing, W. H., Hugh, R. and Johnson, J. G.: Studies on *Aeromonas* Group. Atlanta, CDC, U.S. Public Health Service, 1961. Reproduced courtesy of the authors.

Hugh, R. and Gilardi, G. L.: In Lennette, E. H., Balows, A., Hausler, W. J., and Truant, J. P. (Eds.): *Manual of Clinical Microbiology*, 3rd ed., Washington, D.C. American Society for Microbiology, 1980, Chapture 22.

Hendrie, M. S., and Shewan, J. M.: In Gibbs, B. M., and Skinner, F. A. (Eds.): *Identification Methods for Microbiologists*. New York, Academic Press, 1966, Part A, p. 1.

Bergan, T.: In Starr, M. P., et al. (Eds.): *The Prokaryotes*, Vol. 1. New York, Springer-Verlay, 1981, Chapter 59, p. 666.

Aeromonas and Plesiomonas

Ewing, W. H., and Hugh, R.: In Lennette, E. H., Spaulding, E. H., and Traunt, J. P. (Eds.): *Manual of Clinical Microbiology*, 2nd ed. Washington, D.C., American Society for Microbiology, 1974, Chapter 20.

Ewing, W. H., Hugh, R., and Johnson, J. G.: *Studies on the Aeromonas Group*. Atlanta, Center for Disease Control, U.S. Public Health Service.

Von Graevenitz, A.: In Lennette, E. H., Balows, A., Hausler, W. J., and Truant, J. P. (Eds.): *Manual of Clinical Microbiology*. Washington, D.C. American Society for Microbiology, 1980, Chapter 17.

Chapter 8

BORDETELLA AND ALCALIGENES

M EMBERS of these genera are small, motile, aerobic, Gram-negative rods. Those species occurring in animals are catalase and oxidase positive and nonfermentative.

These genera are placed together for convenience and because they share some characteristics. Only one species of the genus *Bordetella* is of significance in diseases in animals, viz. *Bordetella bronchiseptica. Alcaligenes faecalis* is a nonpathogenic organism that is occasionally recovered from the feces of animals.

Two other species are included in the genus *Bordetella*:

Bord. pertussis. Natural host is humans. It is the cause of whooping cough.

Bord. parapertussis. Natural host is humans. It causes a mild form of whooping cough.

BORDETELLA BRONCHISEPTICA
Synonyms: *Brucella bronchiseptica, Alcaligenes bronchiseptica*

PATHOGENICITY. Dog—common secondary invader in distemper; also associated with respiratory infections including kennel cough (1). Swine—pneumonia; considered to be one of several possible causes of atrophic rhinitis. Respiratory infections in guinea pigs, rabbits (snuffles), cats, ferrets, and other animals. Switzer et al. (2) recovered *Bord. bronchiseptica* from a skunk, opossums, foxes, raccoons, and rats. Gallagher (3) recovered strains from horses with respiratory diseases. The association of this organism with a chronic conjunctivitis in the horse was reported by Miller (4). Infrequent cause of human respiratory infections. It has been recovered in the author's laboratory along with other bacteria from respiratory infections in the horse.

ISOLATION PROCEDURES. The organism grows well on blood agar at 37°C. In forty-eight hours, very small, circular, dewdroplike colonies appear. On further incubation, colonies enlarge, becoming flat and glistening.

MacConkey's agar with 1% glucose added is recommended as a selective medium for the isolation of *Bord. bronchiseptica* from the nasal passages of pigs (5). Small, grayish tan colonies with dark centers are produced in forty-eight hours on this selective medium. Additional tests (see below) are required for definitive identification. This medium was later improved by the addition of furaltadone and nystatin (6). An additional selective medium has been developed for the isolation of *Bord. bronchiseptica* from pigs, which appears to be superior to the media described above (7). This medium is described in Appendix B. A selective medium has also been described for the isolation of *Pasteurella multocida* and *Bord. bronchiseptica* from rabbits (8).

IDENTIFICATION. Smears from colonies disclose small, Gram-negative rods (Fig. 8-1). The organism produces oxidase and catalase and grows on MacConkey's agar. An alkaline reaction is obtained on triple sugar iron agar.

Other significant features include the following:
- Motile.
- Indole and hydrogen sulfide are not produced.
- Litmus milk: alkaline reaction; in five to ten days the medium darkens markedly.
- Urea is hydrolyzed (*Al. faecalis* is negative).
- Carbohydrates are not fermented.
- TSI agar: alkaline.
- Citrate: positive.

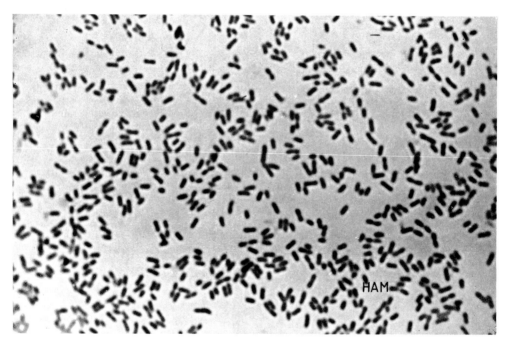

Figure 8-1. *Bordetella bronchiseptica* from a blood agar plate culture. Gram's stain, ×2500 (H. A. McAllister).

The differential characteristics of a number of nonfermentative organisms, including *Al. faecalis*, are listed in Table 8-I.

TABLE 8-I
DIFFERENTIAL CHARACTERISTICS OF SOME NONFERMENTATIVE BACTERIA, EF-4 AND II-j

	Hemolysis	Motile (37°C)	Glucose Oxidation	Oxidase	Nitrate Reduction	Gelatinase	Urease	Phenylalanine Deaminase	Growth on MacConkey's	Growth on SS agar	Citrate	Indole	H₂S (paper)	Growth (42°C)	Pigment
A. lwoffii	-*	-	-	-	-	-	-*	-	+	-*	V	-	-	+	-
A. calcoaceticus	-*	+	+	-	-	V	V	-	+	-*	+*	-	-	+	-
Al. faecalis	-	+	-	+	+*	-	-	U	+	+	+*	-	-	+	-
Bord. bronchiseptica	+*	+	-	+	+	-	+	U	+	+	+	-	-	V	-
M. bovis	+*	-	-	+	-*	+	-	-	-	-	-	-	-*	-	-
M. lacunata	-	-	-	+	+	+	-	-	-	-	-	-	-	-	-
M. nonliquefaciens	-	-	-	+	+*	-	-	-	V	-	-	-	-	V	-
M. phenylpyruvica	-	-	-	+	+	-	+	+	+	-	-	-	-	-	-
M. osloensis	-	-	-	+	V	-	-*	-	V	-	-	-	-*	+	-
M. urethralis	-	-	-	+	-	-	-	-	+	-	+	-	weak	-	U
M-5	-	-	-	+	-	-	-	U	+*	-	-	-	+	U	-
II-j	-	-	-	+	-	+	+	U	-	-	-	+	V	-*	-
EF-4	-	-	-	+	+	V	-	U	V	-	-	-	(-)	V	-or sl. yellow

U = Information unavailable   * = Some exceptions   V = Variable

ALCALIGENES FAECALIS

This organism is found in the intestinal tract of humans and animals. Its recovery from animals except poultry is infrequent, and it is not ordinarily considered pathogenic; however, it has recently been implicated as the probable cause of rhinotracheitis of turkeys and chickens (9). In the early phase of this disease, *Al. faecalis* can be recovered from the mucosa of nasal turbonates and the upper third of the trachea in pure culture. Later in the disease, mixed cultures are obtained, and it helps to employ MacConkey's medium with 4% agar to prevent spreading by *Proteus* (10).

It grows well on ordinary media and produces colorless, salmonellalike colonies on MacConkey's and SS agar. The reaction on triple sugar iron agar slants is negative or alkaline. It can be differentiated from *Pseudomonas aeruginosa* on Seller's medium (Table 8-I; see also Table 7-I) and from *Bord. bronchiseptica* by failure to produce urease.

Other important features include the following:
- It does not ferment glucose or other carbohydrates.
- Indole is not produced.
- Litmus milk is unchanged or alkaline.
- Gelatin is not liquefied.
- Nitrate is usually reduced.
- Hydrogen sulfide is not produced.

REFERENCES

1. McClandlish, I. A. P., Thompson, H., Cornwell, H. J. C., and Wright, N. G.: *Vet Rec,* *102*:298, 1978.
2. Switzer, W. P., Maré, C. J., and Hubbard, E. D.: *Am J Vet Res,* 27:1:1134, 1966.
3. Gallagher, G. L.: *Vet Rec,* *77*:632, 1965.
4. Miller, W. C.: *Vet Rec,* *77*:658, 1965.
5. Ross, R. F., Switzer, W. P., and Maré, C. J.: *Vet Med,* *58*:652, 1963.
6. Farrington, D. O., and Switzer, W. P.: *J Am Vet Med Assoc,* *170*:34, 1977.
7. Smith, I. M., and Baskerville, A. J.: *Res Vet Sci,* *27*:187, 1979.
8. Garlinghouse, L. E., Jr., DiaGiacomo, R. F., Hoosier, G. L. Van, Jr., and Condon, J.: *Lab Animal Sci,* *31*:39, 1981.
9. Simmons, D. G., Gray, J. G., Rose, L. P., Dillman, R. C., and Miller, S. E.: *Avian Dis,* *23*:194, 1979.
10. Simmons, D. G.: Personal communication.

SUPPLEMENTARY REFERENCES

Goodnow, R. A.: Biology of *Bordetella bronchiseptica, Bact Rev,* *44*:722, 1980.
Pittman, M., and Wardlaw, A. C.: In Starr, M. P., et al. (Eds.): *The Prokaryotes,* Vol. 1. New York, Springer-Verlag, 1981, Chapter 85, p. 1075.

CHAPTER 9

BRUCELLA

Margaret E. Meyer

Species in this genus are homogeneous in their morphology and staining characteristics. All are Gram-negative coccobacilli but are sufficiently small that they appear to be cocci rather than short rods with rounded ends (see Fig. 9-1). They are nonmotile, nonencapsulated, non-sporeforming, grow aerobically or microaerophilically but will not grow anaerobically.

Species and their natural reservoirs:

Brucella abortus — cattle
Brucella suis — swine
Brucella melitensis — goats and sheep
Brucella ovis — sheep, especially rams
Brucella canis — dogs
Brucella neotomae — wood and sand rats

PATHOGENESIS AND PATHOGENICITY

Each of the species has a decided host preference for its natural reservoir and is not readily transmitted from the preferential host to dissimilar hosts. When a species does infect a nonpreferential host, it is more apt to localize in the mammary gland and reticuloendothelial tissue rather than in the uterus and fetal membranes. The organisms localize in various tissues and organs following a bacteremia.

B. abortus. In cows, uterine infections usually result in abortion, or birth of weak calves. The supramammary lymph nodes are frequently infected, and the organisms are shed in the milk and also may be recovered in mammary secretions of nonlactating animals. In bulls, genital infections may be inapparent and the organism recoverable from semen, or there may be an apparent orchitis.

B. suis. Brucellosis in swine is usually a more generalized and frequently a more chronic disease than bovine brucellosis. Bacteremia may persist for months with or without localization, and organisms also may persist in the uterus causing a lingering metritis. Genital infections are more frequent in boars than in bulls and may cause necrotic orchitis. Infection in swine is characterized by abortion, stillbirths, decreased litter size, weak pigs, and

85

focal abscessation. Spondylitis is an occasional sequela, especially in sows, and arthritis may occur even in young, weanling pigs.

B. melitensis. The disease in goats and sheep is similar to that in cattle. Sheep are more resistant to infection with this species than are goats.

B. ovis. The primary manifestation of infection with this species is epididymitis accompanied by lesions of the tunica vaginals and sometimes of the testicle. In ewes there is a placentitis and occasionally abortions. Infection without abortion frequently results in weak lambs of low birth weight.

B. canis. Dogs are the only known host for this species. Cases in free roaming dogs and household pets are uncommon. Under kennel conditions, it spreads rapidly both to males and females. It is characterized by abortions and infertility in the bitch accompanied by an intermittent but prolonged bacteremia. Genital infections in males can be silent or can result in orchitis and epididymitis.

B. neotomae. This species has never been found outside its natural reservoir and remains of little or no veterinary significance.

LABORATORY DIAGNOSIS

Great care should be taken in working with *Brucella*: Humans are highly susceptible to this disease, and laboratory infections are common.

Direct Examination of Smears

Koster's staining procedure long has been used for microscopic detection of intracellular brucellae, particularly in the chorionic epithelium. A scraping from a cotyledon is smeared on a slide, fixed, and stained. The staining procedure is outlined in Appendix A.

Fluorescent Antibody Staining

Smears of smooth *Brucella*, which is their usual phase when in animal tissue or freshly isolated, stain specifically with anti-*Brucella* FA conjugate. This method can be used to distinguish *Brucella* in mixed populations of morphologically similar organisms and also to distinguish *Brucella* in impression smears of tissues. The conjugate, with appropriate instructions for use, is available commercially.

Isolation Procedures

Specimens most commonly submitted for isolation of *Brucella* organisms are placental membranes, aborted fetuses, vaginal swabs, semen samples, swabs of male reproductive organs, milk, lymph tissues, and blood.

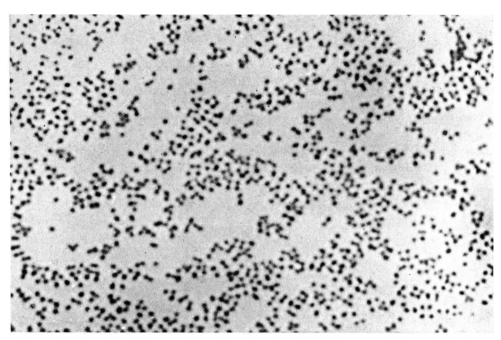

Figure 9-1. *Brucella suis* from a culture. Gram's stain, ×2250. Courtesy of Doctor S. Stanley Schneierson.

The media of choice for isolating *Brucella* from tissue and most body fluids is tryptose agar enriched with 5% seronegative equine or bovine serum. Inoculate duplicate plates for incubation in an atmosphere of air and of 10% carbon dioxide. Most strains of *B. abortus* and all strains of *B. ovis* obligately require carbon dioxide for initial isolation. Since many sources from which brucellae are likely to be isolated are apt to be contaminated, it may be necessary to use a selective medium such as W medium (see Appendix B).

A blood culture bottle containing a diphasic medium is recommended for isolation of brucellae from blood (see Appendix B).

Inoculated plates are incubated at 37°C for at least fifteen days before being discarded as negative.

Examination of Milk for Brucella

Collect 25 to 50 ml of milk from each quarter into a sterile container. Centrifuge at 9000 G for twenty minutes. Decant middle milk, reserving an aliquot of 10 ml, culture a portion of the sediment and a portion of cream on selective media. Mix remaining sediment and cream with reserved aliquot of middle milk and freeze. Should the plates be overgrown with contaminants, or negative when the history of the animal is positive, use frozen reserve for guinea pig inoculation.

Guinea Pig Inoculation

This is the most reliable procedure for isolation of the classical species. Brucellae frequently are present in tissue or fluids but cannot make the transition from growing in animal tissue to growing in an artificial environment. Such strains can be isolated from inoculated guinea pigs. Use this method also for isolation from badly contaminated material. Triturate tissues in nutrient broth with Ten Broeck grinder or mortar and pestle. Guinea pigs are inoculated with up to 5.0 ml doses intraperitoneally for material not badly contaminated. If material is badly contaminated, inoculate subcutaneously or intramuscularly on inside of thigh with. 0.1–0.2 ml dose, on alternate days for three to five doses.

After twenty-four days, blood is drawn by cardiac puncture and an agglutination test is performed to detect *Brucella* antibodies. Keep animals for six to eight weeks. Necropsy animals that die before this time and culture spleen, liver, blood, lymph nodes, and testicles. Necropsy surviving animals at eight weeks, observe for typical lesions, culture, and do agglutination test on blood.

Cultural Characteristics and Identification

Usually after from three to several days incubation, pinpoint, round, smooth, glistening, bluish, translucent colonies appear. As colonies age they increase in size, lose their translucence, and gradually turn from a cream to brownish color. In all instances, media should be incubated fifteen days before being discarded as negative.

In addition to colonial morphology, and organism morphology as determined by Gram's and/or Koster stains, the identity of *Brucella* at the generic level can be ascertained by staining smears with anti-*Brucella* FA conjugate, and by serologic and biochemical methods.

Smooth isolates of questionable identity can be prepared as antigens and used either in tube agglutination or rapid slide agglutination tests. The cell density of the antigen for the tube test is equilibrated to that of the U.S.D.A. Standard Tube Antigen. Rapid slide test antigen is simply a loop full of representative colonies emulsified in a few drops of saline. The tube test is preferable as it is much less likely to give false positive reactions. Organisms of rough morphology require special preparation to use as antigens; thus, rapid serologic methods cannot be used as an aid to identity.

Other characteristics that aid in identification of the genus are the following:
- Nonmotile.
- Catalase production.
- Variable oxidase reaction.
- Production of urease (except *B. ovis*).

- Nitrate reduction.
- No change in litmus milk.
- MR and VP negative.
- Indole negative.
- Gelatin hydrolyzation negative.

The three classical species can be differentiated by their patterns of growth on appropriate concentrations of basic fuchsin and thionin and by other characteristics listed in Table 9-I.

Identification of the biotypes within the species is usually carried out in reference laboratories. Workers interested in biotype identification should consult the Supplementary References on the genus *Brucella*.

Lysis by bacteriophage also is useful for identification purposes. Phages that lyse members of the genus *Brucella* do not lyse members of other genera, which can be useful in generic identification. There are several phages active or partially active upon various members of this genus. The most dependable and useful is the Tbilisi strain (also known as *B. abortus* phage, strain 3). At the RTD (routine test dilution), this phage lyses only smooth strains of *B. abortus*. At 10,000 × RTD it causes lysis from without on most strains of *B. suis*. Typing by other phage strains is usually done by a reference laboratory.

ANTIGENIC STRUCTURE. The agglutination reactions observed with the classical species are considered due to surface antigenic determinants. These determinants are designated as "A" for *B. abortus* antigen and "M" for *B. melitensis* antigen. These two antigens are distributed quantitatively in differing proportions among typical representatives (biotype 1) of the smooth species and among the various biotypes. By use of the technique of agglutinin absorption, it is possible to assay for the proportions of the two antigens in any smooth strain. This technique is of little value for routine identification purposes. In rare instances when it is essential to know the antigenic proportions, it is best to have the procedure done in a reference laboratory.

Organisms that may be difficult to distinguish from *Brucella* by the array of cultural, morphologic, biochemical, or serologic features include *Bordetella bronchiseptica*, *Mima polymorpha*, and *Yersinia enterocolitica*. Organisms that are motile (*Bordetella bronchiseptica*), or are hemolytic, or ferment lactose, can be disqualified as *Brucella*.

Serological Diagnosis of Brucellosis

There are a variety of tests available for serologic diagnosis of brucellosis in farm animals and in dogs. Special antigens must be used for serologic diagnosis of the nonsmooth species of *B. canis* and *B. ovis*. Workers interested in obtaining details of serologic procedures and interpretation of test results may obtain information from the Agricultural Research Service, United

TABLE 9-I
DIFFERENTIATION OF *BRUCELLA* SPECIES AND THEIR BIOTYPES

Species	Urease	Oxidase	Biotype	CO₂ Requirement	H₂S Production	Thionin A	Thionin B	Thionin C	Basic Fuchsin B	Basic Fuchsin C	Agglutination A	Agglutination M	Agglutination R
Br. melitensis	V	+	1	–	–	–	+	+	+	+	–	+	–
			2	–	–	–	+	+	+	+	+	+	–
			3	–	–	–	+	+	+	+	+	+	–
Br. abortus	+	+	1	V	+	–	–	–	+	+	+	–	–
			2	+	+	–	–	–	–	–	+	–	–
			3	V	+	+	+	+	+	+	+	–	–
			4	V	+	–	+	–	+	+	–	+	–
			5	–	–	–	+	+	+	+	–	+	–
			6	–	V	–	+	+	+	+	+	–	–
			7	–	V	–	+	+	+	+	+	+	–
			9	V	+	–	+	+	+	+	–	+	–
Br. suis	+	+	1	–	+	+	+	+	–	–	+	–	–
			2	–	–	–	+	+	–	+	+	–	–
			3	–	–	+	+	+	+	+	+	–	–
			4	–	–	+	+	+	+	±	–	+	–
Br. canis	+	+		–	–	+	+	+	–	+	–	–	+
Br. ovis	–	–		+	–	–	+	+	+	+	–	–	+
Br. neotomae	+	+		–	+	–	–	+	–	–	+	–	–

*Species differentiation is obtained on tryptose agar with the following graded concentrations of dyes; thionin concentration (A) 1:25,000, (B) 1:50,000, (C) 1:100,000: basic fuchsin concentration (B) 1:50,000, (C) 1:100,000.

Modified from G.G. Alton, L.M. Jones, and D.E. Pietz, *Laboratory Techniques in Brucellosis*, 2nd ed., 1975. Courtesy of the World Health Organization, Geneva.

States Department of Agriculture. Additional sources of information are listed in the References.

REFERENCES

Genus Brucella

Alton, G. G., Jones, L. M., and Pietz, D. E.: *Laboratory Techniques in Brucellosis*, 2nd ed. World Health Organ, Monograph Ser., 55, 1975.

Brinley-Morgan, W. J., and Gower, S. G. M.: In Gibss, B. M., and Skinner, F. A. (Eds.): *Identification Methods for Microbiologists* (Technical Series No. 1, Part A). New York, Academic Press, 1966, p. 35.

Meyer, Margaret E.: In Starr, M. P., Stolp, H., Truper, H. G., Balows, A., and Schlegel, H. G. (Eds.): *The Prokaryotes: A Handbook on Habitats, Isolation and Identification of Bacteria.* New York, Springer-Verlag, 1981, Chapter 84.

Brucella canis

Alton, G. G., Jones, L. M., and Pietz, D. E.: *Laboratory Techniques in Brucellosis*, 2nd ed. World Health Organ, Monograph Ser., 55, 1975.

Carmichael, L. E.: *Proceedings of the 71st Annual Meeting US Livestock San Association*, 1967, p. 517.

Brucella ovis

Buddle, M. B. and Boyes, B. W.: *Aust Vet, 29*:145, 1953.

Meyer, Margaret E.: In Blobel, H. (Ed.): *Handbook of Bacterial Infections in Animals.* Jena, Gustav Fischer, 1982, p. 305.

Serologic Diagnosis

United States Department of Agriculture, Agricultural Research Service, National Animal Disease Center, Diagnostic Reagents Division. Manual Nos. 64A, B, C, and D. Ames, Iowa, 1965.

Chapter 10

ENTEROBACTERIA

MARGERY E. CARTER

The enteric bacteria are Gram-negative, facultatively anaerobic, non-sporeforming rods. All are fermentative, oxidase negative, and catalase positive, and many are motile with peritrichate arrangement of the flagella.

Some members of the family reside in the intestinal tract of animals and humans. Others occur in the environment associated with soil, water, plants, or insects.

The nomenclature of the Enterobacteriaceae has undergone considerable change in recent years. In the 8th edition of *Bergey's Manual* (1), the family was divided into tribes based mainly on biochemical reactions. Brenner (2) in *Bergey's Manual of Systematic Bacteriology* recognizes DNA-relatedness studies, and names twenty genera (including 6 new ones) and more than 100 species. Recently named genera are indicated with an asterisk. The genera contain members that are known pathogens; some that are opportunistic pathogens; others that can be isolated from clinical specimens; and species that have no known pathogenicity. Genera belonging to the first three categories are presented in Table 10-V.

GENERAL COMMENTS AND PATHOGENICITY

Cedecea

The two species *C. davisae* and *C. lapagei* have been isolated from human clinical specimens; half of the isolates were from the respiratory tract. The pathogenicity of the genus is unknown.

Citrobacter

The three *Citrobacter* species, *C. freundii, C. diversus* (*C. intermedius* b), and *C. amalonaticus* (*C. intermedius* a), are considered to be opportunistic pathogens in humans, and this is probably their status in animals. They have been isolated from the feces of animals and humans, and from soil, water, and sewage.

Edwardsiella

E. tarda is occasionally isolated from clinical specimens. It is probably a normal intestinal inhabitant of snakes and has been isolated from the intestinal tract of normal animals and humans.

Enterobacter

Enterobacter is found widely in nature. *E. cloacae* and *E. aerogenes* are opportunistic pathogens. *E. cloacae* causes occasional bacteremia in humans (3), and *E. aerogenes* can be associated with bovine mastitis. *E. sakazakii* is the name given to the yellow pigmented variant of *E. cloacae*; it is isolated from food but only rarely from human clinical specimens. *E. gergoviae* has been recovered from human clinical specimens and from cosmetics and from water. *E. agglomerans* is associated with water, soil, and sewage.

Erwinia

Erwinia species include those that are plant pathogens. None are known to be pathogenic for animals or humans.

Escherichia

The lactose-positive member of the genus, *E. coli*, is a normal intestinal inhabitant of humans and animals but is also recovered from a wide variety of diseases in all domestic animals. It may be a primary or secondary agent. Nursing and young animals, under one week of age, are particularly susceptible. Some important diseases are listed below.

CATTLE. Septicemic calf scours and less severe intestinal infections of calves; mastitis.

SWINE. Scours and diarrhea in young pigs; hemorrhagic enteritis and edema disease.

CHICKENS. Airsacculitis; Hjärre's disease or coligranuloma.

DOGS. Urinary infections.

A variety known as *E. coli*, "inactive," represents the previously named A-D Group. This organism is usually lactose negative, nonmotile, anaerogenic, and does not always form indole. *E. blattae* has been isolated from the hind-gut of cockroaches and is not known to be pathogenic.

Hafnia

The single species *Hafnia alvei* is found in the feces of humans and other animals, sewage, soil, water, and dairy produce. It is not considered to be important as a cause of infections in animals.

Klebsiella

K. ozaenae and *K. rhinoscleromatis* can cause disease in humans and are now considered subspecies of *K. pneumoniae*. *K. pneumoniae* subs. *pneumoniae* capsular types 1, 2, and 3 cause pneumonia in humans. The organism occurs widely in nature, notably in wood products used as bedding for cattle. Severe *Klebsiella* mastitis is more prevalent when cows are kept on such bedding.

K. pneumoniae subs. *pneumoniae* has been recovered from other infections in animals, including cervicitis and metritis in mares, wound infections, septicemia, and pneumonia in the dog.

The indole-positive biogroup of *K. pneumoniae* has been named *K. oxytoca*. This organism can be recovered from the intestinal tract of healthy animals and from the environment.

Kluyvera*

The two species *K. ascorbata* and *K. cryocrescens* have been isolated from human clinical specimens, feces, and food. There is no strong evidence of pathogenicity.

Morganella

Proteus morganii has been renamed *Morganella morganii* as DNA/DNA hybridization studies showed that the organism is not closely related to the *Proteus* species. *Morganella* can be isolated from the normal intestinal tract.

Obesumbacterium*

This organism has been isolated only from beer. It is not known to be pathogenic.

Proteus

The genus now contains three species. *P. mirabilis* occurs most frequently in clinical materials from animals. Both *P. mirabilis* and *P. vulgaris* can cause urinary, gastrointestinal, and other sporadic infections in humans and animals. The newly named species, *P. myxofaciens*, has been isolated only from gypsy moth larvae.

Providencia

The three species, *P. alcalifaciens* (*Proteus inconstans* biogroup A), *P. stuartii* (*Proteus inconstans* biogroup B), and *P. rettgeri* (*Proteus rettgeri*), are not frequently recovered from animal specimens. The *Providencia* species are reputed to cause urinary and intestinal infections in humans but are only rarely incriminated in animal diseases.

Rahnella*

The single species *R. aquatilis* is a natural inhabitant of water. It has been isolated from a human burn wound.

Salmonella

There are more than 2,000 closely related serovars. There have been various proposals for the subdivision of the genus. On the basis of different biochemical characteristics, *Salmonella* was divided into three species (4): *S. typhi*, *S. choleraesuis*, and *S. enteritidis*. In this scheme, all the other salmonellae became serovars of *S. enteritidis* (e.g. *S. enteritidis* serovar *typhimurium*). On the basis of numerical taxonomy and DNA relatedness studies, a recent proposal (5) suggests that the genus should consist of a single species, *S. choleraesuis*. This includes all the organisms formerly known as "Arizona". The salmonellae have been placed into five subgenera. Almost all the serovars that are isolated from clinical specimens are in subgenus I. The "Arizona" serovars are included in subgenus III.

The Kauffmann-White scheme, based on the somatic, capsular, and flagella antigens, is still acceptable because of its wide diagnostic usage. With a few exceptions, the antigenic formulae of "Arizona" serovars may be translated into *Salmonella* formulae and included in the scheme.

The *Salmonella* serovars can be subdivided into biovars, which are strains, within a serovar, having different biochemical patterns (e.g. *S. choleraesuis* biovar Kunzendorf). Phagovars are determined by the sensitivity of cultures to a series of bacteriophages.

Salmonellosis assumes one of the following forms: peracute septicemia, acute enteritis, chronic enteritis, or a subclinical carrier state. Some important syndromes caused by *Salmonella* serovars are listed below (grouping is based on O antigen composition):

Group 02 (A)	*S. paratyphi A:*	Paratyphoid fever in man.
Group 04 (B)	*S. schottmuelleri:*	Paratyphoid fever in man.
	S. typhimurium:	Gastroenteritis in man; most prevalent species causing infections in various animal species.
	S. abortus-equi:	Abortion in mares and jennets.
	S. abortus-bovis:	Abortion in cattle.
	S. abortus-ovis:	Abortion in sheep.
Group 06, 7 (C$_1$)	*S. choleraesius:*	Enteritis in pigs; frequent secondary invader in hog cholera; infections in man.
	S. typhisuis:	Infections in young pigs.

Group 06, 8 (C_2)	*S. newport:*	Infections in man, various animals, and especially cattle.
Group 09, 12 (D_1)	*S. enteritidis:*	Infections in various animals; gastroenteritis in man.
	S. gallinarum:	Fowl typhoid, an acute intestinal disease of young chickens and turkeys.
	S. pullorum:	Severe intestinal infections of chicks and poults (pullorum); chronic infections in older fowl.
	S. typhi:	Typhoid fever in man.
	S. dublin:	Severe infections in calves.
Group 03, 10 (E_1)	*S. anatum:*	Keel disease in ducklings; infections in man and animals.

The "Arizona" serovars occur widely in nature and on occasions cause severe or fatal infections in poultry (especially in turkey poults), humans, dogs, cats, and other animal species. They are frequently recovered from snakes and lizards.

Serratia

Of the seven species, only *S. marcescens* is considered a significant pathogen of humans and animals. This organism causes mastitis in cows and has been recovered from pneumonia and septicemia in humans. Roussel, Lucas, and Bouley (6) refer to infections in geckos and tortoises and septicemia in the chicken.

S. odorifera can be isolated from plants and food and may be a rare opportunistic pathogen in humans. *S. liquefaciens* is the most prevalent species in the environment and has been recovered from plants and rodents' intestines.

Shigella

In this genus, the four species cause intestinal infections and dysentery in humans and primates. Their involvement as causes of disease in the domestic animals is rare.

Tatumella*

The single species *T. ptyseos* has been recovered from human clinical specimens, mainly from the respiratory tract. It may be an opportunistic pathogen.

Xenorhabdus*

The two species are parasitic in nematode larvae. They are not known to be animal pathogens. The preferred incubation temperature is 25°C.

Yersinia

This genus is discussed in Chapter 11.

ISOLATION PROCEDURES

SPECIMENS. (A) General specimens include fecal swabs; samples of feces; tissues from a recently dead carcass, such as intestinal sections, liver, kidney, spleen, or bone marrow; and milk samples from mastitis cases. Recovery of enteric bacteria from fresh tissues, in a heavy and fairly pure growth, is suggestive of a bacteremia and may be significant. The routine procedure followed in the isolation and cultivation of enteric bacteria is outlined in Table 10-I.

(B) Samples for the recovery of salmonellae may be varied, and some need additional isolation procedures.

1. Samples from diseased or dead animals can be plated directly onto selective media and an aliquot placed in enrichment broth.
2. Meat and offal samples being checked for salmonellae are usually placed only in enrichment broth. There are often too few organisms present to warrant direct plating.
3. Samples from meat, egg, or milk products that may have undergone heat in processing are first placed in a preenrichment broth. After twenty-four hours incubation, a portion of the nutrient broth can be placed into an enrichment broth.
4. Salmonellae may be recovered from water samples in the following ways:
   a. About 50–100 ml of the water can be added to an equal quantity of double-strength tetrathionate broth.
   b. Pass about 100 ml of water sample through a 0.45 μm membrane filter. The filter can be placed into 10 ml of enrichment broth.
   c. A successful method of isolating salmonellae from streams or large bodies of water is with the aid of a pad of cotton wrapped in cheese-cloth and tied securely with a long piece of string. This is sterilized, and when required it is thrown into the water and attached to the bank by the string. The pad is left *in situ* for forty-eight hours, then recovered and placed in about 100 ml of enrichment broth.

Enrichment Broths and Selective Media for Salmonellae.

There are many enrichment broths, and modifications of these, for the isolation of salmonellae. Some, however, can be toxic for some serovars.

TABLE 10-I
PROCEDURES FOR THE ISOLATION AND IDENTIFICATION
OF THE ENTERIC BACTERIA

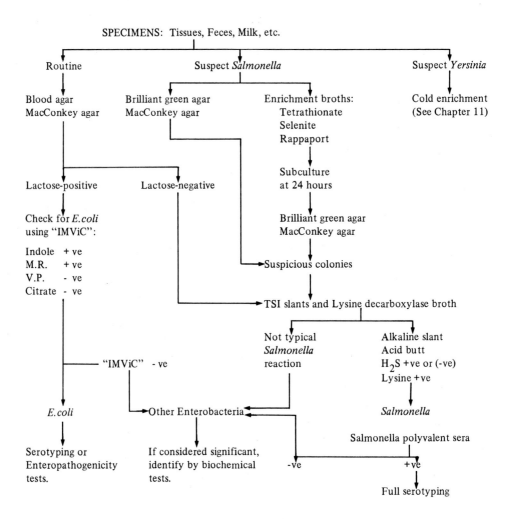

Strains of *S. typhisuis, S. choleraesuis, S. pullorum,* and *S. gallinarum* are inhibited by selenite and tetrathionate broths. Rappaport enrichment broth (7) success-fully encourages the growth of these serovars. GN broth (BBL) supported the growth of all *S. typhisuis* strains tested (8).

The efficiency of selective enrichment broths is influenced by the type of specimen being examined, the proportion of inoculum to broth, and the length and temperature of incubation used. The effectiveness of selenite and tetrathionate are considerably reduced by the addition of certain products such as egg albumen. Food products containing in excess of 1% dextrose should be diluted so that the final concentration of dextrose in the broth is less than 1%.

Excessive amounts of inocula inhibit rather than enhance selectivity of the enrichment broth. Fecal samples should be added in 5–10% quantities, and food products should not exceed 15% in the broth.

Subcultures from enrichment broths onto selective media can be made after twenty-four hours and again after forty-eight hours incubation. It has been found that one subculture at twenty-four hours is sufficient to isolate the great majority of the salmonellae.

Tetrathionate broth can be incubated at 40°C. This temperature inhibits most of the enteric organisms but permits the growth of salmonellae. Care must be taken as at 42–43°C the tetrathionate broth may prove toxic to the salmonellae. The other enrichment broths are incubated at 37°C.

Inhibition of *Proteus* species in tetrathionate broth can be obtained by the addition of 40 µg of novobiocin/ml or 0.125 mg of sodium sulfathiazole per 100 ml of the enrichment broth.

Brilliant green agar, MacConkey agar, and Salmonella-Shigella (SS) agar are among the most generally used selective media, which are also designed to be indicator media.

Brilliant green agar can be inhibitory to *S. pullorum*, *S. gallinarum*, *S. typhi*, *S. choleraesuis*, and *S. typhisuis*. Modified brilliant green agar (Oxoid) was found to support growth of *S. typhisuis* (8). *S. choleraesuis* strains will grow on MacConkey and XLD agars, while some strains of *S. pullorum* were found to be inhibited by all the common selective media except for MacConkey agar.

Sodium sulfadiazine at 8–16 mg/100 ml of brilliant green agar will inhibit *Proteus* and *Pseudomonas* species, but will not affect the growth of salmonellae. The two former organisms tend to have colonies similar in appearance to *Salmonella* on brilliant green agar, that is, pinkish white with an alkaline reaction (red) in the surrounding medium.

Cultural Characteristics

The colonies of the various enteric bacteria on blood agar are not sufficiently distinctive to aid appreciably in their identification, except for the propensity of *P. vulgaris* and *P. mirabilis* to swarm; *Serratia marscens* to produce an orange red pigment, although this is often not seen at 37°C; *Enterobacter sakazakii* to produce a yellow pigment; and *Klebsiella, Enterobacter*, and some strains of *E. coli* to have very mucoid colonies. However, the selective and indicator media, referred to in Table 10-II, do aid in the presumptive identification of some genera, especially salmonellae.

"Mucoidness" or hemolysis produced by some strains of *E. coli* are not now considered to correlate with pathogenicity.

TABLE 10-II

GROWTH OF THE ENTEROBACTERIACEAE ON SELECTIVE AGAR MEDIA

A. RELATIVE GROWTH

Genera	Brilliant Green Agar	S S Agar	MacConkey's Agar
Klebsiella Enterobacter Escherichia	Limited	Very limited	Abundant. *Klebsiella* and *Enterobacter* form large mucoid colonies.
Shigella	Very limited or none	Abundant	Abundant
Edwardsiella Citrobacter Salmonella Arizona Providencia	Abundant	Moderate	Abundant
Proteus	Moderate, will not spread	Moderate, does not usually spread	Abundant, spreads

B. COLONY APPEARANCE

Genera	Brilliant Green Agar	S S Agar	MacConkey's Agar
Klebsiella Enterobacter Escherichia	Yellow green	Red or pink	Red
Providencia Shigella	Pinkish white with a red halo	Whitish or colorless	Whitish or colorless
Edwardsiella Citrobacter Salmonella Arizona	Usually pinkish white with a red halo; yellow green if either lactose or sucrose fermented	Whitish or colorless; if H_2S is produced, a dark central spot forms within 48 hours	Whitish or colorless
Proteus	Yellow green if it is a sucrose fermenter; pinkish white with a red halo if sucrose-neg.	Whitish or colorless, with or without a dark central spot	Whitish or colorless

*Modified from H.A. McAllister, *Procedures for the Identification of Microorganisms from the Higher Animals, 1970.* Courtesy of Lucas Brothers Publishers, Columbia, Missouri.

Identification

The appearance of the colonies on selective/indicator media; the reactions in triple sugar iron (TSI) agar slants; and other biochemical reactions using either conventional or miniature systems (e.g. API or Minitek) are used to identify the Enterobacteriaceae. Brenner (2) warns that as there are now over 100 species of enterobacteria it is risky to identify these organisms to a specific level on the basis of a few tests. Each laboratory must decide what nonroutine tests to use, as these will be necessary in some cases in order to award a specific name; whether it is necessary to speciate in each case; and which species to ignore, as some are not known to be pathogenic for animals. A decision must also be made on the significance of the isolates, based on the handling and age of the samples. The organisms are ubiquitous, and post-mortem invasion of carcasses by the enteric bacteria is rapid.

The appearance of some of the enteric bacteria on selective media is given in Table 10-II. The use of TSI combined with lysine decarboxylase broth can give a presumptive identification of *Salmonella* serovars and is useful for the other enteric bacteria and Gram-negative bacteria generally. The reactions are given in Tables 10-III and 10-IV. It must be remembered that not all salmonellae produce hydrogen sulfide (for example, *S. choleraesuis* does not do so), but the biovar Kunzendorf is positive. *S. typhisuis* varies in the production of H_2S and is lysine negative. The majority of the *Salmonella* serovars, however, are lysine-decarboxylase positive.

TABLE 10-III
REACTIONS NOTED IN TRIPLE SUGAR IRON AGAR SLANTS

Appearance	*Reactions*
Acid butt: yellow, alkaline slant: red	Glucose fermented
Acid throughout medium: butt and slant yellow	Glucose, and sucrose and/or lactose fermented
Gas bubbles in butt and medium frequently split	Gas production
Butt shows blackening	Hydrogen sulfide produced
Unchanged or alkaline butt and slant: medium red throughout	None of the three sugars fermented

No visible change in TSI is observed with *Moraxella* spp., *Pseudomonas aeruginosa*, *Alcaligenes faecalis*, *Bordetella* spp., *Flavobacterium* spp., *Acinetobacter* spp., and most *Neisseria* species.

Organisms Other Than Members of the Enterobacteriaceae Recovered from Feces

The organisms listed below are encountered on the common enteric media and blood agar inoculated from feces.

Pseudomonas aeruginosa (frequent)
Acinetobacter calcoaceticus (frequent)
Aeromonas (frequent)
Yersinia enterocolitica (infrequent)
Alcaligenes faecalis (infrequent)

TABLE 10-IV

REACTIONS OF GRAM–NEGATIVE BACTERIA
ON TSI SLANTS AND IN LYSINE DECARBOXYLASE BROTH

Reactions		Organisms expected and Lysine Reaction		
		Lysine +ve	Lysine −ve	Lysine variable
Acid slant		*Serratia* spp.	*Aeromonas hydrophila*	*Pasteurella* spp.
Acid butt	may be	*Aeromonas hydrophila*	ss. *anaerogenes*	*E. coli* "inactive"
No H₂S	weak	ss. *proteolytica*	*Erwinia* spp.	*Pseudomonas* spp.
No gas			*Actinobacillus* spp.	(some)
			Shigella sonnei	
			Pseudomonas pseudo-	
			mallei	
			Tatumella spp.	
			Yersinia enterocolitica	
Acid slant		*E. coli*	*Cedecea* spp.	*Kluyvera* spp.
Acid butt	strong	*Enterobacter aerogenes*	*Rahnella* spp.	
No H₂S	reaction	*Enterobacter gergoviae*	*Aeromonas hydrophila*	
Gas in butt		*Klebsiella* spp.	ss. *hydrophila*	
			Enterobacter cloacae	
			Providencia (some)	
Acid slant		"Arizona" (some)	*Citrobacter* (some)	
Acid butt			*Proteus vulgaris*	
H₂S produced				
Gas in butt				
Alkaline slant		*Obesumbacterium* spp.	*Shigella* spp.	
Acid butt			*Providencia* spp.	
No H₂S			*Yersinia pseudo-*	
No gas			*tuberculosis*	
			Xenorhabdus spp.	
			Pasteurella multocida	
			(some)	
Alkaline slant		*Salmonella choleraesuis*	*Salmonella typhisuis*	
Acid butt		*Salmonella sendai*	(some)	
No H₂S		*Salmonella abortus-equi*	*Salmonella paratyphi* A	
Gas in butt		*Hafnia alvei*	*Morganella morganii*	
Alkaline slant		*Salmonella pullorum*		
		(some)		
Acid butt		*Salmonella gallinarum*		
H₂S produced		*Salmonella typhi*		
No gas				
Alkaline slant		*Salmonella* serovars	*Citrobacter* spp.	
Acid butt		(most)	*Proteus vulgaris* (some)	
H₂S produced		"Arizona" serovars	*Proteus mirabilis*	
Gas in butt		(most)	*Salmonella typhisuis*	
		Edwardsiella tarda	(some)	

TABLE 10-V

DIFFERENTIATION OF SOME ENTERIC BACTERIA BY BIOCHEMICAL TESTS

- − = 0–10% of strains positive
- (−) = 11–25% of strains positive
- d = 26–75% of strains positive
- (+) = 76–89% of strains positive
- + = 90–100% of strains positive

	Indole production	Methyl red	Voges-Proskauer	Citrate (Simmons')	Hydrogen sulfide (TSI)	Urease (Christensen's)	Phenylalanine	Lysine decarboxylase	Arginine dihydrolase	Ornithine decarboxylase	KCN (growth in)	Motility	Gelatin liquefaction	Malonate	Glucose (gas)	Lactose	Sucrose	Mannitol	Dulcitol	Salicin	Adonitol	Inositol	Sorbitol	Arabinose	Raffinose	Rhamnose	Esculin hydrolysis	ONPG (β-Galactosidase)
Citrobacter amalonaticus	+	+	−	(+)	−	(+)	−	−	(+)	+	+	+	−	(−)	+	d	(−)	+	d	d	−	−	+	+	−	+	−	+
C. diversus	+	+	−	(+)	−	(+)	−	−	d	+	−	+	−	(−)	+	d	(−)	+	d	(−)	+	−	+	+	d	+	d	+
C. freundii	−	+	−	+	(+)	d	−	−	d	(−)	+	+	−	(−)	+	d	d	+	d	−	−	−	+	+	d	+	+	+
Edwardsiella tarda	+	+	−	−	+	−	−	+	−	+	−	+	−	−	+	−	−	−	−	−	−	−	−	−	−	−	−	−
E. tarda biogroup 1	+	+	−	−	−	−	−	+	−	+	−	+	−	−	d	−	+	+	−	−	−	−	−	+	−	−	−	−
Enterobacter aerogenes	−	−	+	+	−	−	−	+	−	+	+	+	−	+	+	+	+	+	−	+	+	+	+	+	+	+	+	+
E. cloacae	−	−	+	+	−	d	−	−	+	+	+	+	−	(+)	+	+	+	+	d	(+)	(−)	(−)	(+)	+	+	+	d	+
E. gergoviae	−	d	+	+	−	+	−	+	−	+	+	+	−	+	+/−	+	+	+	+	(−)	(−)	−	+	+	+	+	+	+
Escherichia coli	+	+	−	−	−	−	−	+	(+)(−)	d	−	d	−	−	+	+	d	+	d	d	−	−	+	+	d	(+)	d	+
E. coli (inactive)	(+)	+	−	−	−	−	−	d	−	(−)	−	(−)	−	−	+/−	(−)	(−)	+	d	−	−	−	(+)	(+)(−)	(−)	d	−	d
Hafnia alvei	−	d	d	−	−	−	−	+	−	+	+	+	−	d	+	−	−	+	−	(−)	−	−	−	+	−	+	−	+
Klebsiella oxytoca	+	d	+	+	−	+	−	+	−	−	+	−	−	+	+	+	+	+	d	+	+	+	+	+	+	+	+	+
K. pneumoniae ss. *pneumoniae*	−	(−)	+	+	−	+	−	+	−	−	+	−	−	+	+	+	+	+	d	+	+	+	+	+	+	+	+	+
Morganella morganii	+	+	−	−	−	+	+	−	−	+	+	+	−	−	(+)	−	−	−	−	−	−	−	−	−	−	−	−	−
Proteus mirabilis	−	+	(−)	d	+	+	+	−	−	+	+	+	+	−	(+)	+	(−)	−	−	−	−	−	−	−	−	d	d	−
P. vulgaris	+	+	−	(−)	+	+	+	−	−	−	+	+	+	−	(+)	+	+	−	−	(+)	−	−	−	−	−	+	(+)	+
Providencia alcalifaciens	−	+	−	+	−	+	+	−	−	+	+	+	−	−	(+)	−	(−)	−	−	d	+	−	−	−	−	d	d	−
P. rettgeri	+	+	−	+	−	+	+	−	−	+	+	+	−	−	−	−	(−)	+	−	d	+	+	−	−	−	d	d	−
P. stuartii	+	+	−	+	−	d	+	+	−	−	(+)	+	−	+	−	−	d	(−)	−	−	−	+	−	−	−	−	−	+
Salmonella subgenus 1	−	+	−	d	+	−	−	+	d	+	−	+	−	−	+	−	−	+	+	−	−	d	+	+	−	+	−	−
subgenus 111 ("Arizona")	−	+	−	+	+	−	−	+	(+)	+	−	+	+	+	+	d	−	+	−	−	d	−	+	+	−	+	+	+
Serratia liquefaciens	−	(+)	(+)	+	−	−	−	+	−	+	+	+	+	−	d	−	+	+	−	+	−	d	(+)	+	+	(−)	+	+
S. marcescens	−	(−)	+	+	−	(−)	−	+	−	+	+	+	+	−	d	−	+	+	−	d	d	(+)	+	−	d	d	+	+
S. odorifera	d	(+)	(+)	+	−	−	−	+	d	d	d	+	d	−	−	+	d	+	−	d	d	+	+	+	d	+	d	+

All grow on MacConkey agar. Readers are referred to the separate discussions of these organisms for additional information.

E. COLI

In general, the serologic classifications of the enteric bacteria parallel the Kauffman-White scheme of the *Salmonella*. Identification of the serotypes of this species is not carried out routinely in most veterinary diagnostic laboratories. The antigens used to designate serotypes are as follows:

Somatic or O antigens: designated by arabic numerals, e.g. 0133.

K (surface or envelope) antigens: these thermolabile antigens designated by the letters L, B, or A with an arabic number, e.g. K4 (B).

H or flagellar antigens: designated by H followed by an arabic number, e.g. H2.

An example of a complete designation would be O111: K4 (B), H2.

Some of the serotypes and O groups and the animal species with which they have been associated are listed in Table 10-VI.

T-10-6

TABLE 10-VI

SOME SEROTYPES AND O GROUPS OF *E. COLI* ASSOCIATED WITH DISEASE IN DOMESTIC ANIMALS

Animal Species	Serotypes and O Groups
Swine	0139:K82(B)
	0141:K85a, b (B)
	0141:K88(L), 0141:K85 a,
	c (B), 0149:K91
	0138:K81 (B); 08:K87 (B?)
	08:K88(L); 018?:K?;
	045:K?
Calves	08, 09, 015, 017, 020, 021,
	026, 035, 055, 078, 0119,
	0114, 015 ,0126, 0137
	(many others reported)
Chickens	02, 078, 01, 08, 011, 022
Lambs	020, 078:K80, 024:K?
	078:K80(B), 08, 09, 015,
	026, 035, 086, 0101, 0137.

The procedure followed in serotyping is as follows:
1. Polyvalent O B antisera are employed in a slide test for grouping.
2. Then, tube tests are carried out with specific O antisera covering those types in the group in order to determine the O group.

Sera are not available commercially for most of the serotypes of importance in animal disease.

It is now known that some enteropathogenic strains of *E. coli* produce one or two enterotoxins. They are referred to as the heat-stable toxin (ST) and the heat-labile toxin (LT). Procedures are now available for the demonstration of the toxins from the enterotoxic (ETEC) strains. These procedures include a suckling mouse test (9) and a rabbit ileal loop test (10) for ST, and a test for LT in chinese hamster ovary (11) and mouse adrenal tissue culture cells (12). These tests are somewhat involved and have limitations for animal isolates of *E. coli*. Immunological assays are being developed, such as an ELISA test for LT, and these give some hope for a suitable diagnostic test for detecting the ETEC strains. Because the capacity to produce enterotoxin can be transmitted by plasmids from one strain of *E. coli* to another, it would follow that in assessing the pathogenic significance of strains isolated from cases of diarrheic disease, it is more important to determine if a given culture produces enterotoxin than to determine its serologic identity. However, the K88 and K99 antigens are known as colonization factors and equate with virulence in calves and pigs. *E. coli* strains with the K99 antigen are also pathogenic for lambs.

Invasiveness has recently been identified as a distinct, but less common, pathogenic mechanism. Enteroinvasive *E. coli* strains (EIEC) can penetrate the intestinal epithelium, mainly that of the large intestine. Invasiveness can be detected by a strain's ability to produce keratoconjunctivitis in the eye of a guinea pig (Sereny test [13]), or by its capacity to penetrate cells in tissue culture (14).

EDWARDSIELLA

On blood agar, strains of *E. tarda* resemble hemolytic *E. coli*. In contrast to the latter, they have the appearance of a salmonella on TSI (see Table 10-IV).

SHIGELLA

Polyvalent *Shigella* sera are available for presumptive identification of strains. Workers are referred to Martin and Washington (4) for the identification of species. Because shigella strains are only rarely isolated in veterinary diagnostic laboratories, detailed serologic examination is not usually carried out.

SALMONELLA

The antigens of *Salmonella* serovars:

SOMATIC OR O ANTIGENS. Thermostable and designated by arabic numerals.

FLAGELLAR ANTIGENS: PHASE 1. Designated by small letters of the alphabet; more or less specific for the salmonellae.

PHASE 2. Designated by Arabic numerals; less specific and duplicated in other bacterial species.

K ANTIGENS (CAPSULAR OR ENVELOPE). Vi antigen, M antigen, and 5 antigen; M or mucoid antigen. These antigens may interfere with agglutinability of O antisera. Boiling of suspensions for ten to twenty minutes destroys the Vi antigen.

The antigenic makeup of some *Salmonella* spp. commonly isolated from animals and animal sources (15) is given in Table 10-VII.

In most veterinary diagnostic laboratories, the salmonella isolates are examined serologically in order to determine the group to which they belong. Group identification is based on the possession of certain somatic antigens. *Salmonella* O antisera are available commercially covering groups A through I. The procedure is a simple slide agglutination test. It is usual to test an isolate first against a polyvalent O serum covering groups A to I. If this is positive, then tests are conducted with the individual group sera.

The culture of *Salmonella* to be serotyped can be taken from a TSI slant or from nutrient agar. Growth from selective media is often unsuitable for typing.

Proteus species may react with the polyvalent salmonella sera; however, *Proteus* species are urease positive and lysine decarboxylase negative, which distinguishes them from the salmonellae.

Further identification to the serovar is carried out with flagella or H antisera. Many serovars are diphasic, having flagella antigens in phase 1 ("specific") and phase 2 ("group" or nonspecific). A culture of *Salmonella* may have organisms in both phases or in just one of the phases. In the latter case, the culture, although capable of giving rise to the alternative phase, usually maintains a constant phase over several generations. Both flagella phases must be identified to obtain the complete antigenic formula of the *Salmonella* and hence determine the serovar. This requires techniques of phase changing, which are rather involved and usually only carried out in reference laboratories.

Lysogenization by certain converting-phages may produce changes in the O-antigenic formulae of the salmonellae. In antigenic groups A, B, and D, the presence of O-antigen 1 (factor 1) is associated with lysogenization, but the presence or absence of this factor in strains of these groups does not alter the name of the serovar. However, in group E, phage E15 alters the O-antigen 3,10 to 3,15, thus making *S. anatum* become *S. newington*. Other serovars in group E are also involved (5).

Variations can occur in salmonellae. To prevent smooth-rough (S→R) dissociation, the freshly isolated strains should be maintained on media without added carbohydrate. Rough strains may autoagglutinate in saline and are unsuitable for typing. Flagellated serovars may give rise to non-

TABLE 10-VII

SOME IMPORTANT *SALMONELLA* SEROVARS
RECOVERED FROM ANIMALS, BIRDS, AND ANIMAL SOURCES (15, 16)

Serovars	Group	Somatic antigens	Flagella antigens Phase 1	Flagella antigens Phase 2	Chickens, Turkeys	Swine	Cattle	Equines	All other birds and animals	Reptiles	Feeds and all other sources
S. typhimurium	B	1,4,[5],12	i	1,2	+	+	+	+	+	•	+
S. abortus-equi*	B	4,12	-	e,n,x				•			
S. abortus-ovis*	B	4,12	c	1,6				•			
S. bredeney	B	1,4,12,27	l,v	1,7	+	•		•	•	•	•
S. derby	B	1,4,[5],12	g,f	–	+	+	•	•	•	•	•
S. agona	B	4,12	f,g,s	–	+	+	•	+	+	•	•
S. saint paul	B	1,4,[5],12	e,h	1,2	+			•	•	+	
S. reading	B	1,4,[5],12	e,h	1,5	+			•	•	+	
S. heidelberg	B	1,4,[5],12	r	1,2	+	•	•		•	+	•
S. san diego	B	4,[5],12	e,h	e,n,z_{15}	+	•	•			•	•
S. typhisuis	C_1	6,7	c	1,5		•	•				
S.choleraesuis	C_1	6,7	c	1,5		+					
S.choleraesuis biovar Kunzendorf	C_1	6,7	[c]	1,5		+					
S. infantis	C_1	6,7,14	r	1,5	+	•	•	•	•	•	•
S.oranienburg	C_1	6,7	m,t	–	•	•	•	•	•	•	•
S.montevideo	C_1	6,7	g,m,[p,s	-	+	•	•	•	•	•	•
S.newport	C_2	6,8	e,h	1,2	•	•	•	•	•	•	•
S.blockley	C_2	6,8	k	1,5	•		•	•		•	•
S.muenchen	C_2	6,8	d	1,2	•	•	•	•		•	•
S.manhattan	C_2	6,8	d	1,5	•	+		•	•	•	+
S.bovismorbificans*	C_2	6,8	r	1,5		•		•		•	
S.kentucky	C_3	8,20	g,s,t	–	•		+			•	
S.panama	D_1	19,12	l,v	1,5	•	+		•		•	
S.gallinarum*	D_1	1,9,12	–	–	•						
S.pullorum	D_1	9,12	–	–	+			•			
S.enteritidis	D_1	1,9,12	g,m	[1,7]	•	•	•	•		•	
S.dublin	D_1	1,9,12,[Vi]	g,p	–		•	+		•		•
S.anatum	E_1	3,10	e,h	1,6	+	•	•	•	•	+	•
S.london	E_1	3,10	l,v	1,6	•	+	•	•	•		•
S.meleagridis	E_1	3,10	e,h	1,w	•	•	•		•		•
S.give	E_1	3,10	l,v	1,7	•	•	•	•	•	•	•
S.newington	E_2	3,15	e,h	1,6	•	•	•			•	•
S.senftenberg	E_4	1,3,19	g,[s],t	–	•	•	•		•		•
S.worthington	G_2	1,13,23	z	1,w	•	•		•	•		•
S.cubana	G_2	1,13,23	z_{29}	$[z_{37}]$			+				•

1 = 0 antigen whose presence is due to phage conversion; [] = antigen present of absent; + = commonly isolated; • = occasionally isolated; * = not listed as commonly occurring in U.S.A.

flagellated variants (OH→O variation). This change tends to be irreversible. Some serovars, such as *S. pullorum*, are permanently without flagella.

The antigenic schema for salmonellae in subgenus 111 ("Arizona") is based on O and H antigens. More than 300 serovars have been identified. The antigenic formulae of most of these "Arizona" serovars have been translated into *Salmonella* formulae and included in the Kauffmann-White scheme (5). The key differential characteristics between salmonella subgenus 1 and salmonella subgenus 111 ("Arizona") are given in Table 10-VIII.

TABLE 10-VIII

MAIN CHARACTERISTICS DIFFERENTIATING
SALMONELLA SUBGENERA 1 AND 111.

	Salmonella subgenus 1	*Salmonella* subgenus 111 ("Arizona")
Malonate	−	+
Lactose	−	v
ONPG	−	+
Dulcitol	+	−

v = variable reaction

SALMONELLA BIOVARS. Biovars have the same antigenic formula, but they may differ in the disease syndrome that they cause and in certain biochemical reactions. The differential biochemical characteristics for some important veterinary biovars are given in Table 10-IX.

TABLE 10-IX

BIOCHEMICAL DIFFERENTIATION OF *SALMONELLA* BIOVARS

	S. choleraesuis	*S. choleraesuis* biovar Kunzendorf	*S. typhisuis*	*Salmonella* (most serovars)
H_2S (TSI)	−	+	v	+
Lysine	+	+	−	+
Citrate (Simmons')	+	+	−	+
Mannitol	+	+	−	+
Inositol	−	−	+	v
Sorbitol	(+)	(+)	−	+

	S. pullorum	*S. gallinarum*	*Salmonella* (most serovars)
Glucose (gas)	(+)	−	+
Dulcitol	−	+	+
Maltose	−	+	+
Ornithine	+	−	+
Rhamnose	+	−	+

v = variable reactions
(+) = most strains positive.

KLEBSIELLA

More than seventy capsular types of *Klebsiella* have been identified. Typing is carried out by some reference laboratories.

PROTEUS, PROVIDENCIA, AND *MORGANELLA*

Species are identified on the basis of characteristics listed in Table 10-X. All deaminate phenylalanine. The main differentiating characteristics are shown in Table 10-V.

TABLE 10-X

DIFFERENTIATION OF THE GENERA *PROTEUS, PROVIDENCIA,* AND *MORGANELLA*

	Proteus	*Providencia*	*Morganella*
Swarming	+	−	−
H$_2$S (TSI)	+	−	−
Gelatin liquefaction	+	−	−
Citrate (Simmons')	v	+	−
Mannose	−	+	+
Maltose	v	−	−
Ornithine	v	−	+
Urease (Christensen's)	+	v	+

v = variable

OTHER ENTERIC ORGANISMS

Biochemical reactions for *Citrobacter, Enterobacter, Hafnia,* and *Serratia* species are given in Table 10-V. The pathogenicity for animals is unknown for the recently named genera *Cedecea, Kluyvera, Rahnella, Tatumella, Obesumbacterium,* and *Xenorhabdus.* Some characteristics for these genera are given in Table 10-XI. For full identification see *Bergey's Manual of Systematic Bacteriology* (2).

REFERENCES

1. Buchanan, R. E., and Gibbons, N. E. (Eds.): *Bergey's Manual of Determinative Bacteriology,* 8th ed. Baltimore, Williams & Wilkins, 1974.
2. Brenner, D. J.: The family Enterobacteriaceae. In Krieg, N. R. (Ed.): *Bergey's Manual of Systematic Bacteriology,* 1st ed., vol 1. Baltimore, Williams & Wilkins, 1983.
3. John, J. F., Jr., Sharbaugh, R. J., and Bannister, E. R.: *Reviews Infect Dis,* 4:13, 1982.
4. Martin, W. J., and Washington, J. A., II: In Lennette, E. H., Balows, A., Hausler, W. J., and Truant, J. P. (Eds.): *Manual of Clinical Microbiology,* 3rd ed. Washington, D.C., American Society for Microbiology, 1980, Chapter 16.
5. Le Minor, L.: Salmonella. In Krieg, N. R. (Ed.): *Bergey's Manual of Systematic Bacteriology,* 1st ed., vol 1. Baltimore, Williams & Wilkins, 1983.
6. Roussel, A., Lucas, A., and Bouley, G.: *Rev Path Comp,* 37(2):27–29, 1969.

TABLE 10-XI

SOME CHARACTERISTICS OF THE RECENTLY NAMED GENERA

	Glucose (gas)	Motility	Reaction on TSI (Slant/butt)	H_2S (TSI)	Lysine	Indole	Lactose	Sucrose	Esculin hydrolysis	Isolation from human clinical specimens.
Cedecea	+	+	A/A	−	−	−	v	v	v	+
Kluyvera	+	+	A/A	−	v	+	+	+	+	+
Rahnella	+	−	A/A	−	−	−	+	+	+	+
Tatumella	−	−	A/A	−	−	−	−	+	−	+
Obesumbacterium	−	−	Ak/A	−	+	−	−	−	−	−
Xenorhabdus	−	+	Ak/A	−	−	v	−	−	−	−

v = variable
Ak = alkaline
A = acid reaction

REFERENCES (cont.)

7. Iveson, J. B., and Kovacs, N.: *J Clin Path, 20*:290, 1967.
8. Blessman, B. H., Morse, E. V., Midla, D. A., and Swaminathan, B.: *Am Assoc Veterinary Laboratory Diagnosticians*, 24th Ann Proc. pp. 1–10, 1981.
9. Dean, A. G., Ching, Y., Williams, R. G., and Harden, L. B.: *J Infect Dis, 125*:407, 1972.
10. Annapurna, E., and Sanyal, S. C.: *J Med Microbiol, 10*:317, 1977.
11. Guerrant, R. L., Brunton, L. L., Schnaitman, T. C., Rebhun, L. I., and Gilman, A. G.: *Infect Immun, 10*:320, 1974.
12. Donta, A. G., Moon, H. W., and Whipp, S. C.: *Science, 182*:334, 1974.
13. Sereny, B.: *Acta Microbiol Acad Sci Hung, 2*:293, 1955.
14. Sonnenwirth, A. C.: In Davis, B. D., Dulbecco, R., Eisen, H. N., and Ginsberg, H. S. (Eds.): *Microbiology*, 3rd ed. Hagerstown, Harper and Row, 1980, Chapter 31.
15. Anon.: *CDC Salmonella Surveillance*, Annual Summary 1979, HHS Publication No (CDC) 82-8219, 1981.
16. McWhorter, A. C., Fife-Asbury, M. A., Huntley-Carter, G. P., Brenner, D. J.: *Modified Kauffmann-White Schema for Salmonella and Arizona*, HEW Publication, No (CDC) 78-8363, 1977.

Chapter 11

PASTEURELLA, YERSINIA, AND FRANCISELLA

IT HAS BEEN widely recognized that the species formerly called *Pasteurella pestis* and *P. pseudotuberculosis* are sufficiently different from *Pasteurella multocida* to warrant the formation of a new genus. The name that has been recommended and widely accepted is *Yersinia*. In *Bergey's Manual, Yersinia* is included in the Enterobacteriaceae. In addition to having an optimum growth at temperatures below 37°C, *Yersinia* are oxidase negative in contrast to *Pasteurella*.

The species that in recent years has been called *Pasteurella tularensis* likewise bears no close relationship to *Pasteurella multocida* or *P. pestis* and *P. pseudotuberculosis* and consequently has been given the new generic name *Francisella*.

PASTEURELLA

Members of this genus are small, nonmotile, Gram-negative rods or coccobacilli. They are facultatively anaerobic, non-sporeforming, fermentative (except *P. anatipestifer*), and oxidase positive.

P. anatipestifer is a nonfermenter and clearly does not belong in the genus *Pasteurella*. Its proper taxonomic place is still uncertain.

PASTEURELLA MULTOCIDA
Synonyms: *Pasteurella septica, P. avicida, P. gallicida*

This is a rather heterogeneous species that may eventually be divided into a number of different biotypes based upon differences in pathogenicity, serological nature, and cultural, morphological, and biochemical characteristics. A number of different serotypes have been identified based upon capsular and somatic antigens (1).

PATHOGENICITY. The species occurs widely in the upper respiratory and digestive tracts of a wide range of birds and mammals.

The diseases with which it is associated are too numerous to list comprehensively. It is a frequent secondary invader or opportunist in a number of pathologic processes.

It is a primary or more frequently a secondary invader in pneumonia of cattle, swine, sheep, goats, and other species, and as well it is involved frequently in the bovine "shipping fever complex" and in enzootic pneumonia of pigs. It is considered the primary cause of fowl cholera and epizootic hemorrhagic septicemia of cattle, bison, and water buffaloes. It is one of the causes of the pleuropneumonia form of "snuffles" in rabbits and a cause of severe mastitis of cattle and sheep.

Because dogs and cats harbor these organisms in their mouths, as commensals, bites inflicted upon humans and other animals are frequently infected with *P. multocida*. A wide variety of infections have been reported in humans.

PASTEURELLA HAEMOLYTICA

Two different biotypes of *P. haemolytica*, designated A and T, have been described based upon differences in fermentative activity, serological characteristics, and pathogenicity (2). Different serotypes have been identified based on capsular and somatic antigens (2).

PATHOGENICITY. This organism is frequently involved as a primary or secondary agent in pneumonias of cattle, sheep, goats, and swine. It is commonly recovered from the bronchopneumonic lungs of cattle with shipping fever. Other important diseases in which it is involved are mastitis of ewes and septicemia of lambs. Strains are sometimes recovered from respiratory infections and salpingitis of fowl.

AVIAN PASTEURELLA HAEMOLYTICA

Organisms isolated from chickens and turkeys that have been called *P. haemolytica* differ in several characteristics from the ruminant strains of *P. haemolytica*. They have larger zones of hemolysis, and unlike ovine and bovine strains of *P. haemolytica* they do not usually ferment dextrin and maltose (3). Avian strains usually ferment trehalose while bovine strains do not (3). Similar strains have been recovered from swine and horses (2).

PATHOGENICITY. This organism has only a low capacity for causing disease. It has been associated with salpingitis and respiratory infections. Infections are usually secondary to another disease or some predisposing condition.

PASTEURELLA PNEUMOTROPICA

Frederiksen (4) refers to the *P. pneumotropica* Complex, which includes at least two different biotypes as well as other varieties with different biochemical characteristics. The Henriksen biotype, which includes strains called *Pasteurella* "Gas," is recovered from dogs and cats. The Jawetz biotype is most often recovered from rodents. Readers are referred to Frederiksen's paper for the identification of these biotypes and varieties.

PATHOGENICITY. This organism, which is a frequent inhabitant of the upper respiratory tract of rodents, has been recovered from pneumonia and other infections, including abscesses in guinea pigs, rats, and hamsters. It is found occasionally as a commensal in the nasopharynx of dogs and cats. Human infections associated with bites are uncommon.

PASTEURELLA AEROGENES

This gas-producing organism, which is probably worldwide in distribution, has most frequently been isolated from the feces of swine (5). There have been recent reports of several isolations from cattle and rabbits (6). It appears to have little potential for causing disease except as a rare secondary invader or opportunist.

PASTEURELLA GALLINARUM

This organism, which is now considered an official species, is found as a commensal in the upper respiratory tract of chickens and turkeys (7). Its potential for causing disease is low, and in respiratory disease processes it is usually present as a secondary invader. Mraz et al. (8) have described the characteristics of this species in some detail.

PASTEURELLA UREAE

This organism occurs infrequently in the nasopharynx of normal humans as a commensal. It has been associated with chronic bronchitis and low-grade infections of the upper respiratory tract of human beings. There are reports of its isolation from the reproductive tract of mice (9) and from swine fetuses (10).

PASTEURELLA ANATIPESTIFER
Synonyms: *Moraxella anatipestifer, Pfeifferella anatipestifer*

As mentioned previously, this organism is a nonfermenter and should not be included in the genus *Pasteurella*.

PATHOGENICITY. It is the cause of an acute or chronic septicemic disease (infectious serositis) of one- to eight-week-old ducklings. Among the principal lesions are those of a fibrinous polyserositis (3). Infections have also been reported in pheasants, quail, waterfowl, and turkeys.

GROUP EF-4

Strains designated EF-4 resemble somewhat *P. multocida*. They are commensals of the oral cavity or respiratory tract of dogs and cats in which they do not usually cause disease (11). They are most often encountered in human infections of dog and cat bite wounds.

PASTEURELLA MULTOCIDA-LIKE ORGANISMS

Occasional cultures are encountered from dogs, cats, poultry, and other animals that resemble *P. multocida* but differ in one or several of the characteristics ascribed to typical strains of this species. If the difference is only in the fermentation of one carbohydrate, they can still be considered *P. multocida*. A number of cultures will be indole negative in SIM medium, peptone water, or tryptone broth. The author has found that heart infusion broth (Difco) supplemented with 2% tryptone is highly satisfactory for indole production by *P. multocida*. Clemons and Gadberry (12) described a medium supplemented with 2% peptone that gave excellent results.

Isolation Procedures

The specimens from which pasteurellas are isolated are quite varied. The preferred medium for the recovery of *Pasteurella* spp. is blood agar. Selective media for the isolation of *P. multocida* and *P. haemolytica* have been described (13).

Cultural Characteristics

P. multocida. Colonies appear after incubation for twenty-four hours at 37°C. They are usually of moderate size, round, and grayish. Some strains produce large mucoid colonies. Fresh cultures have a characteristic musty odor.

P. haemolytica. Satisfactory growth is obtained after twenty-four hours incubation. Colonies are round, grayish, and somewhat smaller than those of *P. multocida*. They are usually surrounded by a zone of beta-hemolysis. This zone varies considerably and may be no larger than the colony and thus not apparent unless the colony is removed. Antibodies in media may inhibit hemolysis. Bovine blood is more suitable than that of the sheep or horse for the demonstration of hemolysis.

P. pneumotropica and *P. gallinarum.* The colonies of these species are indistinguishable from those of *P. multocida*.

P. anatipestifer. This organism grows best on blood or serum agar in an atmosphere of 5–10% carbon dioxide. A candle jar is satisfactory. Small dewdroplike colonies appear within forty-eight hours.

P. aerogenes. The colonies of this species resemble those of nonmucoid *P. multocida* except that they are somewhat smaller. On MacConkey's agar, colonies resemble those of *Salmonella* in twenty-four hours, but in thirty-six hours, they develop a faint pinkish color.

P. ureae. The colonies of this organism resemble those of *P. haemolytica*.

Identification

Gram-stained smears from colonies of the species referred to above reveal small Gram-negative rods or coccobacilli (Fig. 11-1). The reactions on TSI are represented in Table 10-IV. Definitive identification is based upon differential characteristics listed in Table 11-I. Differential characteristics of the two biotypes of *P. haemolytica* are given in Table 11-II. *P. anatipestifer* differs from the other pasteurellas in producing gelatinase, being nonfermentative, and not reducing nitrate.

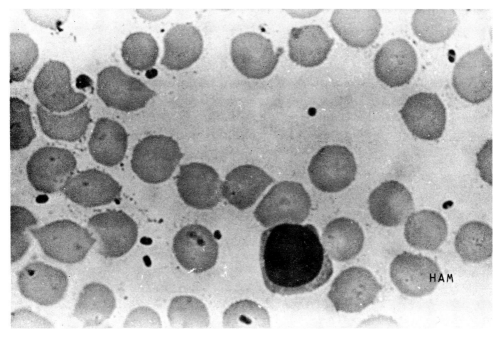

Figure 11-1. *Pasteurella multocida* in blood smear from experimentally infected mouse. Note bipolarity. Wright's stain, ×2250 (H. A. McAllister).

Carter (14) has put forward a proposal for five different biotypes of *P. multocida* on the basis of difference in hyaluronidase decapsulation, acriflavine flocculation, colonial irridescence, action on several carbohydrates, mouse pathogenicity, and serum protection tests. The biotype names proposed were "the mucoid," "the hemorrhagic septicemia," "the porcine," "the canine," and "the feline."

Animal Inoculation

Pure cultures of many strains of *P. multocida* can be obtained by inoculating clinical materials into mice or rabbits subcutaneously.

YERSINIA

Members of this genus are Gram-negative, small rods or coccobacilli. They are facultatively anaerobic, non-sporeforming, fermentative, and oxidase negative. Species of this genus possess characteristics of the Enterobacteriaceae and are now placed in this family.

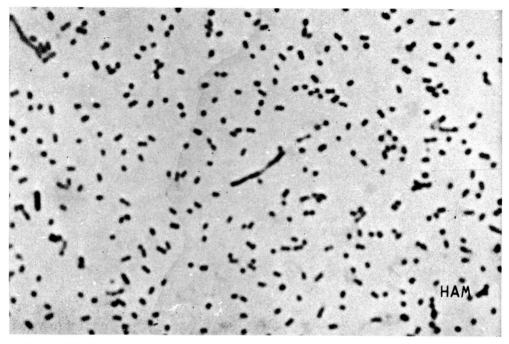

Figure 11-2. *Pasteurella multocida* from a blood agar plate culture. Gram's stain, × 3000 (H. A. McAllister).

TABLE 11-I

DIFFERENTIAL CHARACTERISTICS OF *PASTEURELLA* SPECIES

	MacConkey's Agar	Indole	Urease	Fermentation: Glucose	Lactose	Sucrose	Mannitol
P. multocida	—*	+	—	A	(—)	A	(A)
P. haemolytica	+	—	—*	A	(A)	A	A
P. pneumotropica	—*	+	+	A	(A)	A	—
P. gallinarum	—	—	—	A	—	A	—
P. anatipestifer	—	—	—	—	—	—	—
P. ureae	—	—	+	A	—	A	A
P. aerogenes	+	—	+	A/G	—	AG	—

*Some exceptions

Pathogenicity

Yersinia pestis. Synonym: *Pasteurella pestis.* Plague is basically a disease of rats and many wild rodents, including squirrels and marmots. Humans are

Table 11-II

Differential Characteristics of Biotypes of *Pasteurella haemolytica*

	Biotype A	Biotype T
Fermentation:		
arabinose	+	−
trehalose	−	+
salacin	−	+
xylose	+	−
lactose	+	−
Susceptibility to	(except serotype 2)	
penicillin:	high	low
Serotypes:	1,2,5,6,7,8,9,11,12	3,4,10
Principla location		
in normal host:	nasopharynx	tonsils
Principal disease		
association:	Pneumonia in cattle	septicemia in
	and sheep	feeder lambs
	septicemia in	
	nursing lambs.	

*With permission from Biberstein, E. L. In Bergan, T. (Ed.): *Methods in Microbiology*, Vol. 10, 19xx. Copyright Academic Press, Inc. (London) Ltd., 1978.

considered an accidental host. The organism may kill certain rodents in widespread outbreaks or survive in a latent form in others. Natural infections have been reported in dogs and cats (15).

Yersinia pseudotuberculosis. Synonym: *Pasteurella pseudotuberculosis.* It is the cause of pseudotuberculosis of wild and laboratory rodents. This disease is characterized in its chronic form by the presence of small necrotic nodules in mesenteric lymph nodes, liver, spleen, and lungs. An acute septicemic form is also encountered.

Other species infected include chinchilla, turkeys, rams (epididymoorchitis), and swine. The disease in turkeys may reach epizootic proportions. Severe infections simulating typhoid and appendicitis occur in humans.

Yersinia enterocolitica. Synonym: *Bacterium enterocoliticum.* This widespread organism has been isolated with increasing frequency from humans and animals in recent years. It has been reported as a cause of a pseudotuberculosis-type disease in swine and in chinchilla. Sporadic infections have been observed in hares, birds, cats, minks, dogs, a bushbaby (Galago), sheep, rats, and guinea pigs. Isolations have also been made from cows, horses, feces, water, and milk.

Infections in humans resemble those caused by *Y. pseudotuberculosis.* In addition, the organism has been recovered from cutaneous lesions, from cerebrospinal fluid in meningitis, from the blood in septicemia and bacteremia, from liver abscesses, and from feces in cases of enterocolitis.

OTHER *Yersinia* SPECIES. Three additional species, *Y. intermedia, Y. frederikesii*, and *Y. ruckeri*, have been described (16). They do not appear to have yet been associated with disease in animals or enteric disease in humans.

Isolation Procedures

Organisms of this genus are readily isolated and cultivated on nutrient, tryptose, trypticase, and blood agar. They also grow on such selective and enrichment media as MacConkey's, SS, and desoxycholate agar, and also in selenite and Rappaport broth. Room temperature is preferred to 37°C. *Y. enterocolitica* has been recovered frequently from animal feces. Large inocula and storage in the refrigerator as with listeria have been recommended for the isolation of *Y. pseudotuberculosis*. If one of these species is suspected, plates should be incubated at 22–25°C (room temperature) and at 37°C for not less than forty-eight hours. Longer incubation may be required for enrichment media for the recovery of *Y. enterocolitica*.

The best procedure for the recovery of *Y. enterocolitica* from feces is by cold enrichment as follows: Place some feces, approximately 5% by volume, in 1/15 phosphate buffered saline (PBS) and hold in the refrigerator (4°C) for three weeks. Plate onto MacConkey and SS agar after seven, fourteen, and twenty-one days of enrichment. See Appendix B for preparation of the PBS.

Cultural Characteristics

Y. pestis. After twenty-four hours, small mucoid colonies are visible along the initial streak lines. On further incubation, colonies enlarge and assume a beaten copper appearance. They are raised in the center and flat at the periphery, with an umbonate edge.

Y. pseudotuberculosis. Colonies are round, finely granular, grayish yellow, translucent, and centrally opaque, with a flat periphery showing striations.

Y. enterocolitica. After twenty-four hours incubation on blood agar, colonies are small, round, and gray. They enlarge somewhat on longer incubation. Small, round, pale colonies are produced on SS agar; round, pale colonies with slightly darkened centers are seen on MacConkey's agar after prolonged incubation; and growth on brilliant green agar is sparse, and colonies are green.

IDENTIFICATION. This is based upon the differential features listed in Table 11-III. Workers are referred to the *Manual of Clinical Microbiology* (16) for additional information on differential characteristics.

FRANCISELLA

This genus contains two species, *Francisella tularensis*, formerly *Pasteurella tularensis*, and *F. novicida*. The latter species was isolated from a water sample

TABLE 11-III

DIFFERENTIAL CHARACTERISTICS OF *YERSINIA* SPECIES

	Y. pestis	*Y. pseudotuberculosis*	*Y. enterocolitica*
Motility	—	+25°C	+25°C
Oxidase	—	—	—
Urease	—	+	+
Nitrate reduction	—	+	+*
Sucrose (Acid)	—	—	+*
Indole	—	—	—*
Rhamnose (Acid)	—	+	—
Melibiose (Acid)	—	+	—
Cellobiose (Acid)	—	—	+
Ornithine decarboxylation	—	—	+*

*Some exceptions

and is mainly of academic interest. Both species are small, Gram-negative rods or coccobacilli that require cystine for growth. *F. tularensis*, a facultative intracellular parasite, is an important obligate pathogen with a wide geographic distribution.

Pathogenicity

Tularemia is principally a disease of wild animals. Humans as well as some domestic animals and fowl are susceptible. In nature, the disease is transmitted by insect vectors. Most human infections are acquired as a result of handling and dressing infected rabbits. Infections have been reported in many animal species, including squirrel, opossum, woodchuck, muskrat, skunk, coyote, fox, cat, sheep, deer, game birds, and domestic fowl.

The characteristic lesions observed in wild rabbits and in other animals are small necrotic foci in the liver, spleen, and lymph nodes.

Isolation Procedures

Great care should be exercised in working with infectious material and cultures. A considerable number of serious and occasionally fatal laboratory infections have occurred. A medium containing cystine is essential for cultivation. Cystine heart agar to which is added blood or hemoglobin is satisfactory. Material is inoculated liberally onto this medium, and onto blood agar, and incubated at 37°C for three to five days. Incubation in a candle jar or in 10% carbon dioxide is advantageous.

Frequently, specimens yield a variety of bacteria, and in order to recover *F. tularensis*, it is necessary to inoculate guinea pigs or mice. Approximately a 10% tissue suspension in broth is prepared with a tissue grinder. Graded doses are inoculated subcutaneously. Infections are generally fatal within a week to ten days. The organism can then be isolated from the liver or spleen.

In the case of guinea pigs, blood samples may be taken one week postinoculation and tested for agglutinins to *F. tularensis*. A positive test constitutes a diagnosis of tularemia.

DIRECT EXAMINATION

The best procedure for the rapid and specific identification is direct or indirect fluorescent-antibody staining of smears from exudates and affected tissue such as the necrotic foci in the liver and spleen. Preparations from formalin fixed tissues can also be used. The FA conjugate is available commercially (Difco).

Cultural Characteristics and Identification

Colonies are initially small and dewdroplike. They enlarge with longer incubation and tend to be confluent. Discrete colonies may be difficult to obtain on subculture. Especially notable is the marked greening around colonies on blood agar.

The organism grows poorly in media used for biochemical tests. It is best identified by an agglutination test using a specific antiserum or by the use of specific fluorescent antibody. *F. tularensis* antigen is available commercially (Difco) for the detection of specific antibody in the sera of infected laboratory animals.

REFERENCES

1. Carter, G. R., and Chengappa, M. M.: *Am Assoc Veterinary Laboratory Diagnosticians,* Twenty-fourth Ann Proc. 1981, p. 37.
2. Biberstein, E. L. In Bergan, T. (Ed.): *Methods in Microbiology,* Vol. 10. New York, Academic Press, 1978. p. 253.
3. Rhoades, K. R., and Heddleston, K. L. In Hitchner, S. B., Domermuth, C. H., Purchase, H. G., and Williams, J. E. (Eds.): *Isolation and Identification of Avian Pathogens,* 2nd ed. College Station, Texas, American Association of Avian Pathologists, 1980, Chapter 3.
4. Frederiksen, W.: In Kelian, M., Frederiksen, W., and Biberstein, E. L. (Eds.): *Haemophilus, Pasteurella and Actinobacillus.* New York, Academic Press, 1981, p. 186.
5. McAllister, H. A., and Carter, G. R.: *Am J Vet Res, 35*:917, 1974.
6. Perreau, P., Breard, A., Bercovier, H., and Le Menec, M.: *Bull Acad Vet France, 53*:413, 1980.
7. Hall, W. J., Heddleston, K. L., Legenhausen, D. H., and Hughes, R. W.: *Am J Vet Res, 16*:598, 1955.
8. Miaz, O., Jelen, P. and Bohacek, J.: *Acta Vet Bino, 46*:135, 1977.
9. Ackerman, J. I., and Fox, J. G.: *J Clin Microbiol, 13*:1049, 1981.
10. Suzuki, T. et al.: *J Japan Vet Med Assoc, 33*:219, 1980. (Original not seen; Abstract—*Vet Bull.*)

11. Holmes, B., and Ahmed, M. S.: In Kilian, M., Frederiksen, W., and Biberstein, E. L. (Eds.): *Haemophilus, Pasteurella and Actinobacillus.* New York, Academic Press, 1981, pp. 161–174.
12. Clemons, K. U. and Gadberry, J. L.: *J Clin Microbiol, 15*:731, 1982.
13. Mooris, E. J.: *J Gen Microbiol, 19*:305, 1958.
14. Carter, G. R.: *19th Annual Proceedings American Association of Veterinary Laboratory Diagnosticians.* Madison, Wisconsin, American Association of Veterinary Laboratory Diagnosticians, 1976, p. 189.
15. Rollag, O. J., Skeels, M. R., Nims, L. J., Thelstead, J. P., and Mann, J. M.: *J Am Vet Med Assoc, 179*:1381, 1981.
16. Martin, W. J., and Washington II, J. A.: In Lennette, E. H. (Ed.-in-Chief): *Manual of Clinical Microbiology,* 3rd ed. Washington, D.C., American Society for Microbiology, 1980, Chapter 16.

SUPPLEMENTARY REFERENCES

Pasteurella

Kilian, M., Frederiksen, W., and Biberstein, E. L.: *Haemophilus, Pasteurella and Actinobacillus.* New York, Academic Press, 1981.

Carter, G. R.: In Starr, M. P., et al. (Eds.): *The Prokaryotes,* New York, Springer-Verlag, Vol. 11, Chapter 110, p. 1383.

Carter, G. R. In Blobel, H., and Schieber, T. (Eds.): *Handbuch der bakteriellen Infektionen bei Tieren,* Vol. III. Jena, Gustav Fischer, 1981. p. 557.

Carter, G. R.: In Balows, A., and Hausler, Jr., W. S. (Eds.): *Diagnostic Procedures for Bacterial, Mycotic and Parasitic Infections,* 6th ed. Washington, D.C., American Public Health Association, Inc., 1981, Chapter 34, p. 551.

Yersinia

Quan, T. J., Barnes, A. M., and Poland, J. D.: In Balows, A., and Hausler, Jr., W. J. (Eds.): *Diagnostic Procedures for Bacterial, Mycotic and Parasitic Infections,* 6th ed. Washington, D.C., American Public Health Association, Inc., 1981, Chapter 43.

Francisella

Stewart, S. J.: In Balows, A., and Hausler, Jr., W. J. (Eds.): *Diagnostic Procedures for Bacterial, Mycotic and Parasitic Infections,* 6th ed. Washington, D.C., American Public Health Association, Inc., 1981, Chapter 41.

Chapter 12

ACTINOBACILLÙS

THE ORGANISMS of this genus are small, nonmotile, non-sporeforming, Gram-negative rods and coccobacilli. They grow aerobically and anaerobically and ferment carbohydrates. Three species are widely recognized in animals, viz. *A. lignièresii*, *A. equuli*, and *A. suis*. All occur as commensals in the alimentary, respiratory, and genital tracts of animals.

The following species are listed in the *Approved Lists of Bacterial Names:* *A. lignieresii*, *A. equuli*, *A. suis*, *A. actinomycetemcomitans*, and *A. capsulatus*. *A. actinoides* described earlier appears to be identical to *Haemophilus somnus; A. seminis* according to recent taxonomic studies does not belong in the genus *Actinobacillus*.

PATHOGENICITY

A. lignièresii. This organism is found in infections of principally the head and neck of cattle, sheep, and pigs characterized by granulomatous processes, suppuration, and encapsulation. This disease differs from that caused by *Actinomyces bovis* in that actinobacillosis is spread by way of the lymphatic vessels and the "granules" produced are smaller and best seen with the aid of the microscope. *A. lignièresi* shows a predeliction for the soft tissues as opposed to the affinity for bone shown by *A. bovis*. The infection in swine most frequently involves the sow's udder. There are reports of infrequent infections in the dog, horse, and humans.

A. equuli. This organism, a commensal in the intestine, is a frequent cause of pre– and postnatal infections of foals. Among the manifestations of this infection are enteritis, purulent nephritis with abscesses, suppurative arthritis and pneumonia, abortion and septicemia in mares, and infection of verminous aneurysms in the horse. It has been isolated on occasions from the arthritic joints of swine.

A. suis. Most isolations of this organism are from young pigs with a septicemia that often takes a rapid and fatal course (2). Other manifestations are arthritis, pneumonia, pericarditis, nephritis, and subcutaneous abscesses. It occasionally produces infection in older pigs, and there are reports of its isolation from infections in horses.

A. seminis. This organism has been recovered from cases of ovine epididymitis in rams (2, 3). The disease is clinically indistinguishable from the epididymitis in rams due to *Br. ovis.*

A. actinomycetemcomitans. This is an organism of doubtful pathogenicity, which is usually associated with humans. There is a report of its recovery from cases of epididymitis in rams (4).

A. capsulatus. Arthritis in rabbits is the only disease that has been ascribed to this organism (5). Ross et al. (6) have described an *actinobacillus*-like organism from the vagina of sows.

DIRECT EXAMINATION

This procedure is of value in the presumptive diagnosis of actinobacillosis caused by *A. lignièresii.* Place pus or caseous material in a Petri dish and wash with distilled water. The small, gray white granules characteristic of actino-bacillosis can be seen with a hand lens or the low power of the microscope. Transfer several granules to a slide and crush gently with a coverslip. Examine under low power for club-shaped structures. Remove the coverslip and spread in order to make a thin smear. Stain by Gram's method. The presence of small Gram-negative rods suggests actinobacillosis.

ISOLATION PROCEDURES

Actinobacilli grow well on blood or serum agar in an atmosphere of 5–10% carbon dioxide or in a candle jar. If actinobacillosis is suspected, anaerobic plates should also be inoculated in that the causative agent of the clinically similar disease actinomycosis is caused by the anaerobic organism *Actinomyces bovis.*

CULTURAL CHARACTERISTICS

A. lignièresii. Colonies appear in twenty-four hours and are small, bluish white, smooth, and glistening. They resemble colonies of *Past. haemolytica* and *Past. multocida.*

A. equuli. Colonies are round, raised, gray, and translucent. They attain a size of 3–6 mm and are characteristically viscid or "sticky."

A. suis. Colonies resemble those of *A. lignièresii.* They hemolyze and differ from those of *A. equuli* in not being viscid or "sticky."

A. seminis. Small, pinpoint, round, grayish white, nonhemolytic colonies are produced on blood agar after twenty-four hours incubation. They enlarge considerably after additional incubation.

The colonies of the remaining species resemble in general those described for the more common occurring species. *A. capsulatus* possesses a distinct capsule as do some of the other species.

IDENTIFICATION

All of the species in this genus appear as small Gram-negative rods or coccobacilli (Fig. 12-1).

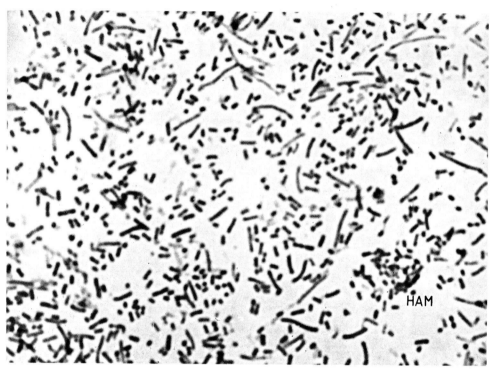

Figure 12-1. *Actinobacillus equuli* from a blood agar plate culture. Gram's stain, ×2500 (H. A. McAllister).

The three most frequently recognized species, *A. lignièresii, A. equuli,* and *A. suis,* have the following characteristics: nonmotile; grow on MacConkey's agar; oxidase and urease positive; reduce nitrates; indole negative; produce acid from glucose.

A. lignièresii. This organism is distinguished from *A. equuli* and *A. suis* on the basis of the criteria listed in Table 12-I. The small gray-white granules containing Gram-negative rods and the suppurative involvement of soft tissues of the tongue or head and neck suggest actinobacillosis; however, some fungous diseases and nocardia infections can produce similar lesions.

A. equuli. The criteria for the identification of this species are listed in Table 12-I. It is closely related to *A. lignièresii* and *A. suis.* It frequently differs from *A. lignièresii* in its capacity for hemolysis. The viscid or mucoid character of the colonies of *A. equuli* is characteristic.

Table 12-I

Differential Characteristics of *Actinobacillus* Species

	A. lignieresii	*A. equuli*	*A. suis*	*A. capsulatus*	*A. seminis*
Hemolysis	−	−	+	−	−
Indole	−	−	−	−	−
Urease	+	+	+	+	−
Catalase	+	−	+	+	−
Hydrogen sulfide	+	v	−	−	−
Trehalose	−	+	+	+	−
Melibiose	−	+	+	+	−
Cellobiose	−	−	+	+	−
Arabinose	−	−	+	+	−
Salacin	−	−	+	+	−
Sorbitol	−	−	−	+	−
Mannitol	+	+	−	+	−

A. suis. Differentiation from the other important *Actinobacillus* spp. is accomplished by the criteria listed in Table 12-I.

A. seminis. Some characteristics of differential significance are the following:
• Nonmotile.
• Litmus milk is unchanged.
• Carbohydrates are not fermented.
• Urea is not decomposed.
• Indole is not produced.

See also Table 12-I, and for further details, readers are referred to the study of Baynes and Simmons (2).

A. capsulatus. Readers are referred to *Bergey's Manual* for a description of this uncommon organism recovered from the tarsal joints of three rabbits. See also Table 12-I.

REFERENCES

1. Mair, N. S., Randall, C. J., Thomas, G. W., Harbourne, J. F., McCrea, C. T., and Cowl, K. P.: *J Comp Path, 84*:113, 1974.
2. Baynes, I. D., and Simmons, G. C.: *Aust Vet J, 36*:454, 1960.
3. Livingston, C. W., and Hardy, W. T.: *Am J Vet Res, 25*:660, 1964.
4. DeLong, W. J., Waldhalm, D. G., and Hall, R. F.: *J Am Vet Med Assoc, 40*:101, 1979.
5. Arseculeratne, S. N.: *Ceylon Vet J, 9*: 5, 1961.
6. Ross, R. F., Hall, J. E., Orning, A. P., and Dale, S. E.: *Int J Syst Bacteriol, 22*: 39, 1972.

SUPPLEMENTARY REFERENCES

Kilian, M., Frederiksen, W., and Biberstein, E. L.: *Haemophilus, Pasteurella* and *Actinobacillus*. New York, Academic Press, 1981.

Chapter 13

HAEMOPHILUS

ERNST L. BIBERSTEIN

The genus *Haemophilus* consists of small, nonmotile, facultatively anaerobic, Gram-negative pleomorphic rods or coccobacilli (Fig. 13-1), which require one or both of two defined growth factors for propagation. In all accepted species these growth factors are X and V, commonly supplied in the form of hemin and nicotinamide adenine dinucleotide (NAD), respectively. Some species require one or the other, some both supplements (see Table 13-I). All recognized species reduce nitrate to nitrites or beyond and, with the exceptions of some strains of *H. ducreyi*, a parasite of humans only, ferment glucose and other carbohydrates (1).

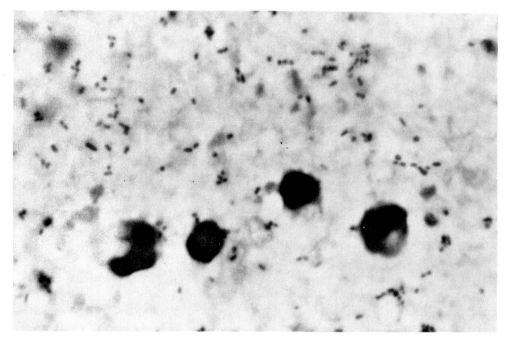

Figure 13-1. *Haemophilus pleuropneumoniae.* Impression smear from porcine lung. Gram stain. × 2250.

Table 13-I

Differential Features of *Haemophilus* spp. and "*Haemophilus*-like"
Agents from Animals (1,15,16,17,18,40,41)

	H. parasuis	*H. pleuropneumoniae*	*H. paragallinarum*	*H. avium*	*H. paracuniculus*	*H. aphrophilus*	*H. haemoglobinophilus*	*H. ovis*	*H. influenzaemurium*	*H. somnus*	*H. agni*	*H. equigenitalis*
X factor required	–	–	–	–	–	+	+	+	+	–	–	–
(Porphyrin test neg.)												
V factor required	+	+	+	+	+	–	–	–	–	–	–	–
Indole	–	–	–	–	+	–	+	–	–	+	–	–
Urease	–	+	–	–	+	–	–	–	–	–	–	–
Ornithine decarboxylase	–	–	–	–	+	–	–	O	O	–	d	–
Arginine dihydrolase	–	–	–	–	+	–	–	O	O	–	–	–
Hemolysis	–	+	–	–	–	–	–	–	–	d	–	–
CAMP reaction	–	+	–	–	–	–	–	–	–	–	–	–
Gas from glucose	–	–	–	–	–	+	–	O	O	–	–	–
Acid from glucose	+	+	+	+	+	+	+	+	+	+	+	–
fructose	+	+	+	+	+	+	–	O	+	+	+	–
sucrose	+	+	+	+	+	+	+	–	+	–	–	–
lactose	d	d	d	d	–	+	–	+	–	–	–	–
d-xylose	–	+	+	d	–	–	+	+	–	+	+	–
d-ribose	+	+	+	O	O	+	d	O	O	O	O	–
d-mannitol	–	+	+	d	–	–	+	d	–	+	–	–
d-sorbitol	–	–	+	d	–	–	–	O	–	+	–	–
Catalase	+	+	–	+	+	–	+	±	+	–	–	+
Oxidase	–	d	–	–	+	–	+	+	–	+	+	+
CO₂ enhances growth	d	–	+	–	+	+	–	–	–	+	+	+
Nitrates reduced	+	+	+	+	+	+	+	+	+	+	+	–
Nitrites reduced	–	+	–	–	O	+	–	O	O	O	O	–

+ = positive
– = negative
± = doubtful
d = less than 90% of strains positive or negative
O = no information

PATHOGENICITY

Sixteen species have been formally described and named, and agents corresponding to the description of *Haemophilus* spp. have been observed in many host species without having been given a specific name. Most are commensals on the mucous membranes of the upper respiratory and genital tracts. Some have a potential for pathogenicity while a few are consistently pathogenic.

The following have been encountered in animals:

Haemophilus parasuis (2). A common V-factor-requiring commensal of the porcine upper respiratory tract, *H. parasuis* becomes involved in swine influenza and enzootic pneumonia as a secondary invader and is an apparently primary cause of Glässer's disease (infectious polyserositis) in young pigs (3, 4). A much rarer bacterium, which occupies the same habitat, is all but indistinguishable from this agent except for its additional requirement for X-factor. Such an organism was first described as *H. suis* in connection with the diseases now associated with *H. parasuis* (5), but recent isolates requiring X-factor lacked a clearcut pathogenic role, and their identity with the original *H. suis* is under dispute. The name has therefore been discontinued for the time being.

H. pleuropneumoniae (6). This organism was commonly called *H. parahaemolyticus* until 1978 (7) and is probably identical with an agent once described as *H. parainfluenzae* (8). It is capable of causing severe epidemics of respiratory disease in swine and, although sometimes present in clinically normal (recovered?) animals, is never considered part of the normal flora.

H. (para)gallinarum (2) is the agent of infectious coryza, an upper respiratory infection of chickens. Originally characterized as requiring both X- and V-factors (9), the current representatives of *H. gallinarum* are invariably V-requiring only and have therefore been named *H. paragallinarum*.

H. pleuropneumoniae and *H. paragallinarum* are the only two *Haemophilus* species of animals that are consistent pathogens, although the infections they cause may sometimes be subclinical, and they may persist and be shed long after cessation of clinical signs.

H. haemoglobinophilus (10), also known as *H. canis*, appears to be a commensal parasite of the canine lower genital tract, particularly in males. On rare occasions it has been associated with cystitis in bitches.

H. avium (11) is a nonpathogenic inhabitant of the upper respiratory tract of chickens. It needs to be carefully differentiated from the pathogenic *H. paragallinarum*, with which it may coexist in a given sample.

H. paracuniculus (12) was isolated from the gut of rabbits suffering from mucoid enteritis. Its significance in this or any other disease is not known.

H. aphrophilus (13) is a frequent member of the oral flora of humans and occasionally associated with infections, including endocarditis and brain abscesses. There has been a report of its occurrence in the pharynx of dogs (14).

The following two agents were once reported as animal pathogens, but, as the original strains were lost and the organisms not encountered for many years, the names were dropped from official lists. Recently the agents were thought to have been rediscovered, but no formal move for reinstatement of the old names has yet been made.

H. ovis was described in 1925 as the cause of a bronchopneumonia of sheep (15) and not seen again until 1978 when an organism resembling it was reported from Britain as an isolate from the nasal passage of normal sheep (16).

H. influenzae-murium (17) has been isolated from the nose and pharynx of mice and credited with causing epidemics of upper respiratory infection and conjunctivitis in mouse colonies (18).

The following three organisms are not *Haemophilus* spp. by the current species definition. They have been described under this designation, however, and are not known by any other name.

H. agni was originally identified as the cause of a septicemic disease of feeder lambs and was subsequently found also associated with pneumonia and mastitis (19). It may be identical with *Histophilus ovis*, reported earlier from similar diseases in Australia (20). It also resembles closely *H. somnus*.

H. somnus (21) causes a septicemia with meningoencephalitis in cattle, a condition sometimes called thromboembolic meningoencephalitis (TEME) (22). It is also involved in bovine respiratory and genital infections (23, 24, 25).

H. equigenitalis (26) is the agent of contagious equine metritis (CEM), a venereal infection of horses, with clinical effects limited to mares (27).

ISOLATION AND IDENTIFICATION

For true *Haemophilus* spp., X- or V-factors or both must be provided in any isolation medium. X-factor is minimally protoporphyrin IX or protoheme and is generally equated with hemin. It is heat-stable and present in adequate amounts in the usual 5% blood agar. V-factor is the heat-labile nicotinamide adenine dinucleotide (NAD, formerly DPN), NAD phosphate (NADP, formerly TPN), or one of its mononucleoside precursors. Although NAD is present in blood, its intracellular location and the presence of NADase in most bloods make blood agar an unsuitable source of V for *Haemophilus* spp. (28).

Both X- and V-factors are adequate and available in chocolate agar or can be replaced by catalase-positive feeder organisms such as *Staphylococcus aureus* in otherwise deficient media. *Haemophilus* spp. will grow as satellite colonies in the immediate vicinity of such bacterial feeder colonies. The feeder organisms are inoculated across the area planted with the specimen so as to encourage satellitic growth (Fig. 13-2). Catalase-negative feeders (e.g. *Streptococcus faecalis*) may supply V- but not X-factor (29).

Mycoplasma agar (PPLO agar, Difco) containing yeast extract or hydrolysate (1–10%) and horse serum (1–10%) is a satisfactory isolation medium for most V-factor-requiring *Haemophilus* spp. (30). Hemoglobin agar (Difco) supplemented with 1% IsoVitaleX® (BBL) satisfies both X- and V-requirements. If X- and V-factors are supplied as hemin and NAD, respectively, 10 µg of

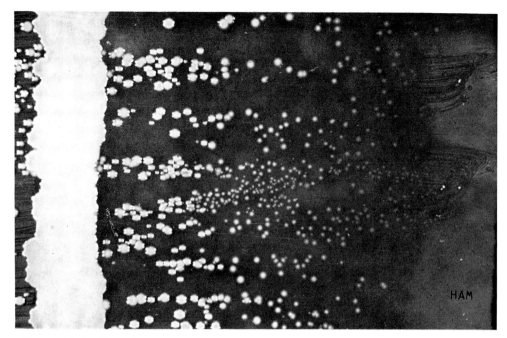

Figure 13-2. Satellite colonies of *Haemophilus* sp. in proximity to the streak growth of *Staphylococcus aureus* (H. A. McAllister).

each per ml of medium is ample for all the organisms listed (31).*

For isolation purposes, media incorporating the critical growth factors in adequate amounts are preferable to reliance on feeder streaks, since the success of the latter method depends on the chance landing of colony-forming units from the specimen close enough to the streak to be benefited by the diffusion of growth factors. The most convenient and satisfactory isolation medium for animal haemophili is chocolate agar, i.e. blood agar prepared by addition of the blood to the melted agar base when this is still at about 75–80°C and held at that temperature for several minutes before the mixture is poured into plates. *H. (para)suis* and *H. (para)gallinarum* do not grow well in the absence of serum, which is adequately supplied in chocolate and mycoplasma agars.

Incubation is at 35–37.5°C, and there should be evidence of growth, though not maximal, within twenty-four hours. Raised carbon dioxide tension (about 5%) is beneficial for the growth of some *Haemophilus* spp., particularly *H. (para)suis, H. (para)gallinarum, H. paracuniculus,* and *H. aphrophilus* (1). It is a requirement for most strains of the nonhaemophili *H. agni, H. somnus,* and *H. equigenitalis* (19, 22, 26).

---

*NAD is freely soluble in water. Hemin can be dissolved in triethanolamine (Eastman) at the rate of 50 mg/ml. Further dilutions can then be made in water.

H. agni and *H. somnus* are readily isolated on either blood or chocolate agar. Little or no growth occurs on unenriched media, but there is no response to X- or V-factor(s). Chances of isolation are improved with uncontaminated specimens by passage of suspect material (tissue, exudate, blood) through an enrichment medium, e.g. infusion broth (22). Embryonated eggs have also been recommended (32).

Swabs from equine genitalia for *H. equigenitalis* are inoculated on Eugon (BBL) chocolate agar plates prepared with horse blood. Since the inoculum is rarely free from contaminants and *H. equigenitalium* is usually resistant to high concentrations of streptomycin, this antimicrobic, at the rate of 200 µg/ml, is included in the chocolate agar. A duplicate set of streptomycin-free media must be inoculated, however, to permit isolation of streptomycin-susceptible strains. Both sets contain amphotericin B (Fungizone, Squibb; 5 µg/ml) and crystal violet (1 µg/ml) for the control of fungal and Gram-positive contaminants, respectively (33).

The following routine has been established as prerequisite for the certification of stallions as CEM free (34): The stallion is bred to two clean test mares. Samples on swabs are then taken from each of the following sites in each mare: cervix, clitoral sinus, and clitoral fossa. A total of six sets of these three swabbings are collected from each mare: at estrus, seven days later, days 2, 4, and 7 after breeding, and at next estrus.

The swabs are placed into Amies transport medium (with charcoal; Difco) immediately upon collection and forwarded to the laboratory. They are to be in transit no longer than forty-eight hours and kept at a temperature of 4–6°C during shipment.

At a laboratory, each swab is inoculated on streptomycin-free and streptomycin-containing horse blood chocolate agar media. These should be in 10 cm glass Petri dishes. Cultures are incubated at 37°C in 5–10% carbon dioxide for forty-eight to seventy-two hours before being first examined. If negative, they should be reincubated up to seven days before being discarded as negative.

CULTURAL CHARACTERISTICS AND IDENTIFICATION

All members of the genus growing on blood agar form very small colonies of less than 1 mm diameter after twenty-four hours of incubation. On richer media, e.g. chocolate agar, colonies up to 2 mm in size occur. They tend to be smooth gray and not obviously pigmented, although *H. somnus* colonies, especially when heaped up with a loop, often appear yellowish. Some strains of *H. pleuropneumoniae* form hard "waxy" colonies (1, 7).

Hemolysis is the rule only with *H. pleuropneumoniae* growing on beef or sheep blood agar. When growing in an area of staphylococcal beta toxin activity, this organism produces the Camp phenomenon (7). Of the other organisms listed, only *H. somnus* has been reported on occasion as being hemolytic, especially when propagated on Columbia-base sheep blood agar (35).

The diagnostic characteristics of animal haemophili are shown in Table 13-I. The requirement for X-factor is best determined by the porphyrin test (36): A loopful of colonial growth from a young culture is suspended in 0.5 ml of a 2 mM solution of delta-amino-levulinic acid (ALA) hydrochloride (Sigma) and 0.8 mM magnesium sulfate in 0.1 M Sørensen phosphate buffer, pH 6.9. The suspension is incubated for at least four hours at 37°C and examined under a Wood's light (about 380 nm). If the porphyrin precursor ALA has been converted to porphyrin, a bright red fluorescence will be observed within four to twenty-four hours. Such a positive test indicates absence of X-factor requirement. Positive and negative controls should be included.

An alternate method calls for the addition of 0.5 ml of Kovacs' reagent to the reaction mixture following incubation. The mixture is shaken vigorously and allowed to separate into two phases. With an X-factor-independent strain, the bottom phase will be red indicating the presence of porphobilinogen, the next intermediate after ALA on the pathway of porphyrin synthesis.

V-factor need is adequately demonstrated by satellitic growth of suspect strains on blood agar with feeder organisms. Since the amounts of V-factor in blood agar are variable and may be sufficient to obscure satellitism, media in which all ingredients have been thoroughly autoclaved and therefore devoid of the heat-labile V-factor are preferable. A common and convenient method of demonstrating the growth factor requirements is the use of three disks or strips impregnated with each and both factors, respectively, and placed on an agar plate that lacks both factors (e.g. proteose peptone) and has been inoculated for the production of confluent growth. Colonies will be clustered around the disc(s) or strip(s) supplying the appropriate supplement(s) (see Fig. 13-3). The method has its pitfalls: cofactor needs, X in particular, may be obscured by "carryover" of an excess of the critical ingredient from growth on a rich medium—a problem that does not complicate the other methods. Further, the presence of any contaminant colonies on the test plates, because of their potential feeder activity, often invalidates the test.

Several of the biochemical tests are most suitably performed by micromethods employing heavy suspensions of bacteria in a small volume (0.5 ml) of substrate solutions. Indole production, urease, ornithine decarboxylase (ODC), and arginine dihydrolase (ADH) activities can be determined in this manner. Results may be positive within four hours, but tubes should be held for twenty-four hours before being discarded as negative (1, 31).

For the indole test, a 0.1% solution of 1-tryptophan in M/15 Sørensen

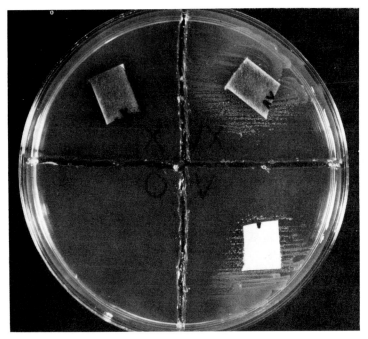

Figure 13-3. Test for cofactor requirements of *Haemophilus* spp. The plate was heavily inoculated prior to application of X-, V-, and XV-factor strips. A quadrant is reserved as negative control (0), and the quadrants are physically separated from each other by incision as a precaution against cross diffusion of growth factors. This organism was *H. pleuropneumoniae* and grew only near the strips containing V-factor. (Reproduced by permission of the *American Journal of Veterinary Research*.)

buffer, pH 6.8, is used. Indole is tested for by the addition of 0.5 ml of Kovacs' reagent (37).

The substrate for the urease test is

KH_2PO_4	0.1 g
K_2HPO_4	0.1 g
NaCl	0.5 g
Phenol red (0.2%)	0.5 ml*
Distilled water	100 ml

Adjust pH to 7.0 with 5N NaOH, autoclave, and add 10.4 ml of a 20% filter-sterilized urea solution.

Urea hydrolysis is signaled by development of a red color (38).

ODC and ADH can be tested for in conventional Møller's medium (39).

---

*0.2% Phenol red solution:

Phenol red crystals	0.2 g
Distilled water	92 ml
1N NaOH	8 ml

Fermentation of carbohydrates is demonstrated in phenol red broth containing 1% of the substrate and 10 µg of filter-sterilized NAD and hemin per ml (31). Serum (1%) may have to be added for adequate growth of *H. parasuis* and *H. paragallinarum*, and a drop of defibrinated blood per 5 ml for that of *H. somnus* and *H. agni*. In these cases, it is wise to include a set of uninoculated controls and a tube lacking the fermentable substrate. *H. paracuniculus* reportedly does not grow in phenol red broth and should be tested in (bromcresol) purple broth instead (12).

REFERENCES

1. Kilian M., and Biberstein, E. L.: Genus *Haemophilus*. In Krieg, N. R. (Ed.): *Bergey's Manual of Determinative Bacteriology*, 9th ed., vol. 1. In press.
2. Biberstein, E. L., and White, D. C.: *J Med Microbiol, 2*:75, 1969.
3. Shope, R. E.: *J Exp Med, 54*:373, 1931.
4. Hjärre, A., and Wramby, G.: *Skand Vet Tidskr, 32*:257, 1942.
5. Lewis, P. A., and Shope, R. E.: *J Exp Med, 54*:361, 1931.
6. Shope, R. E.: *J Exp Med, 119*:357, 1964.
7. Kilian, M., Nicolet, J., and Biberstein, E. L.: *Int J Syst Bacteriol, 28*:20, 1978.
8. Matthews, P. R. J., and Pattison, I. H.: *J Comp Pathol, 71*:44, 1961.
9. DeBlieck, L.: *Vet J, 88*:9, 1932.
10. Friedberger, E.: *Zentralbl Bakteriol Parasitenk Infektionskr Hyg Abt I Orig, 33*:401, 1902/3.
11. Hinz, K. H., and Kunjara, C.: *Int J Syst Bacteriol, 27*:324, 1977.
12. Targowski, S., and Targowski, H.: *J Clin Microbiol, 9*:33, 1979.
13. Khairat, O.: *J Path Bacteriol, 50*:497, 1940.
14. Isom, J. B., Gordy, P. D., Selner, J. C., Brown, L. J., and Willis, M.: *N Engl J Med, 271*:1059, 1964.
15. Mitchell, C. A.: *J Am Vet Med Assoc, 68*:8, 1925.
16. Little, T. W. A., Pritchard, D. G., and Shreeve, J. E.: *Res Vet Sci, 29*:41, 1980.
17. Kairies, A., and Schwartzer, K.: *Zentralbl Bakteriol Parasitenk Infektionskr Hyg Abt I Orig, 137*:351, 1936.
18. Csukás, Z.: *Acta Microbiol Acad Scient Hung, 23*:89, 1976.
19. Kennedy, P. C., Frazier, L. M., Theilen, G. H., and Biberstein, E. L.: *Am J Vet Res, 19*:645, 1958.
20. Roberts, D. S.: *Austral Vet J, 33*:330, 1956.
21. Bailie, W. E.: Characterization of *Haemophilus somnus*, new species, a microorganism isolated from infectious thromboembolic meningoencephalitis of cattle. Ph.D. Dissertation, Kansas State University, Manhattan, Ks, 1969.
22. Kennedy, P. C., Biberstein, E. L., Howarth, J. A., Frazier, L. M., and Dungworth, D. L.: *Am J Vet Res, 21*:403, 1960.
23. Corstvet, R. E., Panciera, R. J., Rinker, H. B., Starks, B. L., and Howard, C.: *J Am Vet Med Assoc, 163*:870, 1973.
24. Corboz, L., and Nicolet, J.: *Schweiz Arch Tierheilk, 117*:493, 1975.
25. Chladek, D. W.: *Am J Vet Res, 36*:1041, 1975.
26. Taylor, C. E. D., Rosenthal, R. O., Brown, D. E. J., Lapage, S. P., Hill, L. R., and Legros, R. M.: *Equine Vet J, 10*:136, 1978.
27. Crowhurst, R. C.: *Vet Rec, 100*:476, 1977.
28. Krumwiede, E., and Kuttner, A. G.: *J Exp Med, 67*:429, 1938.

29. Pickett, M. J., and Stewart, R. M.: *Am J Clin Path, 23*:713, 1953.
30. Nicolet, J.: *Pathol Microbiol, 31*:215, 1968.
31. Kilian, M.: *J Gen Microbiol, 93*:9, 1976.
32. Panciera, R. J., Dahlgren, R. R., and Rinker, H. B.: *Pathol Vet, 5*:212, 1968.
33. Hook, K. R.: *Veterinary Services Memorandum 555.5.* Washington D.C., U.S. Department of Agriculture, APHIS, May 18, 1981.
34. Anonymous: *Federal Register, 45,* No. 3:1003, 1980.
35. Mannheim, W., Pohl, S., and Holländer, R.: *Zentralbl Bakteriol Hyg I Abt Orig A, 246*:512, 1980.
36. Kilian, M.: *Acta Path Microbiol Scand Sec B, 82*:835, 1974.
37. Clarke, P. H., and Cowan, S. T.: *J Gen Microbiol, 6*:187, 1952.
38. Lautrop, H.: *Bull WHO, 23*:15, 1960.
39. Møller, V.: *Acta Path Microbiol Scand, 36*:158, 1955.
40. Biberstein, E. L.: In Kilian, M., et al. (Eds.): *Haemophilus, Pasteurella, and Actinobacillus.* London, Academic Press, 1981, p. 125.
41. Morse, J., and Biberstein, E. L.: Unpublished observations.

SUPPLEMENTARY REFERENCES

Kilian, M., Frederiksen, W., and Biberstein, E. L. (Eds.): *Haemophilus, Pasteurella and Actinobacillus.* London, Academic Press, 1981.

Chapter 14

MORAXELLA AND ACINETOBACTER

THE TAXONOMY of these genera and their species is in a state of confusion. They have been included in the same chapter because they have a number of characteristics in common. *Acinetobacter* resemble *Moraxella* more than they do other Gram-negative bacteria. They are nonfermentative, Gram-negative bacteria, and in this respect they resemble such genera as *Alcaligenes*, *Bordetella*, and *Pseudomonas*.

One suspects these nonfermentative bacteria when TSI and Kligler's present an alkaline or unchanged slant and a neutral butt without gas and hydrogen sulfide production (see Table 10-IV).

MORAXELLA

These are small, aerobic, nonmotile, Gram-negative rods or coccobacilli. They produce catalase and oxidase but do not attack sugars.

The following species of particular veterinary interest are listed in the *Approved Lists of Bacterial Names*: *Moraxella bovis*, *M. equi*, *M. ovis*, *M. lacunata*, and *M. nonliquefaciens*. The only species of appreciable pathogenic significance is *M. bovis*. The name *M. equi* was suggested by Hughes and Pugh (1) for moraxellae recovered from horses and ponies with conjunctivitis. It is not considered to be the primary cause of the disease in horses.

An organism originally described as a *Neisseria* was recovered from sheep with "infectious keratoconjunctivitis" and has since been given the name *Moraxella ovis* (2). Its primary causal relationship to conjunctivitis in sheep has not yet been demonstrated. *M. nonliquefaciens* has been isolated from aborted equine fetuses (3), from septicemia in a goat (3), and from various infectious processes in swine and dogs (3). *M. liquefaciens* (*lacunata*) has been recovered from goats with viral pneumonia and encephalitis (3). There are also reports of unspeciated *Moraxella* associated with conjunctivitis and keratitis in sheep (4) and goats (5) and from various porcine clinical specimens (6).

Several organisms, not yet given species names, have been described that resemble *Moraxella*. They are occasionally recovered from human clinical specimens and are sometimes considered to be responsible for low-grade infections. They have been designated as groups M-3, M-4, EF-4, M-5, and M-6. Only EF-4 and M-5 have been recovered from animals. The former has been isolated from several animal species, and the latter has been recovered from the mouths of dogs (7) and consequently may result in bite infections.

The principal differential characteristics of *Moraxella* species groups EF-4 and M-5 are listed in Table 8-I. Readers are referred to the tables of Weaver et al. (8) on *Moraxella* and related organisms and to Tatum et al. (9) for further information. Readers should consult the references (1,2) for the identification of *M. equi* and *M. ovis*.

MORAXELLA BOVIS

PATHOGENICITY. This organism is consistently recovered from cattle with infectious keratoconjunctivitis or pinkeye. There is still some question as to its primary role in this disease.

ISOLATION PROCEDURES. This species is usually isolated from swabs taken from affected eyes. The organism requires a medium enriched with blood or serum, and blood agar is routinely used.

CULTURAL CHARACTERISTICS AND IDENTIFICATION. Small, round, grayish white colonies with narrow zones of beta-hemolysis are produced in twenty-four hours. Gram-stained smears disclose small, Gram-negative rods and coccoid forms.

The following are important differential features (see also Table 8-I):
- Carbohydrates are not attacked.
- No growth on MacConkey's agar.
- Oxidase positive and catalase negative.
- Nonmotile.
- Litmus milk becomes alkaline with the development of three zones: Upper—deep blue; Middle—lighter colored soft curd; Lower—white with coagulation.
- Nitrates are not reduced.
- Indole is not produced.
- Gelatin is liquefied slowly.

ACINETOBACTER

Two species of *Acinetobacter* are listed in the *Approved Lists of Bacterial Names*, viz., *Acinetobacter calcoaceticus* and *A. lwoffii*. Their equivalents in the earlier terminology are as follows:

A. *calcoaceticus* = A. *calcoaceticus* var. *anitratum* = *Herellea vaginicola*

A. *lwoffii* = A. *calcoaceticus* var. *lwoffii* = *Mima polymorpha* (oxidase negative forms)

The organism that was called *Mima polymorpha* var. *oxidans* and that was designated M-4 has been given the name *Moraxella urethralis* (9).

Both species occur widely in nature and are recovered frequently from water, soil, food, and milk.

The greater prevalence in recent years of *Acinetobacter* in pathologic processes is thought to be due in large part to an alteration of the microbial flora as a result of the wide use of antibiotics.

The evidence for the significance of these organisms as causes of disease in animals is based, as it is in humans, on the isolation of pure cultures in association with a disease process or illness. The significance of these species in human infections has been widely recognized for a number of years.

Acinetobacter lwoffii

PATHOGENICITY. This species has been recovered from a variety of clinical specimens from fowl, dog, horses, swine, cattle, sheep, and laboratory and zoo animals (10). Pure cultures have also been recovered from a number of bovine and equine aborted fetuses (10).

Acinetobacter calcoaceticus

PATHOGENICITY. This species, like *A. lwoffii*, has been recovered from various clinical specimens from farm animals, pets, and laboratory and zoo animals (10). Of special interest was the recovery of this organism from the blood of sick dogs (10). Skovdal (11) attributed bronchopneumonia in mink to this organism.

ISOLATION PROCEDURES AND CULTURAL CHARACTERISTICS. These organisms grow well on simple media, but they are usually isolated on blood agar. On this medium, colonies of both species attain a diameter of 2–3 mm in twenty-four hours at 37°C.

The colonies of *A. lwoffii* are usually somewhat larger than those of *C. calcoaceticus*; both are low, convex, butyrous, opaque, and grayish white. Strains of the latter are usually beta-hemolytic on bovine blood agar in forty-eight hours, while cultures of the former are generally nonhemolytic.

These organisms do not usually grow on SS agar; they produce colorless to pinkish colonies on MacConkey's agar and blue colonies on EMB agar. They grow on TSI and KIA but do not alter it except for the production of an alkaline slant by some cultures.

IDENTIFICATION. Both species are Gram-negative and predominantly coccoid (Fig. 14-1). Diplococci and rods are seen, and there is considerable pleomorphism. Morphologically, they resemble the *Neisseria* with which they can be confused.

They are identified on the basis of the characteristics listed in Table 8-I.

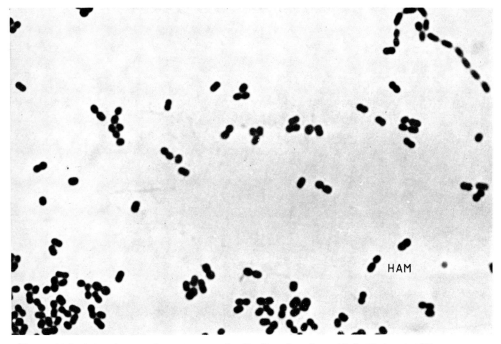

Figure 14-1. *Acinetobacter calcoaceticus* var. *lwoffii*. Gram's stain, ×2250 (H. A. McAllister).

REFERENCES

1. Hughes, D. E., and Pugh, G. W., Jr.: *Am J Vet Res, 31*:457, 1970.
2. Lindqvist, K.: *J Infect Dis, 106*:162, 1960.
3. Aubert, G. D., and Toma, B.: *Acad Vet France, 44*:97, 1972.
4. Baker, J. R., Faul, W. B., and Ward, W. R.: *Vet Rec, 77*:402, 1965.
5. Pande, P. G., and Sekariah, P. C.: *Curr Sci, 29*:267, 1960.
6. Larsen, J. L., Bille, N., and Neilsen, N. C.: *Acta Pathol Microbiol Scand (B), 81*:181, 1973.
7. Saphir, D. A., and Carter, G. R.: *J Clin Microbiol 3*:344, 1976.
8. Weaver, R. E., Tatum, H. W., and Hollis, D. G.: *The Identification of Unusual Pathogenic Gram-negative Bacteria* (Elizabeth O. King), Preliminary Revision, 1972. Published as training materials by the Center for Disease Control, Public Health Service, Department of Health, Education and Welfare, Center for Disease Control, Atlanta, Georgia.
9. Rubin, S. J., Granato, P. A., and Wasilauskas, B. L.: In Lennette, E. H., Balows, H., Hausler, W. J., Jr., and Truant, J. P. (Eds.): *Manual of Clinical Microbiology*, 3rd ed. Washington, D.C., American Society for Microbiology, 1980, Chapter 21.
10. Carter, G. R., Isoun, T. T., and Keahey, K. K.: *J Am Vet Med Assoc 156*:1313, 1970.

SUPPLEMENTARY REFERENCES

Weaver, R. E., Tatum, H. W., and Hollis, D. G.: *The Identification of Unusual Pathogenic Gram-negative Bacteria* (Elizabeth O. King), Preliminary Revision, 1972. Published as training materials by the Center for Disease Control, Public Health Service, Department of Health, Education and Welfare, Center for Disease Control, Atlanta, Georgia.

Rubin, S. J., Granato, P. A., and Wasilauskas, B. L.: In Lennette, E. H., Balows, A., Hausler, W. J., Jr., and Truant, J. P. (Eds.): *Manual of Clinical Microbiology*, 3rd ed. Washington, D.C., American Society for Microbiology, 1980, Chapter 2.

CHAPTER 15

NON–SPOREFORMING ANAEROBIC BACTERIA

J. F. Prescott and M. Chirino-Trejo

Anaerobic bacteria can be isolated from a wide variety of infectious processes in animals, particularly where necrosis is present. Examples of such nonspecific infectious processes that commonly contain a wide range of anaerobic organisms are shown in Table 15-I. Any chronic infection involving necrosis of tissue is liable to contain anaerobic bacteria. The presence of a foul odor is convincing evidence of anaerobic infection, but anaerobes may cause infections without such smells.

The distinction of non-sporeforming anaerobic bacterial infections from infections due to other anaerobes (*Clostridia*—Chapter 19; *Actinomyces*—Chapter 22) is largely artificial since *Clostridia* and *Actinomyces* are commonly found in nonspecific infections of the type shown in Table 15-I. Infections due to these organisms are distinguished from non-sporeforming anaerobic infections because the former are often specific (e.g. blackleg, lumpy-jaw, tetanus), whereas the latter are nonspecific infections, usually arising endogenously. They are caused by bacteria, present as part of the normal flora, which take advantage of impaired host defenses to express their generally weak pathogenic potential.

Nonspecific anaerobic infections associated with endogenous bacterial flora are caused by a variety of different genera and species, shown in Table 15-II. This table shows that *Clostridia* and *Actinomyces* species are not common in these types of infections. Although a wide variety of bacteria have been isolated from nonspecific anaerobic infections in man, in general the majority of infections are caused by a relatively limited number of species. In animals the situation is marginally more complicated because of differences in the potentially pathogenic normal flora in the common animal species. Nevertheless, the great majority of anaerobic bacteria from nonspecific infections in animals can be readily characterized using the criteria developed for clinical isolates from disease in human patients (1–6). Infections with non-sporeforming anaerobes are often "mixed infections"—there may be several anaerobic and aerobic bacteria present. In some cases, there are well-recognized associations of aerobes with anaerobes. For example, *Corynebacterium pyogenes* often implies the presence of either *Fusobacterium*

Table 15-I

Infections likely to involve non-sporeforming anaerobic bacteria

Infectious process	Comment or example
Abscess	Any site, particularly deep.
Head and neck	
Chronic sinusitis	Horse especially.
Periodontal abscess	Tooth root abscess, dog.
Gingivitis	Chronic gingivitis in dogs and cats.
Guttural pouch infection	Horse.
Pleuropulmonary	
Aspiration pneumonia	All species.
Pleuritis, pleural effect	Especially chronic — cat-bite empyaema.
Chronic bronchopneumonia	
Intraabdominal	
Peritonitis	Navel-infection, penetrating foreign body — traumatic reticuloperitonitis of cattle, post-surgical.
Liver abscess	Cattle — *F. necrophorum*.
Female genital tract	
Post-parturient endometritis	All species — mixed with aerobes.
Pyometra	Dog.
Mastitis	Chronic in cattle — with *C. pyogenes*.
	Summer mastitis of cattle.
Soft tissue	
Cellulitis	All species — may be crepitant due to gas.
	Cat-bite cellulitis.
Gas gangrene — deep wounds	Often with *Clostridia*.
Skeletal tissue	
Arthritis	*F. necrophorum* — cattle.
Osteomyelitis	Penetrating wounds, compound fracture, post-surgical.
Foot rot	Sheep — *B. nodosus, F. necrophorum*
	Cattle — *F. necrophorum, B. melaninogenicus*.

Any chronic infection involving necrosis of tissue is liable to involve anaerobic bacteria. Their presence is often indicated by the presence of foul-smelling discharges.

necrophorum or *Peptococcus indolicus* (7,8) and, if appropriate culture techniques are used, these organisms may be recovered. Their presence in mixed infections makes the isolation and identification of non-sporeforming anaerobic bacteria challenging.

In veterinary medicine, certain infections due to non-sporeforming anaerobes can be described as specific. These include foot rot in sheep caused by *Bacteroides nodosus* and liver abscesses in cattle caused by *Fusobacterium necrophorum*. In these cases, the organisms are cultured repeatedly from typical disease. Another example of a specific infection may be "summer mastitis" caused by a mixture of *C. pyogenes, P. indolicus, F. necrophorum,*

Table 15-II

Relative frequency (% total) of anaerobic bacteria from non-specific infections in domestic animals.

Genus, Species	Author			
	1	2	3	4
Bacteroides (total)	46	51	31	40
B. ureolyticus (corrodens)	1	3	—	—
B. fragilis	1	3	4	10
B. melaninogenicus*	19	11	13	17
B. oralis	1	4	—	—
B. ovatus	—	1	—	—
B. ruminicola†	6	3	—	—
Other‡	19	28	14	13
Fusobacterium (total)	6	9	11	19
F. necrophorum	1	4	5	7
F. nucleatum	1	1	—	9
Other‡	4	5	5	3
Clostridium (total)	8	4	30	9
C. perfringens	2	1	19	
C. septicum	1	—	—	
C. sordellii	1	—	—	
Other‡	5	3	11	
Eubacterium (total)	3	2	3	1
E. lentum	1	—	—	
Other‡	1.7	2	—	
Propionibacterium (total)	5	1	4	6
P. acnes	4	—	2	5
Other‡	1	1	2	1
Actinomyces (total)	6	1	9	—
Peptostreptococcus (total)	15	6	7.5	15
P. anaerobius	12	—	7.5	13
P. intermedius	1	—	—	2
Other‡	2.3	6	—	—
Peptococcus (total)	2	12	—	—
P. indolicus	1	5	—	
Other‡	1.1	7	—	
Lactobacilli (total)	1	3	—	2
Other genera	1	5	—	—
Unidentifiable	9.2	1	—	—

*Includes *B. asoccharolyticus, B. intermidius*

† Subspecies *brevis, ruminicola*

‡ Uncommon or unidentifiable species.

[1]Hirsh, Biberstein, Jang: *J Clin Microbiol,* 10:188, 1979.

[2]Prescott: *Can J Comp Med,* 43:194, 1979.

[3]Berg, Hales, Scanlan: *Am J Vet Res,* 40:876, 1979.

[4]Love, Jones, Bailey, Johnson: *J Med Microbiol,* 12:207, 1979.

Bacteroides melaninogenicus subspecies *melaninogenicus*, and a microaerophilic coccus (9).

Anaerobic bacteria may be isolated from some perhaps unexpected sites. For example, non-sporeforming organisms such as *Bacteroides fragilis, P. indolicus*, and *Eubacterium lentum* have been isolated from 12 percent of udders of cattle with subclinical mastitis (10). As veterinary diagnostic laboratories apply the correct techniques for sample transport and for isolation and identification of clinically important anaerobes, then our knowledge of a neglected area of veterinary bacteriology should blossom. Much of the pioneering work has already been done, by medical and a few veterinary bacteriologists, and a series of simple biochemical identification systems are now available that bring anaerobic identification within the realm of small laboratories.

COLLECTION AND TRANSPORT OF SPECIMENS

Proper collection and transport of specimens is essential for laboratory diagnosis of nonspecific anaerobic infections. Samples contaminated with normal flora are not suitable. These include intestinal contents, feces, throat, and mouth swabs. Specimens of the type shown in Table 15-I are all suitable.

Clinical material for anaerobic culture is best taken using a needle and syringe from which the air has been expelled. Pus or other fluid material is aspirated and the needle capped with a rubber bung. If this specimen does not reach the laboratory within thirty minutes, then it must be transported in an anaerobic transport tube. Swabs are less satisfactory but, if used, they must not be the dry cotton wool type because of the lethal effect of both drying and exposure to air. Swabs must be transported in anaerobic transport medium with a minimum of delay, and kept at 4°C if they are not processed immediately. The maintenance of anaerobic conditions for large volumes of anaerobically infected tissue obtained postmortem does not seem to be so critical, so long as the tissue is kept at a reasonably sized volume such that oxygen penetration is minimized. An example is a 6 × 6 × 6 cm cube of lung from a cow with chronic bronchopneumonia. Such specimens should be kept at 4°C and processed as soon as possible. If smaller volumes of tissue are transported, or if transport is prolonged, then optimal isolation results will be obtained if material is transported in a Bio-Bag® (Marion Scientific Corp.). This is a small heat-sealable plastic bag containing a gas-generator, catalyst, and anaerobic indicator. It is also suitable for transporting swabs or syringes.

Anaerobic Transport Containers and Media

Anaerobes must be protected from oxygen during transport. Specimens should be placed in a container with an anaerobic environment. Such containers are available commercially (Anaport® vials — Scott Laboratories; Port-a-cul® — BBL). Commercially available swabs for use with appropriate transport media in an anaerobic environment include the Anaerobic Culturette (Marion Scientific Corp.) and the Vacutainer Anaerobic Transporter (Becton-Dickinson). A suitable transport medium for swabs is prereduced anaerobically sterilized (PRAS) Cary and Blair agar. This is also available commercially (BBL). It can be made as described in Appendix B (11).

Swabs for use with Cary and Blair medium can be prepared by placing in tubes similar to those described in Appendix B, gassing the tubes with N_2, and capping under N_2. Swabs are removed from their airtight anaerobic environment as required and then transported, buried in the Cary and Blair medium, immediately to the laboratory.

Direct Microscopic Examination of Specimens

Direct microscopic examination of all specimens should always precede attempts at isolation. Gram's stain will indicate the number of different morphological types of bacteria to be expected on aerobic and anaerobic culture plates. Many anaerobic bacteria cannot be distinguished morphologically from common aerobic pathogens, although there are exceptions. These include *F. necrophorum*, which is often present in clinical material as a long, filamentous Gram-negative beaded rod (1.0 × 20–100 µm), and *F. nucleatum*, which is a Gram-negative thin rod with tapered ends, often in pairs. *Bacteroides melaninogenicus* and *B. asaccharolyticus* are generally coccobacillary. The distinctive appearance of *Clostridium perfringens* and *Actinomyces* (or *Nocardia*) is described in other chapters.

Fluorescent antisera are available to identify the *B. fragilis* and *B. melaninogenicus* group of organisms (Fluorotec-F, Fluorotec-M — Pfizer Diagnostics). While there is no experience with their use in veterinary diagnostic laboratories, the experience with them in medical laboratories is that they are specific and sensitive (>90%) (12).

ANAEROBIC CULTURE TECHNIQUES

A variety of anaerobic culture techniques are available, and the method of choice will depend on the main purposes of the laboratory. The three major methods are the use of anaerobic jars, the use of Hungate roll-rubes, and the use of anaerobic chambers. The recovery of clinically important non-sporeforming anaerobes is equally good with anaerobic jars when compared to the other techniques and, because this approach is cheapest and is familiar to most diagnostic bacteriologists, is the method recommended. Details on the other methods are available (2). The use of chopped meat broth or thioglycollate broth as the sole primary isolation medium for anaerobes must be avoided as dangerously old-fashioned.

Anaerobic jars are of two types—those that can be evacuated (Torbal, Baird-Tatlock) and those that cannot (GasPak). Both depend on the removal of O_2 by catalytic reaction at room temperature with H_2 to form water. The catalyst used is palladium, coated on alumina, and held in a wire-mesh that comes with the jar. For optimal results the catalysts must be reactivated by heating at 160°C for two hours in a dry oven after every use. They should be stored in an airtight jar with a dessicant.

The anaerobic jar of choice is one that can be evacuated, because this results in rapid anaerobiosis and is significantly cheaper than the envelope system. Air is removed from the jar by evacuating to 25 mmHg and the vacuum replaced with a mixture of 5–10% CO_2, 5–10% H_2, and the balance N_2. The process should be repeated three times, and a catalyst must be included. These gas mixtures can be bought from commercial gas suppliers. The alternative method is to use H_2O and CO_2 generator envelopes (GasPak, Oxoid), which are activated by adding 10 ml of water. In both types of jar a methylene blue anaerobic indicator (BBL) should be included to show that anaerobic conditions are achieved, with thirty minutes in the evacuation system, and two to three hours in the envelope system.

Plates must be maintained in an anaerobic atmosphere both before and after inoculation. If it is not convenient to incubate them anaerobically within ten to twenty minutes of inoculation, plates can be stored in a gentle stream of 100% CO_2, made fully O_2-free by passage over copper filings kept at 300°C (1).

Loops used to inoculate plate or liquid media should be made of platinum or pure nickel wire; nichrome wire is not suitable.

Media for Primary Isolation of Anaerobic Bacteria

Enriched blood (5% v/v) agar plate media should be used. Any of the following are suitable: Brucella, Columbia, or Schaedler, or brain-heart infusion supplemented with 0.5% yeast extract. All these bases should be supplemented with vitamin K_1 (10 μg/ml final concentration) and hemin (5 μg/ml final concentration) (see Appendix). It is generally recommended that plates be used within two weeks of preparation.

A selective medium sometimes used to isolate *Bacteroides* species and to encourage pigmentation by the black-pigmented *Bacteroides* is kanamycin-vancomycin-laked blood agar, made from Brucella agar with 75 μg/ml kanamycin added before autoclaving and vancomycin (7.5 μg/ml), vitamin K, and 5% (v/v) laked blood added after autoclaving. Laked blood is made by freezing and thawing whole blood three times.

Plate media should be prereduced by storage in an anaerobic atmosphere for six to twenty-four hours before use, and inoculated plates should be incubated anaerobically *immediately* after streaking. Prereduced anaerobic plates for daily use can be stored during the day in the storage jar, which is flushed with O_2-free CO_2. If anaerobic conditions cannot be immediately established after inoculation, plates can be placed in a jar that is continuously flushed with O_2-free CO_2, obtained by passing O_2-free CO_2 through the heated copper filings as described earlier. The greatest cause of failure to isolate anaerobes is the failure to immediately provide the anaerobic conditions required.

Many authors recommend the use of liquid media as an adjunct in the isolation of anaerobes, particularly where low numbers are present or antibiotics have been used. They may also provide a fail-safe device in case the anaerobic jars are not set up properly or the GasPaks fail to activate. Suitable liquid media includes thioglycollate medium, without the indicator (BBL–135C), supplemented with 1 mg/ml sodium bicarbonate, 10% (v/v) horse serum or 5% (v/v) Fildes enrichment (BBL), and the vitamin K-hemin concentrations described earlier. In liquid media the trial concentration of vitamin K_1 is 0.1 µg/ml. An alternative medium is cooked meat broth with 1% (w/v) glucose. All liquid media should be placed in a boiling water bath for ten minutes to drive off dissolved oxygen and cooled before use or should be stored with loose caps for forty-eight hours before use in an anaerobic environment.

In addition to anaerobic plate media and conditions, all specimens should be routinely cultured for aerobes on blood and MacConkey's agar.

Incubation of Cultures

Specimens must be streaked in the traditional manner and cultures incubated anaerobically at 37°C as soon as possible. The jars must *not* be opened for forty-eight hours, and plates should be incubated for seven days before discarding. In general, most isolates will grow in forty-eight hours given proper media and methods.

Liquid media should be inoculated near the bottom of the tube with 1–2 drops of material by Pasteur pipette. Broth tubes should be incubated in an anaerobic atmosphere with screw caps loosened, and not discarded for seven days.

Examination of Cultures

Anaerobic plates should be examined after forty-eight hours of incubation. Plates should be exposed to air for minimal time since certain anaerobes will die within minutes; although the majority of clinically important non-sporeforming anaerobes are more robust, all anaerobic plates when not under examination should be stored in an O_2-free CO_2 atmosphere. They must not be left exposed on the bench top.

Colonies should be examined with a dissecting microscope or hand lens, and each colony type described. Nonspecific anaerobic infections are commonly "mixed infections" and contain both anaerobic and aerobic organisms. The different colony types should all be subcultured to aerobic and enriched anaerobic blood agar plates to purify the colonies and to identify which of the colony types are only facultative anaerobes. Certain facultative bacteria may at first only grow anaerobically; examples include *Listeria monocytogenes* and some strains of *E. coli*. The different colonies should be Gram-stained and inoculated into peptone-yeast-glucose (PYG) broth that has been previously heated for ten minutes in a boiling water bath to remove oxygen. Peptone-yeast-glucose broth is used for gas liquid chromatographic (GLC) analysis of volatile and nonvolatile acids. It will be incubated for forty-eight hours in an anaerobic environment with the cap loosened. The method of preparation is given in Appendix B. An alternative to PYG broth is cooked meat carbohydrate (CMC) prepared as described in Appendix B. The GLC profiles for CMC differ slightly from those in PYG; only PYG profiles are described here.

Plates should be examined under UV light for the presence of the brick-red fluorescing colonies of *B. melaninogenicus*, *B. asaccharolyticus*, and *Veillonella parvula*.

The primary plates and subcultures are then reincubated for a further forty-eight hours. At this time the anaerobic organisms will be distinguished from the aerobically growing facultative anaerobes. The black-pigmented *Bacteroides* will usually have developed pigment, and *B. ureolyticus* will usually show pitting of the agar.

Actinomyces species are discussed in Chapter 22. They may grow under 10% CO_2 or even be aerotolerant. *Clostridium perfringens* is discussed in Chapter 19. It is found in nonspecific anaerobic infections and is readily identified. Where the Gram-stain of the original specimen suggests *C. perfringens*, the organism may be isolated after approximately twelve hours of incubation. It is readily identified. On occasion it is also aerotolerant.

If the colony types, recovered on plate media, their Gram reaction, and their shape do not correspond to the morphological types of bacteria seen in the original specimen, then the cooked meat glucose broth or thioglycollate medium should be subcultured to an anaerobic plate.

IDENTIFICATION OF NON-SPOREFORMING ANAEROBIC BACTERIA

Pure cultures of non-sporeformers must be obtained before they can be identified. They should always be subcultured onto the supplemented, enriched, and prereduced anaerobic media described.

The individual laboratory has to decide to what level it will take identification of non-sporeforming anaerobes. Some laboratories will be satisfied with isolation, a description based on Gram reaction and colonial morphology, and antibiotic sensitivity; other laboratories will want better identification and will base it on the use of commercially available "Kit" identification systems or semiconventional media and analysis of metabolic products; a few specialist laboratories will want to use conventional media (prereduced anaerobically sterilized, PRAS) and identify isolates definitively using the classic VPI manual (1). In this description, the characteristics used in the identification of non-sporeforming anaerobes will be based on the descriptions in the *VPI Anaerobe Laboratory Manual.*

Gas Liquid Chromatographic Analysis of Metabolic End Products

Gas liquid chromatographic analysis of metabolic end products from carbohydrates is essential in the identification of genera of non-sporeforming anaerobic bacteria (see Table 15-III). Volatile fatty acids and nonvolatile fermentation products are key characteristics of anaerobic bacteria, are reproducible and reliable, and are so characteristic of some species as to act as "fingerprints" of that species.

The major problem with GLC analysis is that of initial cost of equipment. Running costs of reliable equipment will, however, be low. There are several GLC machines that have been designed specifically for work with the identification of anaerobes. They currently cost $6–8,000. Examples are the Capco Anaerobe Identification System (Capco Instruments, P.O. Box 9093, Sunnyvale, Ca 94086) or the Antek Series 460A Anaerobe Identification System (Antek Instruments Inc., 6005 North Leeway, Houston, Tx 77076). Any laboratory attempting to identify anaerobic bacteria must have GLC facilities available to it.

Details of procedures used to extract volatile fatty acids or nonvolatile acids (e.g. pyruvic, succinic, lactic acids) from PYG or CMC are given elsewhere (1, 2, 3). The type of columns to be used, their packing material, and details on operating conditions of the GLC equipment are all described for anaerobic bacteriology (1, 2, 3). They will not be repeated here. The methods used must conform with those of the *VPI Anaerobe Laboratory Manual* (1), and the profiles of metabolic products obtained must be compared with those shown in that book and identified and quantified by use of control solutions. Different manuals of anaerobic bacteriology (1, 2, 3) may show minor differences in metabolic products from carbohydrates; these minor differences relate to the media used and to the conditions of analysis of metabolic products. It is important to follow one set of methods and not to interchange them.

The GLC equipment designed for identification of anaerobes is easy to operate and has a long lifetime. Without such equipment anaerobe identification becomes haphazard, although the Minitek (BBL) system can to some extent be used without GLC analysis.

Table 15-III

Genera of clinically important non-sporeforming anaerobic bacteria

Morphology	Gram-Reaction	Major metabolic acids produced	Other Characteristic	Genus
Rods	Positive	Propionic, acetic	Catalase generally positive	*Propionibacterium*
			Catalase negative	*Arachnia*
		Ratio lactic: acetic > 1:1		
		Lactic acid only		*Lactobacillus*
		Succinic acid major product		*Actinomyces*
		Ratio lactic: acetic < 1:1		*Bifidobacterium,*
		Butyric acid not produced		*Eubacterium,*
		Produce butyric and other acids,		*Lachnospira*
		or no major acids		
Rods	Negative	Butyric acid (± minor isobutyric, isovaleric)	Nonmotile or peritrichous flagella	*Fusobacterium*
		Lactic acid only	As above	*Leptotrichia buccalis*
		Other	As above	*Bacteroides*
Cocci	Positive	Lactic acid	No fermentable carbohydrate required.	*Streptococcus*
		Not as above	As above	*Peptococcus, Peptostreptococcus*
Cocci	Negative	Propionic, acetic		*Veillonella*

Commercially Available Anaerobe Identification Systems

There are several "Kit" identification systems available commercially for the identification of anaerobic bacteria. All suffer from the drawback that they are designed for medical bacteriology and the common non-sporeforming anaerobes recovered from clinical disease in man. Anaerobes of veterinary importance may not be included in their data base, an example being *Peptococcus indolicus*. The reliability of all the systems is greatly improved if metabolic products of the anaerobes are determined by GLC.

The best system currently available is the Minitek (BBL) system. The system uses filter paper discs impregnated with a range of biochemical substrates. These discs are dispensed into a series of wells within a disposable plastic plate. The wells are filled with an anaerobe broth in which is suspended a dense suspension of the washings of a blood agar plate of the anaerobe, adjusted to a standard turbidity. Thirteen biochemical tests are performed, and all the equipment is supplied. The plastic plates are incubated in an anaerobe jar using GasPak (BBL) envelopes, and the results are recorded after forty-eight hours. The manufacturer's recommendations should be followed throughout. The recorded results give a numerical identification, which is decoded using the book supplied by the manufacturer that, when combined with GLC results, allows the majority of isolates in a veterinary diagnostic laboratory to be identified. The method is easy to use and gives 97–100% correlation with conventional identification systems (13).

The API 20A system does not give such good correlation with conventional identification (13) and has a smaller data base than the Minitek system. It consists of a plastic strip containing twenty cupules each containing dehydrated substrates. The substrates are rehydrated with a suspension of a broth, provided by API, containing a standardized inoculum of the unknown organism. Incubation is for twenty-four and forty-eight hours at 37°C. The major drawbacks are the lack of clear color reactions with some organisms and the limited range of bacteria that can be identified. Supplementation of results with GLC analysis of metabolic products makes it a reasonable system, but inferior to the Minitek system. The strips are cheaper than the Minitek system—both cost about $5.00 to identify a single isolate, but the Minitek system includes the added costs of some initial equipment.

Other systems available currently include the Micro-Media prefrozen multiple biochemical substrate panel system (Micro-Media Systems Inc., 10,000 Falls Road, Potomac, Md. 20854) and the Anaerobe-Tek™ multiple/agar media system (Flow Laboratories Inc.). There is no published information on their use in veterinary laboratories, and it should be borne in mind that they were designed for the identification of common human pathogens.

Individual laboratories have described good results with simple micromethods of their own make (14).

Identification Using Conventional Media

Conventional media not designed to support the growth of anaerobes cannot be used. The definitive identification of anaerobic bacteria by the diagnostic laboratory is based on the use of the methods and manual of the *VPI Anaerobe Laboratory Manual.* This involves the use of PRAS media, which is, however, difficult to prepare in the average laboratory. It is available commercially, but is expensive (Scott Laboratoreis, Inc., Fiskeville, R.I. 02823). An alternative method more acceptable to most laboratories, because of the time and expense involved in using PRAS media, is to use the conventional tubed liquid and semisolid media described by Dowell and coworkers at the Centers for Disease Control (15). These media are also available commercially (Carr-Scarborough Microbiologicals, Stone Mountain, Ga.; Nolan Biological Laboratories, Tucker, Ga.). This basal medium can be prepared in conventional media kitchens from the commercial dried product (CHO medium base, Difco) or prepared from its constituents. It is a semisolid medium containing peptone, yeast extract, L-cystine, sodium thioglycollate, NaCl, ascorbic acid, and bromothymol blue (see Appendix B). CHO medium base tubes with carbohydrate are inoculated near their base with a few drops of a well-grown culture from the chopped meat carbohydrate or the supplemented thioglycollate medium described. The tubes should be incubated in an anaerobe jar with loosened caps and are inspected for acid production or other parameter on day 1, 2, and 7. The CHO medium contains bromothymol blue that turns yellow at pH $\leq$ 6.0, when the 0.6% (w/v) carbohydrates are fermented. PRAS carbohydrate medium (1) is inoculated through the rubber seal and the pH measured when good growth has occurred.

In addition to the fermentation tests described, the following tests should be done:

a. Esculin hydrolysis. Determined in esculin broth (see Appendix) with 0.1% esculin, after the addition of 1% ferric ammonium chloride.

b. Indole production. Determined in cooked meat broth without carbohydrate after extraction with 1 ml of xylene, adding 1 ml of Ehrlich's reagent and observing the rapid development of the pink color.

c. Nitrate reduction. Determined after good growth in indole-nitrate broth (see Appendix) in the traditional way.

d. Gelatin hydrolysis. Use Thiogel medium (see Appendix) and test in the usual manner after seven days.

e. Urea hydrolysis. A rapid test uses a heavy suspension from a blood agar plate in 0.5 ml sterile urea broth (Difco), incubating aerobically for twenty-four hours. A red color is positive. If the indicator is reduced, add Nessler's reagent to show ammonia production.

f. Growth in 20% bile. Add 20% bile (w/v) to supplemented thioglycollate broth with 1% (w/v) glucose. Record after forty-eight hours as stimulated, inhibited, or no effect by comparing to a similar tube without the bile.

g. Catalase. Remove some of the culture from a blood plate to a slide, add 3% H_2O_2. Ensure no medium is carried over, since blood contains catalase.

h. Gram reaction. The Gram reaction of anaerobic bacteria is sometimes difficult to determine by traditional staining procedures, which often give false Gram-negative results. The KOH test is useful in cases of doubt. Add 2 drops of 3% (w/v) KOH to a slide and add a loopful of a forty-eight-hour culture on plate media. The loop is stirred in a circular motion and occasionally raised 1–2 cm from the slide surface. If the suspension becomes stringy and can be lifted 1–2 cm from the plate within thirty seconds of mixing, then the organism is Gram-negative.

Gram-Negative Anaerobic Rods: Bacteroides and Fusobacterium

The identification of common clinically important *Bacteroides* is shown in Table 15-IV.

The black-pigmented species of *Bacteroides* may show pigmentation on kanamycin-vancomycin-laked blood agar earlier than on enriched blood agar plates, where it may take seven days to develop. All these species will show brick-red fluorescence under long wavelength UV light before the black pigment develops. Cellular morphology is usually coccobacillary.

The *B. fragilis* group described in Table 15-IV consists of *B. thetaiotaomicron*, *B. distasonis*, *B. vulgatus*, and *B. uniformis*, all uncommon bile-resistant clinical isolates.

Bacteroides ureolyticus (*B. corrodens*) forms transparent colonies that usually pit the surface of the agar plates. Most strains grow poorly in broth unless supplemented with sodium formate and fumaric acid.

In general, little difficulty will be experienced in the isolation and identification of the *Bacteroides* species shown in Table 15-IV, provided the methods described are followed.

A species requiring special attention in veterinary laboratories is *B. nodosus*, the cause of foot rot in sheep. Gradin and Schmitz (16) have described a selective medium, which consists of Eugon agar (BBL), with 0.2% (w/v) yeast extract (Pfizer Inc.), 10% defibrinated horse blood (v/v), and 1 µg/ml (w/v) lincomycin. Media must be prereduced before inoculation, and the plates must be incubated anaerobically for five or six days and not discarded for seven days. An alternative medium has been described by Thorley (17). It consists of 1% (w/v) proteose peptone (Difco), 1% (w/v) trypsin 1:250 (Difco), 1% (w/v) liver digest L27 (Oxoid), and 0.5% (w/v) NaCl. To prevent precipita-

Table 15-IV

Characteristics of Bacteroides species

Species	Growth in 20% bile	Esculin hydrolysis	Catalase	Urease	Indole	Black pigment	Arabinose	Glucose	Lactose	Maltose	Rhamnose	Salicin	Sucrose	Trehalose	Fatty acids from peptone-yeast-glucose*
													Fermentation of:		
B. fragilis	+	+	+⁻	-	-	-	-	+	+	+	-	-	+	+	A,S,p,(lb,iv,l,f)
B. ovatus	+	+	-,+	-	-	-	+	+	+	+	+	-	+	+	A,S,p,(ib,iv,l)
B. fragilis group	+	V	V	-	V	-	+	+	+	+	V	V	+	V	A,S,(p,ib,iv,l,f)
B. eggerthii	+	+	-	-	+	-	+	+	+	+	+⁻	-	-	-	A,S,P,ib,b,iv,(l)
B. splanchnicus	+	+	-	-	+	-	+	+	+	-	+	-	-	-	A,S,P,ib,b,iv,(l)
B. bivius	-	-	-	-	-	-	-	+	+	+	-	-	-	-	A,S,iv,(ib,f).
B. capillosus	-	+	-	-	-	-	-	+⁻	-⁺	+⁻	-	-	-	-	a,s,(l,f,p)
B. disiens	-	-	-	-	-	-	-	+	+	+	-	-	-	-	A,S,(ib,iv,f,p)
B. putredinis	-	-	+⁻	-	+	-	-	-	-	-	-	-	-	-	a,p,ib,b,iv,(l,f)
B. ruminicola†															
subsp. brevis	-	+	-	-	-	-	+	+	+	+	V	+	+	-	A,S,F,(p,ib,iv)
subsp. ruminicola	-	+	-	-	-	-	+	+	-	+⁻	-⁺	-⁺	+⁻	-	A,S,f,p,(ib,b,iv,l)
B. ureolyticus	-	-	-	+	-	-	-	-	-	-	-	-	-	-	a,s,(fpl)
B. asaccharolyticus	-	-	-	-	+	+⁻	-	-	-	-	-	-	-	-	A,p,ib,B,iv,(l,s)
B. intermedius	-	-	-	-	+	+⁻	-	+	-⁺	+⁻	-	-	V	-	A,S,ib,iv,(l,p)
B. melaninogenicus															
subsp. melaninogenicus	-	+⁻	-	-	-	+	-	+	+	+	-⁺	-	+⁻	-	A,S,(p,ib,iv,l,f)
subsp. levii	-	-	-	-	-	+	-	+	+	-	-	-	-	-	A,P,B,iV,ib,S
B. nodosus	-⁺	-	-	-	-	-	-	-	-	-	-	-	-	-	a,s,p,(ib,b,iv)

Key:

+ = positive
- = negative
+⁻ = most strains positive
-⁺ = most strains negative
V = variable.

*Fatty acids: capital letters indicate major quantities, lower case minor amounts, in parentheses indicates variable production. A = acetic; B = butyric; C = caprylic; IB = isobutyric; IC = isocaproic; IV = isovaleric; P = propionic; V = valeric; F = formic; L = lactic; S = succinic.

†*B. ruminicola* subsp. *ruminicola* is xylose positive, subsp. *brevis* is xylose negative.

tion during autoclaving, the dissolved materials are brought to boiling at pH 8.5, filtered and cooled, the pH adjusted to 7.4, and then autoclaved. The agar concentration used is 5.0% (w/v), which high concentration prevents swarming. Before pouring plates, 0.05% (w/v) of a cysteine-HCl solution is added. Plates should be stored anaerobically after pouring.

On the medium described by Gradin and Schmitz, certain strains produce colonies that are small and have a "snowflake" appearance. Colonies of primary isolates on the medium described by Thorley are low in profile and show central papular studding, fusing near the colony center to give a series of shiny reticulations, or a low peak. There are also clear internal bands. The outer zone consists of a fine, granular, and diffusely spreading edge made up of migrating microcolonies. Details on the colonial variation seen on subculture are available (17, 18). Virulent isolates can be tested for proteolytic activity in a variety of ways (19, 20), and tests to determine serotypes are done as described (17, 21).

Isolations of *B. nodosus* are most likely to be made where Gram staining shows the presence in active foot rot lesions of *B. nodosus*. Typically, this is a large Gram-negative rod (0.6–0.8 × 3–10 μm) with terminal enlargements at one or both ends and a general curvature to the cell, which in many cases is surrounded by radially disposed Gram-negative bacilli (Fig. 15-1).

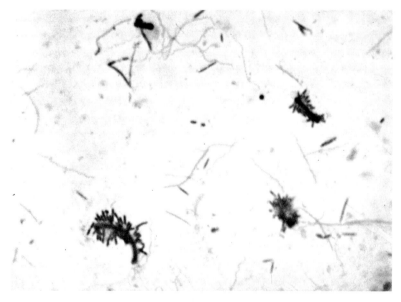

Figure 15-1. Smear from a case of foot rot showing *Bacteroides nodosus* surrounded by small bacteria. Also seen are a number of spirochetes. Courtesy of W. I. B. Beveridge.

Material for culture should be taken from necrotic material at the junction of healthy with separating tissue from the feet of sheep, and transported in Cary and Blair anaerobic medium to the laboratory. Those portions of necrotic material showing greatest numbers of *B. nodosus* on smears should be shaken vigorously for thirty seconds in 10 ml 0.25 M sterile sucrose and the suspension then seeded on anaerobic media (22).

The identification of clinically important *Fusobacterium* species is shown in Table 15-V.

Table 15-V
Characteristics of Fusobacterium species

Species	Growth in 20% bile	Indole	Esculin hydrolysis	Propionate from lactate	Propionate from threonine	Fermentation of Glucose	Levulose	Mannose	Fatty acids from peptone-yeast-glucose*
F. gonidiaformans	−+	+	−	−	+	+	−	−	B,a,p,(l,f,s)
F. naviforme	−	+	−	−	−	−+	−	−	B,a,p,L,(f,s)
F. necrophorum	−	+	−	+	+	−	−	−	B,a,p,(L,f,s)
F. nucleatum	−	+	−	−	+	−	+	−	B,a,p,s,(L,f)
F. russii	−+	−	−	−	−	−	−	−	B,a,(L,f,s)
F. mortiferum	+	−	+	−	+	+	+	+	B,A,p,(L,F,iv,s)
F. varium	+	+−	−	−	+	+	+	+	B,A,L,(S,p)

Key:
+ = positive
− = negative
+⁻ = most strains positive
−⁺ = most strains negative
V, variable.
*Fatty acids: capital letters indicate major quantities, lower case minor amounts, in parentheses indicates variable production. A = acetic; B = butyric, C = caprylic; IB = isobutyric; IC = isocaproic; IV = isovaleric; P = propionic; V = valeric; F = formic; L = lactic; S = succinic.

Fusobacterium necrophorum is a Gram-negative pleomorphic bacillus 0.5–1.75 μm in diameter, ranging in length from small coccobacilli to filaments 100 μm in length. The latter will be sometimes seen in clinical material, especially fluids, and are characteristic. Cellular morphology in cultures will vary with age and type of media used. The colonies are commonly beta-hemolytic, round, gray with an entire edge, and are generally 2–3 mm. A biochemical property that distinguishes *F. necrophorum* from other *Fusobacterium* species is its strong lipase activity, readily demonstrable on egg yolk agar. Details of the differentiation of the biovars of *F. necrophorum* are available (23, 30).

Fusobacterium nucleatum has a distinctive morphology, with thin, Gram-negative rods with tapered ends, often in pairs arranged end-to-end. Most stains of *F. necrophorum* and *F. nucleatum* show chartreuse fluorescence under UV light.

Gram-Positive Non-Sporeforming Rods: Actinomyces, Arachnia, Bifidobacterium, Eubacterium

The identification of some of the more common clinically important Gram-positive non-sporeforming rods is shown in Table 15-VI. *Actinomyces* speciation is discussed in Chapter 22.

The cellular morphology of these organisms is highly variable, depending on media and growth conditions used. The morphology varies from small to large rods, with club-shaped or bifurcated ends being sometimes common; at times the organisms may occur as their filaments. At times the organisms may mimic streptococci in appearance or may readily decolorize and appear Gram-negative.

It has been proposed that *Corynebacterium suis*, a cause of pyelonephritis in swine, be redesignated *Eubacterium suis* (29). Apart from the identifying characteristics shown in Table 15-VI, the organism ferments glycogen and starch, but not most other carbohydrate substrates.

Anaerobic Cocci: Peptococcus, Peptostreptococcus, Streptococcus, and Veillonella

The identifying characteristics of these organisms is given in Table 15-VI.

In general, anaerobic cocci are slow-growing, and biochemical tests may require three to five days for sufficient growth to occur before they can be read. *Peptococcus magnus* (1–2 µm diameter) is distinguished from the less common, but biochemically similar, *P. micros* on the basis of size.

Peptococcus indolicus is a common clinical isolate from ruminants, often found with *C. pyogenes*. It is common in cases of summer mastitis of heifers or dry cows. Details of its biochemical characterization are available (24), but a simple diagnostic procedure is to demonstrate its peptocoagulase by clumping of citrated rabbit plasma, similar to the demonstration of staphylococcal clumping factor.

Antimicrobial Susceptibility Testing of Non-Sporeforming Anaerobes

In general, antimicrobial susceptibility testing of non-sporeforming anaerobes is unnecessary since the susceptibility is predictable (2, 25), and resistance of *B. fragilis* isolates to penicillin does not seem to occur in animal isolates, although it is common in human medical isolates. A study by Kimsey and Hirsh (26) showed that 90–95 percent of anaerobic bacteria isolated from a variety of animals were susceptible to levels of ampicillin, chloramphenicol, minocycline, and tetracycline, and 60 percent to penicil-

Table 15-VI

Differential characteristics of common cocci and Gram-positive non-sporeforming anaerobic rods from clinical specimens

Species	Gram reaction / Morphology	Indole	Esculin hydrolysis	Gelatin hydrolysis	Nitrate reduction	Cellobiose	Glucose	Lactose	Maltose	Mannitol	Sucrose	Catalase	Major products in peptone-yeast-glucose*
						(Fermentation of:)							
Actinomyces species	+, rods	–	V	–	V	V	V	+	+⁻	+⁻	+	–+	S;L;a>1:1
Arachnia propionica	+, rod	–	–+	V	+	+	–	+	+⁻	+	+⁻	–	P,A,(f,s,l,iv)
Bifidobacterium eriksonii	+, rod	–	+⁻	–	–	–	+	+	+	+	+	–	A,F,s,L,
Eubacterium lentum	+, rod	–	–	–	+⁻	+⁻	+	–	–	+	–	–	(a,f,s,l)
E. suis	+, rod	–	–	–	–	–	–	–	–	+⁻	–	–	A,F
E. limosum	+, rod	–	+	–+	+⁻	–	+	+	–	–+	–	–	L,a,b,(s,p)
Propionibacterium acnes	+, rod	–	–	+⁻	+	+⁻	+	+⁻	–	–	–+	+⁻	P,A,(f,L,s,iv)
Peptostreptococcus anaerobius	+, coccus	–	–	+⁻	–	–+	+⁻	–+	–	–+	V	–	A,(iv,ib,ic,b)(l,s,p)
P. micros	+, coccus	–	–	–	–	–	–	+	–	–	–	–	A,(l,s,f)
P. parvulus	+, coccus	–	+	–	–	–	+	+	+	–+	–	–	a,L,(s)
P. productus	+, coccus	–	+	–	–	–	+	+	+	+	+⁻	–	A,s,(l,p)
Peptococcus asaccharolyticus	+, coccus	+	–	–	–	–	–	–	–	–	–	–	A,b,(f,l,s)
P. indolicus	+, coccus	+	–	V	+⁻	+⁻	–	+	–	–	–	–	A,B,p
P. magnus	+, coccus	–	–	–	–	–	–	–	–	–	–	–+	A,(f,l,s)
P. saccharolyticus	+, coccus	–	–	–	+⁻	–	+	+	+	–	–	+	F,A
Streptococcus intermedius	+, coccus	–	+	–	–	+⁻	+	+	+	–	+	–	L,(f,a,s)
Veillonella parvula	–, coccus	–	–	–	+	+	–	–	–	–	–	V	a,p,(l)

Key:

+ = positive
– = negative
+⁻ = most strains positive
–+ = most strains negative
V = variable.

*Fatty acids: capital letters indicate major quantities, lower case minor amounts, in parentheses indicates variable production. A = acetic; B = butyric; C = caprylic; IB = isobutyric; IC = isocaproic; IV = isovaleric; P = propionic; V = valeric; F = formic; L = lactic; S = succinic.

lin levels achievable using standard oral or intramuscular dosing. Between 90–95 percent of isolates tested had minimal inhibitory concentrations of 1 µg/ml clindamycin and 2 µg/ml metronidazole, serum levels that in humans would easily be achieved by oral treatment with proper doses. All common aminoglycosides tested, including gentamicin, were ineffective against anaerobes. About 85 percent of human sporeforming anaerobes tested are sensitive to achievable levels of trimethoprim-sulpha combinations. Love and others (27), using a qualitative susceptibility test for anaerobes isolated mainly from cat bite abscesses, found all isolates to be sensitive to chloramphenicol (12 µg/ml), erythromycin (3 µg/ml), and deoxycycline (6 µg/ml) and 99 percent to be sensitive to 2 units/ml penicillin and 2.5 µg/ml amoxycillin.

It should nevertheless be remembered that non-sporeforming anaerobic infections are commonly mixed infections, and that anaerobes may be protected against the "correct" antibiotic, chosen on susceptibility data, by the production of, for example, beta-lactamases or chloramphenicol acetylase by aerobic opportunist pathogens. The presence of the very large numbers of bacteria often found within abscesses may result in inactivation of antibiotics penetrating the outer part of the abscess, with minimal effect on total bacterial numbers, even though the organisms are susceptible to antibiotic levels achieved—there are just too many bacteria.

The routine disc diffusion (Kirby-Bauer) technique used for rapidly growing anaerobes must *not* be used for anaerobe susceptibility testing. The methodology and zone size criteria have not been developed for anaerobes. Until such time as a disc diffusion method becomes available, the small diagnostic laboratory will find the broth-disc method of Wilkins and Thiel (28) the method of choice. A frozen antibiotic MIC panel is produced by MicroMedia Systems Inc. for anaerobic organisms; in our experience, non-sporeforming anaerobes isolated from clinical infections in animals do not always grow in the media used in this system.

Add one or more standard antibiotic sensitivity disc of the antimicrobial agents shown in Table 15-VII to 5 ml of brain-heart infusion broth supplemented with cysteine (0.05% w/v), hemin (0.0005% w/v), menadione (0.002% w/v), and yeast extract (0.5%). These tubes must be placed in a boiling water bath for ten minutes to remove oxygen, cooled, and the discs added. Following inoculation they are incubated in anaerobic conditions with loosened caps. A positive growth control tube, without antibiotics, is always included.

Table 15-VII

Preparation of broth-disc tubes

Antimicrobial agent	Disc Content	Number of discs/tube	Final antimicrobial concentration/ml
Penicillin G	10 U	1	2 U
Ampicillin	10 µg	2	4 µg
Cefoxicin	30 µg	3	18 µg
Carbenicillin	100 µg	5	100 µg
Cephalothin	30 µg	1	6 µg
Erythromycin	15 µg	1	3 µg
Clindamycin	2 µg	8[*]	3.2 µg
Chloramphenicol	30 µg	2	12 µg
Tetracycline	30 µg	1	6 µg

[*] Store overnight anaerobically to remove oxygen from large numbers of discs.

REFERENCES

1. Holdeman, L. V., Cato, E. P., and Moore, W. E. C.: *Anaerobe Laboratory Manual*, 4th ed. Blacksburg, Va., Virginia Polytechnic Institute and State University, 1977.
2. Sutter, V. L., Citron, D. M., and Finegold, S. M.: *Wadsworth Anaerobic Bacteriology Manual.* St. Louis, C. V. Mosby Company, 1980.
3. Allen, S. D., and Siders, J. A.: In Lannette, E. H., Balows, A., Hausler, W. J., and Truant, J. P. (Eds.): *Manual of Clinical Microbiology*, 3rd ed, Washington, D.C., American Society of Microbiology, 1980, pp. 397–417.
4. Berkhoff, G. A., and Redenberger, J. L.: *Am J Vet Res, 38*:1069, 1977.
5. Hirsh, D. C., Biberstein, E. L., and Jang, S. S.: *J Clin Microbiol, 10*:188, 1979.
6. Berg, J. N., Fales, W. H., and Scanlan, C. M.: *Am J Vet Res, 40*:876, 1979.
7. Schwan, O., and Holmberg, O.: *Vet Microbiol, 3*:213, 1978/1979.
8. Roberts, D. S., Graham, N. P. H., Egerton, J. R., et al.: *J Comp Pathol, 78*:1, 1968.
9. Sorensen, G. H.: *Nord Vet Med, 30*:199, 1978.
10. DuPreez, J. H., Greeff, A. S., and Eksteen, N.: *Onderstepoort J Vet Res, 48*:123, 1981.
11. Vera, H. D., and Power, D. A.: In Lannette, E. H., Balows, A., Hausler, W. J., and Truant, J. P. (Eds.): *Manual of Clinical Microbiology*, 3rd ed. Washington, D.C., American Society of Microbiology, 1980, pp. 965–999.
12. Mouton, C., Hammond, P., Slots, J., et al.: *J Clin Microbiol* 11:682, 1980.
13. Hansen, S. L., and Stewart, B. J.: *J Clin Microbiol* 4:227, 1976.
14. Morgan, J. R., P. Y. K. Liu, Smith, J. A.: *J Clin Microbiol* 4:315, 1976.
15. Dowell, V. R. Jr., and Hawkins, T. M.: Laboratory methods in anaerobic bacteriology. *CDC Laboratory Manual.* Atlanta, Ga., Center for Disease Control, 1977.
16. Gradin, J. L., and Schmitz, J. A.: *J Clin Microbiol* 6:298, 1977.
17. Thorley, C. M.: *J Appl Bact* 40:301, 1976.
18. Skerman, T. M., Erasmuson, S. K., and Every, D.: *Infect Immun, 32*:788, 1981.
19. Stewart, D. J.: *Res Vet Sci, 27*:99, 1979.

20. Egerton, J. R., and Parsonson, I. M.: *Aust Vet J, 45*:345, 1969.
21. Schmitz, J. A., and Gradin, J. L. *Can J Comp Med, 33*:440, 1980.
22. Merritt, G. C.: *Aust Vet J, 36*:388, 1960.
23. Shinjo, T., Miyazato, S., Kaneuchi, C., et al.: *Jpn J Vet Sci, 43*:223, 1981.
24. Schwan, O.: *J Clin Microbiol* 9:157, 1979.
25. Sutter, V. L., and Washington, J. A. II: In Lannette, E. H., Balows, A., Hausler, W. J., and Truant, J. P. (Eds.): *Manual of Clinical Microbiology*, 3rd ed. Washington, D.C., American Society of Microbiology, 1980, pp. 475–477.
26. Kimsey, P. B., and Hirsh, D. C.: *J Vet Pharmacol Therap, 1*:63, 1978.
27. Love, D. N., Bailey, M., and Johnson, R. S.: *Aust Vet Pract* Sept. 1980.
28. Wilkins, T. D., and Thiel, T.: *Antimicro Agents Chemother, 3*:350, 1973.
29. Wegienek, J., and Reddy, C. A.: *Int J Syst Bacteriol, 32*:218, 1982.
30. Berg, J. N., and Scanlan, C. M.: *Am J Vet Res, 43*:1580, 1982.

Chapter 16

MICROCOCCUS AND STAPHYLOCOCCUS

JOHN R. COLE, JR.

THE FAMILY Micrococcaceae includes unchained, Gram-positive aerobic or facultatively anaerobic cocci. The following genera are listed in the family: *Micrococcus, Staphylococcus,* and *Planococcus.*

Organisms of the genus *Staphylococcus* are aerobic and facultatively anaerobic, catalase positive, nonmotile, non-sporeforming, fermentative, Gram-positive cocci. Although they are usually seen in clusters and pairs, short chains may be seen in smears from fluid media (Fig. 16-1). They occur commonly as commensals on the skin and mucous membranes of man and animals.

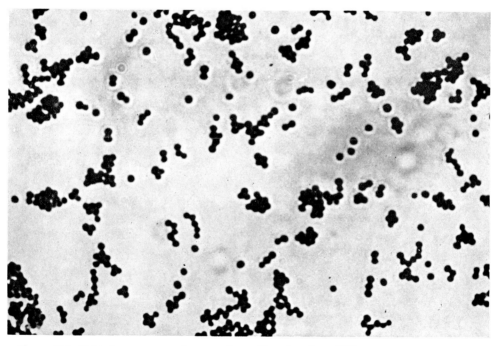

Figure 16-1. *Staphylococcus aureus* from a culture. Gram's stain, ×2250.

There are at least fourteen recognized species of *Staphylococcus*. The principal distinguishing features of the five species considered to be important animal pathogens or frequent isolates are as follows:

161

Staphylococcus aureus (*Staphylococcus pyogenes*). Typical strains are coagulase positive, are beta-hemolytic, ferment maltose and mannitol, and have pigmented colonies. In order to qualify for identification as *S. aureus*, a strain must produce coagulase. The capacity to produce this enzyme may, on occasion, be weak.

Staphylococcus epidermidis. This species is considered an opportunistic pathogen that does not produce coagulase. Members of this species ferment maltose but not mannitol, vary in their capacity to produce hemolysis, and are usually nonpigmented.

Staphylococcus saprophyticus. This opportunistic human pathogen closely resembles *S. epidermidis*, and is differentiated by its resistance to novobiocin. It is coagulase negative, is nonhemolytic, and has been isolated from air, soil, dust, and the surface of animal carcasses.

Staphylococcus intermedius. This coagulase positive, beta-hemolytic, nonpigmented species is primarily cultured from dogs. It has, however, been isolated from horses, pigeons, and milk. This species may weakly ferment maltose.

Staphylococcus hyicus subsp. *hyicus.* Most strains are coagulase positive, produce no hemolysis, and are nonpigmented. It is considered an important pathogen in pigs and has been isolated from cattle and poultry.

ISOLATION PROCEDURES, CULTURAL CHARACTERISTICS, AND PATHOGENICITY

Specimens to be cultured for staphylococci and other cocci are inoculated onto sheep or bovine blood agar, and into brain-heart infusion semisolid medium or thioglycollate broth. The growth from semisolid or liquid media is examined for cocci by staining and is subcultured to blood agar if the plate cultures are negative. Colonies are apparent after incubation for eighteen to twenty-four hours at 37°C.

Several useful selective media are of value for the isolation of staphylococci from heavily contaminated clinical materials. Two of the most frequently used are mannitol salt agar and phenylethyl alcohol agar (see Appendix B).

S. aureus. Colonies are round, glistening, convex, smooth, and opaque. Beta-hemolysis is produced by most strains, and frequently a double zone is apparent in which the central clear zone is surrounded by a band of partial hemolysis. Colonies may possess white, golden, or lemon pigmentation.

Members of this species are involved in a wide variety of disease processes, such as suppurative wound infections, pyemia of lambs, mastitis, pyoderma of the dog, cat, and other animals, abscesses in many animal species, and botryomycosis of horses, cattle, and swine. They are also frequent secondary invaders and opportunists. Staphylococcal enteritis occurs in humans but has not yet been reported in animals.

S. epidermidis. The cultural characteristics of strains of this species are very similar to those of *S. aureus.* Colonies are usually white or colorless, but pigmentation, as with *S. aureus,* is sometimes present. Beta-hemolysis, although not usually seen with these strains, is encountered on occasions.

These bacteria are opportunistic pathogens and have been implicated as causing mastitis in cattle, as well as abscesses and skin infections in various animal species and humans.

S. saprophyticus. Colonies of this species are often pigmented on initial isolation and are nonhemolytic.

Bacteria included in this species are considered opportunistic human pathogens. They have been cultured from a variety of human infections that involve the urinary tract, involve wounds, and are associated with artificial prostheses.

S. intermedius. Colonies are white, circular, smooth, and glistening and do not produce pigment. They are usually beta-hemolytic on blood agar.

Reports of infections due to these bacteria are confined to pyodermas and mastitis in dogs. Their significance in other animal species is not known.

S. hyicus. Pigmentation of the colonies and zones of hemolysis varies with this species. Zones of discoloration may occur around the colony when "aged" blood agar is used, and this should not be confused with beta-hemolysin production.

Infections by this organism occur in pigs and cattle. Exudative dermatitis or greasy pig disease (1,2) and polyarthritis (3) due to *S. hyicus* have been reported in pigs. It has been cultured from skin, udder lesions, and milk in cattle.

IDENTIFICATION

Characteristics that are recommended for use in identification of *Staphylococcus* species are shown in Table 16-I. For a more detailed explanation of these procedures, see articles listed in the Supplementary References.

The staphylococci can be identified not only by the use of conventional procedures but also by a more rapid and readily available commercial system such as the API Staph-Ident System (Analytab Products, Inc., Plainview, N.Y.) (4,5). This miniaturized system will satisfactorily differentiate most of the staphylococci encountered in diagnostic laboratories.

OTHER COCCI

In addition to the staphylococci, other Gram-positive cocci are associated with humans and animals as generally harmless commensals. Most of these are included in the genera listed below. The nomenclature is that of *Bergey's*

TABLE 16-I

Characteristics of Staphylococci and Micrococci

Genus and Species	Coagulase*	Hemolysis (⌐)	Pigment	Catalase	Maltose[+]	PAB[‡]	Mannitol[+]	DNase	Novobiocin[‡]	Glucose[‡‡]	Oxidase[§§]
S. aureus	+	+	+	+	+	+[6]	+	+[**]	S	+	−
S. epidermidis	−	±	±	+	+	+[§]	−	−	S	+	−
S. saprophyticus	−	−	±	+	+	NA	±	−	R	−	−
S. intermedius	+	+	−	+	±	+[//]	±	±	S	+	−
S. hyius subsp. hyius	±	−	−	+	−	−	−	+	S	+	−
Micrococcus sp.	−	−	±	+	−	NA	−	−	S	−	±

*Tube test using rabbit plasma (EDTA)

+Acid produced aerobically

‡Purple Agar Base (Difco Laboratories, Detroit, MI) with 1% maltose for detecting aerobic acid production

§Wide yellow zone around colony

//Slight yellow to yellow-green zone under colony

¶Diffuse alkaline color (deep purple) around colony streak

**Poultry isolates may be negative

−−S = Sensitive, zone diameter μ 16 mm; R = Resistant, 5 μg disc, P Agar medium.

++Anaerobic acid production (O–F Medium)

    Reaction observed in 6% tetramethyl phenylene diamine-hydrochloride in dimethyl sulfoxide

Manual (3). They have little or no significance as causes of disease in animals, but they are encountered on occasion in clinical materials.

Micrococcus. These cocci resemble the staphylococci morphologically and occur singly and in clusters of varying size. They are Gram-positive to Gram-variable, and both motile and nonmotile strains are encountered. Some are not pigmented. Cowan and Steel (4) distinguish staphylococci from some species of micrococci on the basis of the O/F test (see Appendix C). Most staphylococci split sugars by fermentation, while those species of micrococci that break down sugars do so oxidatively (see Table 16-I). The modified oxidase and benzidine tests, resistance to lysozyme and lysostaphin, acid production in glycerol-erythromycin medium, and growth on the selective medium of Schleifer and Kramer are additional tests for differentiating staphylococci from micrococci (8,9). These procedures should be incorporated when definitive identification of the isolate is required.

Planococcus. These are motile, nonpathogenic cocci that occur singly and in pairs, triads, or tetrads. They are found in sea water and are rarely encountered in clinical materials.

The anaerobic genera *Sarcina* and *Peptococcus* are now included in the family *Peptococcaceae*, and will be referred to in Chapter 15 and 17. *Gaffkya* is no longer recognized as a genus (6).

PHAGE TYPING

A large number of phage types from animals and man have been identified. Determination of phage type is frequently of value in the study of the epidemiology of staphylococcal infections. Generally speaking, the phage types found in animals differ from those found in man so that there appears to be host species specificity. Also, there is a tendency for the phage types important in disease outbreaks to change over an extended period of time. Phage typing is performed at the Center for Disease Control, Atlanta, Georgia, and some other reference laboratories. For details of the procedures, consult the Supplementary References.

ENTEROTOXIN

Some strains of *S. aureus*, including those responsible for staphylococcal food poisoning, produce thermostable enterotoxins. These toxins can be detected and identified by agar gel immunodiffusion procedures. Examination of foods and cultures for enterotoxin is carried out in some public health laboratories.

REFERENCES

1. Underdahl, N. R., Grace, O. D., and Twiehaus, M. J.: *Am J Vet Res, 26*:617, 1965.
2. L'Ecuyer, C.: *Can J Comp Med, 31*:243, 1967.
3. Phillips, W. E., King, R. E., and Kloos, W. E.: *Am J Vet Res, 41*:274, 1980.
4. Maddux, R. L., and Koehne, G.: *J Clin Microbiol, 15*:984, 1982.
5. Kloos, W. E., and Wolfshohl, J. F.: *J Clin Microbiol, 16*:509, 1982.
6. Buchanan, R. E., and Gibbons, N. E. (Eds.): *Bergey's Manual of Determinative Bacteriology,* 8th ed. Baltimore, Williams and Wilkins, 1974.
7. Cowan, S. T., and Steel, K. J.: *Manual for the Identification of Medical Bacteria,* 2nd ed. Cambridge, Cambridge U Pr, 1974.
8. Kloos, W. E.: *Clin Microbiol Newsletter, 4*:75, 1982.
9. Faller, A., and Schleifer, K. H.: *J Clin Microbiol, 13*:1031, 1981.

SUPPLEMENTARY REFERENCES

Batty, I., and Walker, P. D.: Colonial morphology and fluorescent labelled antibody staining in the identification of species of the genus *Clostridium. J Appl Bacteriol, 28*:112, 1965.

Botulism In the United States (Contains a Section on Laboratory Methods). U.S. Department of Health, Education and Welfare, Public Health Service. (Available from the Center for Disease Control, Atlanta, Georgia.)

Cruickshank, R., (Ed.): *Medical Microbiology,* 11th ed. Baltimore, Williams & Wilkins, 1965.

Holdeman, L. V., and Moore, W. E. C. (Eds.): *Anaerobe Laboratory Manual,* 4th ed. Blacksburg, Va., V.P.I. Anaerobic Laboratory, Virginia Polytechnic Institute and State University, 1977.

Smith, L. DS.: *The Pathogenic Anaerobic Bacteria,* 2nd ed. Springfield, Thomas, 1975.

Sterne, M., and Batty, I.: *Pathogenic Clostridia.* London, Butterworth Scientific Publications, 1975.

Staphylococcus

Baird-Parker, A. C.: A classification of micrococci and staphylococci based on physiological and biochemical tests. *J Gen Microbiol, 30*:409, 1963.

Devriese, L. A., and Hajek, V.: Identification of pathogenic staphylococci isolated from animals and foods derived from animals. *J Appl Bacteriol, 49*:1, 1980.

Hajek, V.: *Staphylococcus intermedius*, a new species isolated from animals. *Int J Syst Bacteriol, 26*:401, 1976.

Kloos, W. E.: Natural populations of the genus *Staphylococcus. Ann Rev Microbiol, 34*:559, 1980.

Kloos, W. E., and Smith, P. B.: Staphylococci. In Lennette, E. H. et al. (Eds.): *Manual of Clinical Microbiology*, 3rd ed. Washington, D.C., American Society for Microbiology, 1980, Chapter 7.

Phillips, W. E., Jr., and Kloos, W. E.: Identification of coagulase-positive *Staphylococcus intermedius* and *Staphylococcus hyicus* subsp. *hyicus* isolates from veterinary clinical specimens. *J Clin Microbiol, 14*:671, 1981.

Sewell, C. M., et al.: Clinical significance of coagulase-negative staphylococci. *J Clin Microbiol, 16*:236, 1982.

Phage Typing

Gillies, R. R., and Dodds, T. C.: *Bacteriology Illustrated*. Baltimore, Williams and Wilkins, 1965.

Balows, A., and Hausler, W. J., Jr. (Eds.): *Diagnostic Procedures for Bacterial, Mycotic and Parasitic Infections*, 6th ed. Washington, American Public Health Association, Inc., 1981.

Enterotoxin

Bennett, R. W., and McClure, F.: Collaborative study of the serological identification of staphylococcal enterotoxins by the microslide gel double diffusion test. *AOAC, 59*:594, 1976.

Chapter 17

STREPTOCOCCUS AND RELATED COCCI

T HE GENERA of the family Streptococcaceae contain facultatively anaerobic, Gram-positive cocci that are catalase negative and fermentative. They occur singly, in pairs, or in chains. Many occur in nature, and some are commensals in the respiratory, genital, and alimentary tracts and skin of animals and humans. Most of the potential pathogens are in the genus *Streptococcus.*

Bergey's Manual list the following genera in the family Streptococcaceae: *Streptococcus, Leuconostoc, Pediococcus, Aerococcus,* and *Gemella.* The family Peptococcaceae contains the following genera of anaerobic cocci: *Peptococcus, Peptostreptococcus, Ruminococcus,* and *Sarcina.*

STREPTOCOCCUS

Organisms of this genus are characterized by their capacity to produce chains of cocci of varying lengths. They are classified in various ways. Four principal categories can be recognized by the characteristics listed in Table 17-I. The pyogenic category contains the majority of the disease-producing strains. The viridans group is a serologically heterogeneous group that produces a greenish discoloration of blood agar or alpha-hemolysis. The lactic group includes strains recovered from milk, while the habitat of the enterococci is usually the intestine.

TABLE 17-I
PRINCIPAL CATEGORIES OF STREPTOCOCCI

	CAPACITY TO GROW IN THE FOLLOWING:			
Categories	*On MacConkey's Agar*	*At 45°C*	*In 6.5% NaCl Agar or Broth*	*In 0.1% Methylene Blue Milk*
Pyogenic	—	—	—	—
Viridans	—	+	—	—
Lactic	—	—	—	+
Enterococci	+	+	+	+

LANCEFIELD CLASSIFICATION. The streptococci are also placed in the well-known Lancefield groups by means of a precipitin procedure (see

167

Appendix C) based on a group-specific antigen called component C. The groups can be broken down to types on the basis of specific M and T antigens employing agglutination tests. The latter are seldom carried out in veterinary diagnostic laboratories. Species within groups are identified by the criteria listed in Table 17-II.

Pathogenicity

Str. pyogenes (human: group A): Many disease processes in humans. Reported infrequently as a cause of bovine mastitis and other infections in animals.

Str. zooepidemicus (animal pyogenes: group C): Cervicitis, metritis, and abortion in the mare; navel and other infections in foals; cervicitis, metritis, and mastitis in the cow; septic arthritis, abortion, and septicemia in swine.

Str. equisimilis (group C): Strangles, wound infections, genital infections, and mastitis in the equine; frequent cause of a variety of infections in swine and cattle; various infections in the dog and fowl; and human infections.

Str. equi (group C): Principal cause of strangles, genital infections, mastitis, and other infections of the equine.

Gross (1) lists the following species as causing infections in poultry: *Str. zooepidemicus*, highly pathogenic; *Str. faecalis*, highly pathogenic; and *Str. faecium (Str. durans)*, slightly pathogenic.

Streptococci Frequently Causing Bovine Mastitis

Str. agalactiae (group B): Bovine mastitis; rarely from other animals. Group B streptococci resembling closely *Str. agalactiae* have been recovered frequently from various human infections (2).

Str. dysgalactiae (group C): Bovine mastitis and polyarthritis of lambs. Dennis (3) incriminated this species or a closely related organism in antepartum streptococcal infections in lambs.

Str. uberis (viridans group; antigenically heterogeneous): Accounts for a small proportion of bovine mastitis.

Other streptococci are recovered from the udder of cows with mastitis. Hamilton and Stark (4) reported the recovery of group G strains from cases of mastitis.

Enterococci

These belong to group D and include *Str. faecium (Str. durans)*, *Str. bovis*, *Str. equinus*, and *Str. faecalis*. They generally occur as commensals in the intestinal tract of humans and animals and have occasionally been incriminated in sporadic infections. *Str. faecalis* has been recovered from lesions of endocarditis in chickens. Unlike other streptococci, some strains of enterococci are motile.

TABLE 17-II

DIFFERENTIAL CHARACTERISTICS OF IMPORTANT STREPTOCOCCI FROM ANIMALS

Species	Group	Hemo-lysis	Treha-lose	Sorbi-tol	Manni-tol	Sali-cin	Lac-tose	Raffi-nose	Inu-lin	Escu-lin	Sod. Hip-purate*
						Fermentation:	Acid				
Str. pyogenes	A	β	+	−	v	(+)	+	−	−	−	−
Str. zooepidemicus	C	β	−	+	−	+	+	−	−	−	−
Str. equisimilis	C	β	+	−	−	(+)	v	−	−	−	−
Str. equi	C	β	−	−	−	+	−	−	−	−	+
Str. agalactiae	B	α,β,γ	+	−	−	(+)	+	−	−	−	+
Str. dysgalactiae	C	α,β,γ	+	−	−	−	+	−	−	−	−
Str. uberis		α,β,γ	+	+	+	+	+	−	+	+	+
Str. faecalis	D	α,γ	+	+	+	+	+	−	−	+	v
Str. bovis	D	α	v	−	v	+	+	+	+	+	−

*Hydrolysis

Additional Groups

GROUP E: Infections in sheep and swine, including cervical lymphadenitis (jowl abscesses) in pigs.

GROUP E: (large colony type): Throats of humans, dogs, and other animals.

GROUP G: Streptococcal lymphadenitis in cats (5).

GROUP L.: Miscellaneous infections in the dog.

GROUP N: *Str. lactis* — Milk and dairy products.

GROUP Q: *Str. avium* — Chicken feces.

GROUP R: Septicemia and other infections in pigs; meningitis in humans. Most of the porcine infections have been attributed to *Str. suis* type 11 (6).

GROUP S: Throats of pigs; septicemia, meningitis, and arthritis in young pigs.

PROVISIONAL GROUPS (7): Representatives of groups P and T have also been implicated in septicemia, arthritis, and endocarditis in pigs. Alpha-hemolytic streptococci, which probably belong in Group P or another of these groups, have been isolated from nursing pigs with pneumonia and septicemia (8).

Viridans Streptococci

The term *viridans* is used to refer to those serologically heterogeneous streptococci that produce alpha-hemolysis. It is not a satisfactory term in that several serologically homogeneous and frequently pathogenic species may also produce alpha-hemolysis, e.g. *Str. agalactiae.*

The following are well-known viridans streptococci:

Str. acidominimus: Cattle: udder, vagina, and milk products.

Str. thermophilus: Milk and milk products.

Str. uberis: Cattle: vagina, tonsils, and mastitis.

Streptococcus pneumoniae (Diplococcus pneumoniae)

This organism, which resembles some alpha-hemolytic streptococci, is discussed separately at the end of this chapter.

Other Genera of the Family Streptococcaceae

Leuconostoc, Pedicoccus. Species of these genera have no veterinary significance.

Aerococcus. Cocci of organisms of this genus form tetrads. They have formerly been referred to as *Gaffkya. Aerococcus viridans* has been implicated in human urinary infections and endocarditis.

Gemella. The cocci of this genus occur singly or in pairs with adjacent sides flattened. They are recovered from the human respiratory tract.

Anaerobic Cocci

These are included in the family Peptococcaceae and are discussed in greater detail in Chapter 15. The genera of this family as listed in *Bergey's Manual* are referred to below. Little is known about their significance in animal infections. Readers are referred to *Bergey's Manual* (9) and the *Anaerobe Laboratory Manual* (10) for their identification.

Peptococcus. This genus includes anaerobic cocci that occur singly, in pairs, tetrads, or irregular masses. Six species are described, and all are probably commensals of human beings, in which they occasionally cause infections. Limited experience in the author's laboratory suggests that these organisms seldom cause infections in animals.

Peptostreptococcus. Cocci of this genus occur singly, in pairs, and in short or long chains. All require an anaerobic atmosphere for initial isolation; however, on subsequent subculture, some strains will grow microaerophilically or aerobically. Five species are described in *Bergey's Manual*, and all are probably commensals in humans. Some have been implicated in a variety of infections, including wound infections, puerperal fever, pleurisy, appendicitis, sinusitis, dental infections, arthritis, abscesses, and vaginitis. Based on experience in the author's laboratory, these organisms are associated with similar disease processes in animals.

Ruminococcus. These spherical or elongated coccoid organisms are found in the normal rumen of cattle and sheep and in the intestinal tract of rabbits. They have no pathogenic significance.

Sarcina. These spherical cells occur in packets of eight or more. One of the two species described, viz. *Sarcina ventriculi*, has been recovered from rabbit and guinea pig stomach contents. They have not been implicated in disease.

Hemolysis

The kind of hemolysis obtained on or in blood agar as a result of the growth of streptococci is given considerable emphasis, although it is frequently a variable characteristic as indicated in Table 17-II.

For practical purposes, hemolysis is described as follows:

Alpha (α) hemolysis: Greenish (viridans) zone around colony.

Beta (β) hemolysis: Clear zone of hemolysis around colony.

Gamma (γ) or no hemolysis: No hemolysis apparent around colony.

Factors Influencing Hemolysis

Hemolysis may be influenced by (a) the basic medium employed, (b) the kind of blood used, (c) the length of the incubation period, (d) the atmosphere (anaerobiosis may result in a reduction of hemolysis), and (e) position of the colonies. Some difference may be noted between surface and subsurface colonies.

Sheep's blood in trypticase soy agar is recommended for the careful study of hemolysis, but for practical purposes, media containing other kinds of blood are satisfactory.

Isolation Procedures

The procedures are those recommended as routine for aerobes and facultative anaerobes. After inoculation to blood agar, swabs are incubated in BHI semisolid broth or Schaedler broth because some of the streptococci are anaerobic or microaerophilic and they may be accompanied by Gram-negative anaerobic bacteria.

Selective media especially formulated for the isolation and growth of the streptococci are referred to in Appendix B.

Cultural and Morphological Characteristics

Small, grayish colonies that yield Gram-positive cocci are usually evident in twenty-four hours. Considerable colonial variation may be noted, including mucoid, smooth (glossy), and matte (rough) colonies. Cultures of *Str. equi* from strangles and other infections of the horse may be strikingly mucoid. If chains are present on solid media, they are usually short, and considerable pleomorphism may be noted. Distinct chains are produced in liquid media such as serum, BHI, or thioglycollate broth (Fig. 17-I).

IDENTIFICATION

It is advisable to determine the category (see Table 17-I) to which the streptococcus belongs by inoculating a MacConkey's agar slant, a tube of BHI broth for incubation at 45°C, a 6.5% NaCl agar slant, and a tube of 0.1% methylene blue broth. The last is frequently omitted in that lactic streptococci are relatively uncommon.

On the basis of the results of the tests listed in Table 17-I and growth on blood agar, the culture may be reported as a streptococcus, alpha-, beta-, or nonhemolytic, belonging to the pyogenic, viridans, lactic, or enterococcus group.

If it is desired to determine the species, the differential criteria listed in Table 17-II are employed. For the identification of those species not listed, reference should be made to *Bergey's Manual*. The identification of streptococci usually associated with bovine mastitis is dealt with in Table 34-I.

Low concentrations of bacitracin are specifically active against group A streptococci. A sensitivity disc containing 0.02 units of bacitracin* is placed

---

Taxo A discs (BBL).

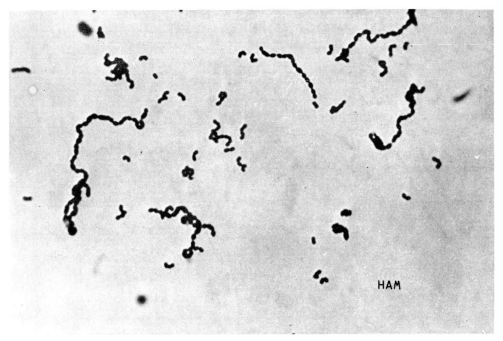

Figure 17-1. *Streptococcus* sp. from a culture. Gram's stain, ×2250 (H. A. McAllister).

on a blood agar plate previously streaked with the streptococcus to be examined. Zones of inhibition of 15–20 mm are obtained with group A strains.

Edward's medium, which is referred to in Chapter 34, is used for the rapid identification of the important mastitis streptococci. The Camp test, which is used for the rapid presumptive identification of *Str. agalactiae*, is described in detail in Appendix C.

Besides the Lancefield groups listed in Table 17-II, *Bergey's Manual* lists groups E, G, L, M, O, and R. Other unofficial Lancefield groups were referred to previously. These groups can only be identified with certainty by a precipitin test of the kind described in Appendix C. Streptococcal groups A, B, C, and G can, with considerable reliability, be identified by slide coagglutination using commercially available reagents* consisting of protein A containing staphylococci coated with antibodies specific for group A, B, C, and G streptococci (11). A latex agglutination procedure is available to identify groups A, B, C, D, F, and G.†

The porcine group E strains produce acid from glucose, mannose, fructose, glycerol, trehalose, cellobiose, salacin, maltose, mannitol, sorbitol, and usu-

---

*Pharmacia Diagnostics, 800 Centennial Avenue, Piscataway, New Jersey, 08854.

†Wellcome Diagnostics, 3030 Cornwallis Road, Research Triangle Park, N.C. 27709.

ally galactose. Xylose, arabinose, rhamnose, melibiose, raffinose, lactose, and inulin are not fermented. Beta-hemolysis is usually evident in forty-eight hours.

Fluorescein-labelled globulins are available for the Lancefield grouping of strains, but, except for the identification of group A and B strains, they are reported to be of questionable value.

STREPTOCOCCUS PNEUMONIAE
Synonym: *Diplococcus pneumoniae*

This species resembles closely the alpha-hemolytic streptococci. It has only recently been included in the genus *Streptococcus*. It is a Gram-positive, capsulated, nonmotile coccus that occurs singly, frequently in pairs, and also as short chains. Its natural habitat is the upper respiratory tract of humans and, to a lesser extent, animals. More than eighty serotypes have been identified on the basis of serologic differences in the capsular substances.

PATHOGENICITY. *Str. pneumoniae* is an important cause of pneumonia, otitis, sinusitis, and meningitis in humans. It is found occasionally in the upper respiratory tract of animals. Respiratory infections have been reported in calves, monkeys, and rabbits. The organism may be established in rat and guinea pig colonies and produce severe pneumonic disease under certain circumstances. Romer (12) has referred to isolations of pneumococci from a number of animal species, including horses, sheep, goats, swine, and a cat.

Str. pneumoniae types III, IV, and XIX have been identified as producing infections in guinea pigs (12). Infections in rats due to types II, III, VIII, XVI, and XIX have been reported (13, 14, 15).

ISOLATION PROCEDURES. The organism grows well on blood agar at 37°C. Growth is favored by incubation in a candle jar or in air containing 10% carbon dioxide. Some cultures will not grow in primary culture without carbon dioxide.

CULTURAL CHARACTERISTICS. Small, round colonies with elevated edges and zones of alpha-hemolysis are produced in eighteen to twenty-four hours. Although most cultures produce colonies with a characteristic concavity, especially evident on prolonged incubation, some strains are moist and mucoid. Stained smears reveal Gram-positive cocci occurring in characteristic pairs, singly, and as short chains.

IDENTIFICATION. This is based upon the following additional criteria:
- Bile soluble.
- Inhibited by optochin* (see Appendix C); alpha-hemolytic streptococci are not inhibited.
- Fermentation of glucose, lactose, sucrose, and inulin.

---

*Taxo P discs (BBL).

- Mannitol is not fermented.

Other less important characteristics are as follows:

- Gelatin is not liquefied.
- Nitrate is not reduced.
- Indole is not formed.
- Litmus milk is acidified and coagulated.

Serotyping is carried out by reference laboratories, and typing sera for the more common serotypes are available commercially (Difco).

REFERENCES

1. Gross, W. B.: In Hofstad, M. S., Calnek, B. W., Helmboldt, C. F., Reid, W. M., and Yoder, H. W., Jr. (Eds.): *Diseases of Poultry*, 6th ed. Ames, Iowa St U Pr, 1972, p. 385.
2. Norcross, N. L., and Oliver, N.: *Cornell Vet, 66*:240, 1976.
3. Dennis, S. M.: *Vet Rec, 82*:403, 1968.
4. Hamilton, C. A., and Stark, D. M.: *Am J Vet Res, 31*:397, 1970.
5. Tillman, P. C., Dodson, N. D., and Indiveri, M.: *J Clin Microbiol, 16*:1057, 1982.
6. Sanford, S. E., and Tilker, M. E.: *J Am Vet Med Assoc, 181*:673, 1982.
7. de Moor, C. E.: *Acta Leiden, 32*:220, 1963 (Abstract in *Vet Bull*).
8. Brown, L. N.: *Proc 73rd Meeting US Animal Hlth Assoc*, 1969.
9. Buchanan, R. E., and Gibbons, N. E. (Eds.): *Bergey's Manual of Determinative Bacteriology*, 8th ed. Baltimore, Williams & Wilkins, 1974.
10. Holdeman, L. V., and Moore, W. E. C.: *Anaerobe Laboratory Manual*, 4th ed. Blacksburg, Va., Virginia Polytechnic Institute, 1977.
11. Hahn, G., and Nyberg, I.: *J Clin Microbiol, 4*:99, 1976.
12. Romer, O.: *Pneumococcal Infections in Animals*. Copenhagen, Mortensen, 1962.
13. Ford, T. M.: *Lab Anim Care, 15*:48, 1965.
14. Baer, H.: *Can J Comp Med, 31*:216, 1967.
15. Weisbroth, S. H., and Freimer, E. H.: *Lab Anim Care, 19*:473, 1969.

SUPPLEMENTARY REFERENCES

Krantz, G. E., and Dunne, H. W.: *Am J Vet Res, 26*:951, 1965.

Skinner, F., and Quesnel, L. (Eds.): *Streptococci*. New York, Academic Press, Inc., 1978.

Wannamaker, L. W., and Matsen, J. M. (Eds.): *Streptococci and Streptococcal Diseases: Recognition, Understanding and Management*. New York, Academic Press, 1972.

Chapter 18

BACILLUS

MEMBERS of this genus are large, aerobic, Gram-positive (old cultures decolorize easily), sporeforming rods. They are catalase positive; many are fermentative, and most are motile.

Bacillus spp. occur widely in nature, being found in the air, water, and soil. They are among the most common laboratory contaminants and are usually ignored when recovered from clinical materials. One should, however, keep in mind the possibility of encountering *B. anthracis.*

The various species of *Bacillus* produce a variety of colonies. With experience, one can frequently identify them by their characteristic morphology and texture. Some general features of colonies follow.

B. cereus. Hemolytic, greenish gray color, and ground glass appearance.

B. mycoides. Moldlike or nebulalike colonies.

B. mesentericus (and other species). Slimy colonies that are sticky and mucoid.

B. mycoides. Many strains are beta-hemolytic, and colonies are rough, opaque, and spreading; slimy colonies are encountered, and yellow, orange, or brown pigmentation is seen. Growth in broth is frequently characterized by the production of a surface pellicle.

Another feature of these organisms that may result in confusion is the Gram variability of a number of species. There are many species of the *Bacillus* genus. The characteristics of some of them are presented in Table 18-II. The Supplementary References provide information on additional strains.

BACILLUS ANTHRACIS

Pathogenicity

Moderate to acute septicemic infections occur in cattle, sheep, caribou, elk, bison, water buffaloes, horses, mink, and other mammals. Fowl are generally resistant.

SWINE. The disease results in acute pharyngitis with extensive swelling and hemorrhage of the throat region.

This animal is less susceptible than cattle. There may be an acute pharyngitis with lesions in the region of the throat and neck.

HORSES. If infection is by ingestion, there is a septicemia with enteritis and colic. Infection may also be through the skin in which there is marked edema and lymphadenitis.

DOGS AND CATS. A rare infection resembling that seen in swine.

HUMANS. Pulmonary anthrax, "malignant carbuncle," or cutaneous anthrax and intestinal anthrax.

Isolation Procedures

All procedures should be carried out in a biological safety hood.

If anthrax is suspected, the carcass should not be opened, and a diagnosis should be attempted on blood obtained from a superficial vein. In swine and horses, fluid or exudate can be aspirated with a syringe from swollen tissues if the organism cannot be demonstrated in blood smears. If a necropsy is performed, great care must be taken not to contaminate the immediate area with spores. Spores are formed when the vegetative forms are exposed to air.

If tissues are submitted, a composite suspension is prepared with a Ten Broeck grinder or mortar and pestle using sterile physiological saline as a diluent. Blood is used as it is.

1. Smears are made and stained by Gram's method. The finding of large, typical Gram-positive rods indicates the likelihood of infection.

It should be kept in mind that clostridial organisms are frequently found in the blood and tissues shortly after death. They can be readily eliminated on the basis of the absence of square-ended capsules and failure to grow aerobically. Sterne (1) recommends Giemsa and Wright's stains for the demonstration of the capsule. By these procedures it is stained a reddish mauve and appears square-ended (Fig. 18-1). Another widely used procedure is to stain heat-fixed or alcohol-fixed (flame rather than let dry) blood smears with 1% polychrome methylene blue for five to ten minutes (see Appendix A). The capsule and capsular material seen among the bacilli stain pink to purple (McFadyean's reaction). After fixation, smears can be immersed in 1:1000 mercuric chloride for thirty minutes in order to kill the spores.

2. Inoculate animals with the tissue suspension and blood if the latter is available.

1 guinea pig:	1.0 ml subcutaneously
1 mouse:	0.1 ml intraperitoneally
1 mouse:	0.3 ml intraperitoneally
1 mouse:	swab blood or suspension over a scarified area at the base of the tail

If anthrax bacilli are present, animals begin to die, usually after twenty-four hours, and show numerous capsulated, square-ended, large rods in stained smears from the spleen and blood. Animals dying in less than

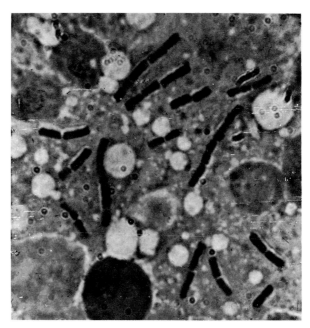

Figure 18-1. Giemsa-stained smear showing the capsule of *Bacillus anthracis*, ×2000. Courtesy of J. W. Huff and T. Vera.

twenty-four hours are examined for other agents, including toxigenic strains of *Bacillus cereus* and clostridial organisms.

3. Blood plates are inoculated from blood and/or tissue suspension and incubated aerobically at 37°C. If rough, flat, gray, nonhemolytic colonies appear, they are transferred to broth. Broth cultures are checked for motility and the characteristic "cotton wool" growth.

4. Cultures of nonhemolytic, nonmotile bacilli are held, and if typical deaths do not occur in animals inoculated with the original material, these cultures are inoculated into animals as in 2.

5. Pure cultures of the same suspicious bacilli are inoculated into salacin and methylene blue broths, litmus milk, and nutrient gelatin. They are also checked for their ability to grow at 45°C. The maximum temperature for the growth of *B. anthracis* is considered to be 43°C.

Strains of *B. cereus* resemble closely the anthrax bacilli. Some differential characteristics are listed in Tables 18-I and 18-II. *B. subtilis, B. mesentericus,* and *B. mycoides* can be differentiated from *B. anthracis* on the basis of their motility. They and other *Bacillus* spp. can be identified by reference to Table 18-II. This table provides some differential characteristics of important *Bacillus* species; however, readers should consult *Bergey's Manual* for the definitive identification of species.

If animals do not die and the reactions do not conform to those of the

TABLE 18-I

DIFFERENTIATION OF *BACILLUS ANTHRACIS* AND *BACILLUS CEREUS*

	B. anthracis	B. cereus
Motility	—	Usually +
Capsulation	+	Noncapsulated*
Salicin (Acid)	Slow or not at all	Rapid
Methylene Blue Reduction	Slow	Rapid
Gelatin Hydrolysis	Slow	Rapid
Litmus Milk	Slowly coagulated and peptonized	Rapidly coagulated and peptonized
Hemolysis	None, or very slightly	Often markedly hemolytic
Penicillin	Sensitive†	Not sensitive

*Other species are.

†Resistant strains have been reported.

anthrax bacillus, the examination for anthrax is considered negative. If the animals die and the criteria listed for *B. anthracis* are fulfilled, a positive diagnosis is warranted.

Other Methods of Identification

Bacteriophage Identification

The use of bacteriophage for the differentiation of this pathogen from the nonpathogenic *Bacillus* spp. is considered reliable (2,3,4,5). The Cherry gamma phage* is used as a 1:10 dilution of a frozen suspension with a titer of 10^4. The phage can be propagated on strain 14, which makes a suitable control. Known cultures of *B. anthracis* and *B. cereus* should be included as controls. The procedure is as follows:

1. On one-half of a blood agar plate, streak a culture of *B. anthracis* as a control. On the other half, streak the culture to be identified.
2. Several droplets of the phage preparation are added to each half of the inoculated blood plate.
3. Incubate the plate in the upright position for eight to twelve hours. Clear zones of lysis will be seen on the control culture and on the suspected culture if it is *B. anthracis*.

String of Pearls Test

This test (6) is based upon the alteration in morphology that strains of *B. anthracis* undergo in the presence of penicillin. A tryptose agar plate containing 0.5 units/ml of penicillin is swabbed from a twenty-four-hour broth culture

*Available from the Center for Disease Control, Atlanta, Georgia.

TABLE 18-II
DIFFERENTIAL CHARACTERISTICS OF SOME *BACILLUS* SPECIES

Species	Motility	Urease	Citrate	Starch Hydrolysis	Gela-tinase	Nitrate Reduction	Growth at 65°C	Acid Production			VP
								Glucose*	Arabinose*	Mannitol*	
B. anthracis	–	–	(+)	+	+	+	–	+	–	–	+
B. subtilis	+	(–)	+	+	+	+	–	+	+	+	+
B. cereus var. mycoides	(+)	(–)	+	+	+	+	–	+	–	–	+
B. megaterium	+	(–)	+	+	(+)	–	–	+	v	+	–
B. stearothermophilus	+	(–)	–	+	+	+	+	±	v	–	–
B. licheniformis	+	v	+	+	+	+	–	±	+	+	+
B. lentus	+	+	–	+	–	–	–	–	–	–	–
B. firmus	+	–	–	+	+	+	–	–	–	–	–
B. coagulans	+	–	–	+	–	v	–	+	v	v	v
B. pumilus	+	–	+	–	+	–	–	+	+	+	+

*Because of ammonia production from peptones, peptone-free media must be used for the testing of carbohydrate utilization. A suitable ammonium salt medium is described by Smith, Gordon and Clark (Aerobic Sporeforming Bacteria, U. S. Dept. Agric., Agriculture Monograph, No. 16, 1952).

of the suspected organism. After incubation for three to six hours, the surface of the agar is examined microscopically under a coverslip. Anthrax bacilli swell and appear as chains of spheres, which rupture on further incubation leaving only a spore. The morphology of the nonpathogenic and saprophytic *Bacillus* spp. remains unchanged. Known cultures of *B. anthracis* and *B. cereus* should be included as controls.

The tests mentioned above should be used along with other criteria such as pathogenicity for experimental animals. There have been reports of strains being resistant to penicillin.

BACILLUS CEREUS

This widespread saprophyte is capable of infecting the bovine udder and producing acute and sometimes fatal gangrenous mastitis. It has also been implicated as a cause of abortion in cows and ewes (7,8). In mastitis, isolation should be attempted during the acute phase of the disease, as the organism may be absent in later samples. For identification, see Tables 18-I and 18-II.

BACILLUS LICHENIFORMIS

This organism has been reported as a cause of abortion in cows (9).

REFERENCES

1. Sterne, M.: Diseases due to bacteria. In Stableforth, A. W., and Galloway, I. A. (Eds.): *Infectious Diseases of Animals*. London, Butterworth Scientific Publications, 1959, vol. 1.
2. Brown, E. R., and Cherry, W. B.: *J Infect Dis, 96*:34, 1955.
3. Brown, E. R., Moody, M. D., Treece, E. L., and Smith, C. W.: *J Bacteriol, 75*:499, 1958.
4. Leise, J. M., Carter, C. H., Friedlander, H., and Freed, S. W.: *J Bacteriol, 77*:655, 1959.
5. Buck, C. A., Anacker, R. L., Newman, F. S., and Eisenstark, A.: *J Bacteriol, 85*:1423, 1963.
6. Jensen, J., and Kleemeyer, H.: *Zentralbl Bakteriol Orig, 159*:494, 1953.
7. Wohlgemuth, K., Bicknell, E. J., and Kirkbride, C. A.: *J Am Vet Med Assoc, 161*:1688, 1972.
8. Wohlgemuth, K., Kirkbride, C. A., Bicknell, E. J., and Ellis, R. P.: *J Am Vet Med Assoc, 161*:1691, 1972.
9. Ryan, A. J.: *Vet Rec, 86*:650, 1970.

SUPPLEMENTARY REFERENCES

Wolf, J., and Barker, A. N.: In Gibbs, B. M., and Shapton, D. A. (Eds.): *Identification Methods for Microbiologists*. New York, Academic Press, 1968, p. 93.

Cowan, S. T.: *Manual For the Identification of Medical Bacteria*, 2nd ed. Cambridge, Cambridge U Pr, 1974.

Feeley, J. C., and Patton, C. M.: In Lennette, E. H. (Ed.-in Chief) *Manual of Clinical Microbiology*, 3rd ed. Washington, D.C., American Society for Microbiology, 1980, Chapter 13.

Gordon, R. E., Haynes, W. C., and Pang, C. N.: *The Genus Bacillus*. Agricultural Handbook No. 427. Washington, D.C., Agricultural Research Service, U.S. Department of Agriculture, 1973.

Chapter 19

CLOSTRIDIUM

O RGANISMS of this genus are large, anaerobic, sporeforming, Gram-positive rods. They are motile (except *Cl. perfringens*), fermentative, and catalase negative.

They reside in the soil and in the intestinal tracts of animals and humans. There are both pathogenic and nonpathogenic species. The latter are only rarely encountered in fresh tissues. The ubiquity of clostridial spores results in their occasional presence as contaminants in clinical materials.

PATHOGENICITY

The principal diseases are tabulated below. Less frequent and important infections, often mixed, occur in all the domesticated species. Clostridial species occur in a quiescent state in a considerable number of apparently normal livers and spleens of cattle, sheep, dogs, and probably other animals.

Clostridium chauvoei (Cl. feseri). Blackleg or blackquarter of cattle and sheep. Isolated on occasion from the apparently normal livers and spleens of cattle and the livers of dogs.

Cl. septicum. Malignant edema of horses, cattle, sheep, and swine. The internal form of the disease in sheep is called braxy. Gangrenous dermatitis of chickens.

Cl. perfringens (Cl. welchii). Wound infections, gangrenous mastitis, and enterotoxemias (dealt with in detail below) and food poisoning (humans). It has been suggested that this organism may have a role in diarrhea in the horse, hemorrhagic enteritis in the dog, colitis X in the horse, and mucoid enteritis in rabbits.

Cl. novyi (Cl. oedematiens) type A. Gas gangrene in cattle, sheep, and humans. Big head in rams. Septicemia in tortoises.

Cl. novyi (Cl. oedematiens, Cl. gigas) type B. Black disease (infectious necrotic hepatitis of sheep and cattle). Isolated on occasion from apparently normal livers of sheep and cattle.

Cl. bubalorum. A cause of osteomyelitis in buffaloes in Indonesia. It is closely related to *Cl. novyi* type B and had been referred to as *Cl. oedematiens* type C.

Cl. novyi. An unknown type has been associated with sudden death in pigs (1).

Cl. haemolyticum (*Cl. oedematiens* type D). Bacillary hemoglobinuria of cattle and sheep resulting from the infection of liver infarcts.

Cl. tetani. Tetanus in a wide range of animal species.

Cl. sordellii. Recovered by itself and occasionally with other clostridia. Its significance is uncertain. Some workers claim that it can produce lesions resembling those of blackleg in cattle and "big head" in rams.

Cl. botulinum. Food poisoning due to the toxins of this species is dealt with in a separate section below. Wound and infant botulism has been described in humans but not in animals.

Cl. colinum. The cause of ulcerative enteritis, originally called "quail disease," an acute disease of game birds, young turkeys, young chickens, quail, grouse, and partridges (1).

Cl. difficile. Pseudomembranous colitis in humans and ileocecitis in laboratory animals (2) on antibiotic regimens.

EXAMINATION FOR CLOSTRIDIA

DIRECT EXAMINATION. Smears are prepared for Gram's and fluorescent antibody staining from the suspected site of infection. Clostridial infections yield large, Gram-positive rods. The number and size of the organisms seen may be quite variable. Long filamentous forms suggest infection with *Cl. septicum* (Fig. 19-1).

ISOLATION PROCEDURES

DIRECT CULTIVATION. A general procedure is outlined in Table 19-I. Media are the following: freshly poured blood agar plates, thioglycollate broth (with glucose), Schaedler broth, cooked meat medium, and the differential media required for the tests referred to in Table 19-II. The procedures for anaerobic cultivation are given in Chapter 15.

C. colinum (3, 4). Primary isolation of this fastidious organism has been made in tryptose-phosphate glucose broth with 8% horse plasma and in the yolk sac of five- to eight-day embryonated chicken eggs. After several passages in these media or in thioglycollate broth with 3–10% horse serum, the organism can be grown on blood agar like other fastidious anaerobes.

CULTURAL CHARACTERISTICS

If clostridia are present, colonies of all but the very fastidious will be apparent after three to four days incubation. The growth of many strains may be apparent after forty-eight hours. The aerobic and anaerobic plates should be compared. If colonies resembling those of the clostridia are seen on the anaerobic plate but not on the aerobic plate, Gram-stained smears should be made from the former. The presence of large Gram-positive rods suggests clostridia (Fig. 19-2). The rather characteristic colonies of *Bacillus*

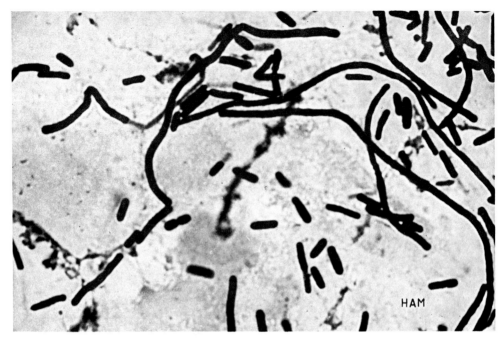

Figure 19-1. Long forms of *Clostridium septicum* seen in guinea pig liver impression smear. Gram's stain, × 2250 (H. A. McAllister).

spp. are usually encountered only on the aerobic plates. *Cl. perfringens* produces characteristic round, grayish, smooth colonies, while the other clostridia infecting animals produce small, ill-defined colonies with irregular peripheries. Colonies usually have zones of beta-hemolysis; many strains of *Cl. perfringens* produce a characteristic zone of double hemolysis.

Colonies of *C. colinum* are 1–3 mm in diameter, have filamentous margins, are grayish, glossy, and semitranslucent. Spores are not readily produced on usual media.

IDENTIFICATION

Cl. chauvoei, Cl. septicum, and *Cl. sordellii* may be identified by the fluorescent antibody staining procedure described below. *Cl. perfringens* and *Cl. botulinum* are the most commonly encountered clostridia causing disease in animals. *Cl. perfringens* (double zone of hemolysis), *Cl. botulinum,* and some other important clostridia can usually be identified according to the criteria listed in Table 19-II. Because this table includes only some of the more important clostridia, it may be necessary to submit cultures that cannot be identified or about which there is doubt as to their identification to a reference laboratory. The identification of fermentation products by gas chromatography has been found to be helpful in identification (5).

TABLE 19-I

A PROCEDURE FOR THE ISOLATION OF CLOSTRIDIA

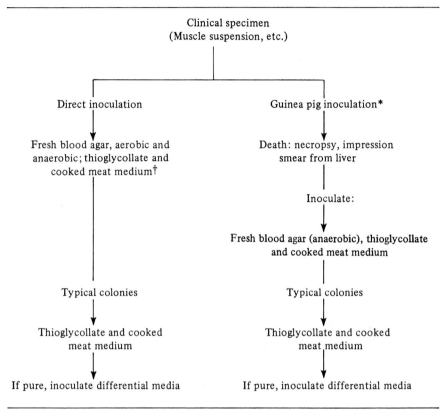

Clinical specimen
(Muscle suspension, etc.)

Direct inoculation   Guinea pig inoculation*

Fresh blood agar, aerobic and   Death: necropsy, impression
anaerobic; thioglycollate and   smear from liver
cooked meat medium†

                                            Inoculate:

                                 Fresh blood agar (anaerobic), thioglycollate
                                       and cooked meat medium

Typical colonies   Typical colonies

Thioglycollate and cooked   Thioglycollate and cooked
meat medium   meat medium

If pure, inoculate differential media   If pure, inoculate differential media

* Young adult guinea pigs are preferred.
† Several tubes should be used. After inoculation place one tube in boiling water for five minutes, another for ten minutes and one is left unheated. These tubes after sufficient incubation are plated to blood agar for subsequent identifications. If materials are fresh, heating may not be necessary.

The differential media and tests listed in Table 19-II are described in Appendices B and C along with an interpretation of the reactions. The shape and position of spores may be helpful in identification. *Cl. perfringens* is exceptional in that spores are seldom produced.

GUINEA PIG INOCULATION

Because *Cl. chauvoei, Cl. septicum, Cl. novyi,* and *Cl. sordellii* can be identified in clinical materials by the fluorescent antibody procedure described below, it is seldom necessary to inoculate guinea pigs.

Tissue weighing about 10 g is placed in a sterile mortar containing 15–20 ml of broth. Mince the tissue finely with sterile scissors, and express the

TABLE 19-II
DIFFERENTIAL CHARACTERISTICS OF IMPORTANT SPECIES OF CLOSTRIDIA*

Species	Spores	Milk Digestion	Gelatinase	Indole	Egg yolk agar LEC	Egg yolk agar LIP	Glucose	Maltose	Lactose	Salicin	Sucrose	Other
Cl. perfringens	OS	+	+	-	+	-	+	+	+	v	+	Double zone hemolysis; spores rare
Cl. septicum	OS	-	+	-	-	-	+	+	+	+	-	
Cl. chauvoei	OS	v	v	-	-	-	+	v	+	-	+	
Cl. novyi, A	OS	-	+	-	+	+	+	v	-	-	-	
Cl. novyi, B	S	+	+	+	+	+	+	+	-	-	-	
Cl. haemolyticum	OS	+	+	+	v	-	+	-	-	-	-	Urease +
Cl. bordellii	OS	+	+	+	+	-	+	+	-	v	-	Urease -
Cl. bifermentans	OS											
Cl. botulinum:												
Group I	OS	+	+	-	-	+	+	+	-	v	-	
Group II	OS	-	+	-	-	+	+	+	-	v	-	
Group III	OS	-	+	v	v	+	+	v	-	-	-	
Cl. tetani	RT	-	+	-	-	+	-	+	-	-	-	
Cl. sporogenes	OS	+	+	-	-	-	+	+	-	v	-	
Cl. histolyticum	OS	+	+	-	-	-	-	-	-	+	-	Microaerophilic
Cl. colinum	OS	-	-	-	-	-	+	+	-	v	+	Spores rare
Cl. difficile	OS	-	v	-	-	-	+	-	-	-	-	

* KEY: OS = oval, subterminal
      RT = round, terminal

PLATE I

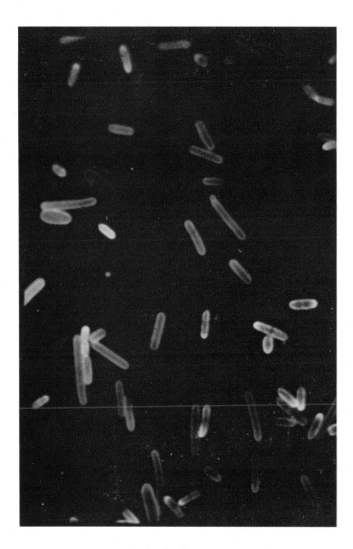

Plate I. A smear of mixed *Clostridium septicum* and *Cl. chauvoei* culture, stained with a mixture of fluorescein isothyocyanate labeled *Cl. septicum* antiserum and lissamine rhodamine B. 200 *Cl. chauvoei* labeled antiserum. Courtesy of Burroughs Wellcome Co. (U.S.A.), Research Triangle Park, North Carolina.

PLATE II

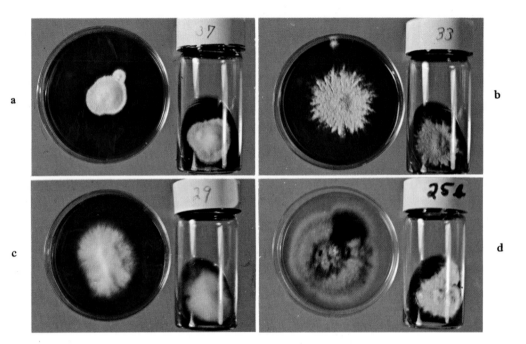

Plate II. Dermatophyte test medium (Fungassay®): *a, Trichophyton mentagrophytes* at fourteen days; *b, Microsporum gypseum* at fourteen days; *c, Microsporum canis* at fourteen days; *d*, (Petri dish) common contaminant with no color change, ten days; (bottle on right) typical growth of *Microsporum gypseum*. Color reproduction courtesy of Pitman-Moore Inc., Washington Crossing, New Jersey.

PLATE III

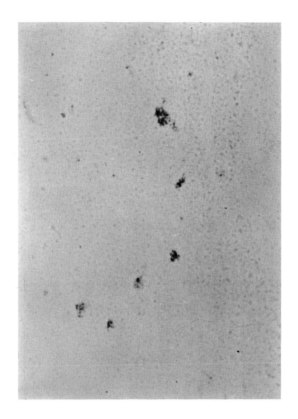

Plate III. *Mycobacterium paratuberculosis* in a smear of rectal mucosa from a cow with Johne's disease. The microorganisms are short rods which appear in clumps. Ziehl-Neelsen stain, X1200. The assistance of Doctor A. G. Karlson, Mayo Graduate School of Medicine, University of Minnesota, in preparation of this photograph is greatly appreciated. Color reproduction courtesy of Smith, Kline and French Laboratories, Philadelphia, Pennsylvania.

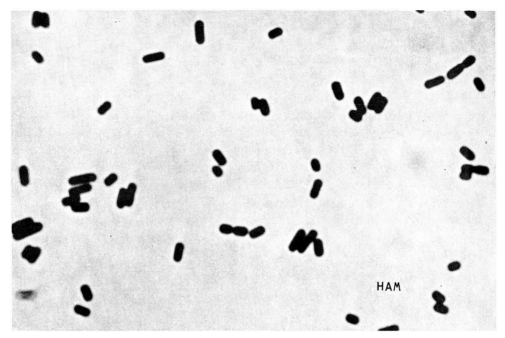

Figure 19-2. *Clostridium perfringens* from blood agar plate culture. Gram's stain, ×2250 (H. A. McAllister).

tissue fluid with the pestle. Pour off the supernatant after the tissue fragments have settled. Two guinea pigs are inoculated intramuscularly in the rear leg, one with 0.5 ml and the other with 1.0 ml of the supernatant. If only one guinea pig is used, employ the larger dose.

IMPORTANT CONSIDERATIONS

1. Plates should be placed in an anaerobic environment shortly after inoculation, as some clostridial species are oxygen sensitive. Twenty minutes exposure may be bactericidal.

2. Heat thioglycollate, cooked meat medium, litmus milk, and differential media in boiling water for ten minutes prior to inoculation to remove oxygen.

3. Motility can be demonstrated in semisolid trypticase agar medium.

4. To inoculate differential media, add a drop of the thioglycollate or cooked meat medium culture to each tube. Stab with the inoculating loop if the medium is semisolid, and incubate until appreciable growth is obtained.

5. Make impression smears from the livers of the dead guinea pigs. Long, filamentous Gram-positive organisms suggest *Cl. septicum*.

6. Clostridial infections are occasionally mixtures of two or more species.

7. The more fastidious clostridia, e.g. *Cl. novyi* strains, *Cl. haemolyticum,*

Cl. sordellii, and *Cl. chauvoei*, require fresh blood agar plates. Small pieces of sterile muscle tissue added to fluid media increase the growth of fastidious anaerobes.

8 *Cl. septicum, Cl. chauvoei, Cl. perfringens, Cl. tetani*, and other clostridia are found normally in the intestinal tracts of farm animals, and potentially pathogenic clostridia can be recovered from the livers of some normal animals.

9. The significance of the isolation from tissues of the clostridia that are found normally in the intestinal tract is often questionable. These organisms may reach the tissues from the intestinal tract within a very short time after death.

10. An oxygen indicator should always be placed in the anaerobic jar. A satisfactory indicator solution consists of a mixture in a cotton-stoppered test tube of 2 ml of each of the solutions listed below. The mixture is boiled until it loses its color, then immediately placed in the anaerobic jar. The light tinge of blue that appears at the top of the indicator will disappear, and the tube will remain clear in a completely anaerobic atmosphere.

Solutions required: (a) 6 ml of 1 N sodium hydroxide, with distilled water added to make 100 ml; (b) 3 ml of 0.5% aqueous methylene blue, with distilled water added to make 100 ml; (c) 6 g glucose, with distilled water added to make 100 ml. Add a crystal of thymol to the last solution as a preservative. A disposable anaerobic indicator is available commercially (BBL–GasPak).

11. All types of *Cl. novyi* are considered pathogenic or potentially pathogenic for guinea pigs except type C (*Cl. bubulorum*). The pathogenicity of the other types can be enhanced if equal amounts of calcium chloride solution (5–10%) and culture are injected into the muscles of the thigh of the guinea pig. The dose recommended is 0.5 ml. Guinea pigs dying as a result of *Cl. haemolyticum* infection may have a hemoglobinuria.

12. For those strains that do not readily produce surface colonies on blood agar, it is advisable to pour over the inoculated plate sufficient freshly melted and cooled agar to make a thin layer. Pour from the lightly inoculated side to the more heavily inoculated side.

13. Failure to cultivate the fastidious anaerobes may be due to peroxide formation in media resulting from heating and storage in the air. For this reason, media should not be oversterilized or overheated. Media that have been placed in boiling water to remove oxygen should not be set aside for later use. Fresh media should be used if at all possible. The adequacy of methods can be tested with known cultures of the fastidious *Cl. novyi* type B and *Cl. haemolyticum*.

14. Spreading of organisms on solid media can be reduced by the addition to the surface of several drops of glycerin, which is spread over the surface

evenly by means of a glass spreader. Excessive dampness of plates can be prevented by placing a round filter paper (11 cm diameter) in the cover of the plate.

15. A useful seal ("vaspar") for fluid anaerobic cultures is prepared by melting together equal amounts by weight of Vaseline® petroleum jelly and paraffin. This is autoclaved at 121°C for thirty minutes and melted prior to adding to culture tubes.

IDENTIFICATION BY ANTISERA

Antisera against *Cl. chauvoei*, *Cl. septicum*, *Cl. novyi* (types A and B), and *Cl. sordellii* are available commercially* for the identification of these species by passive protection tests. The procedures given below are those recommended by the producer.

1. Determine the approximate lethal dose of a thioglycollate broth culture of the strain to be identified by inoculating intramuscularly pairs of guinea pigs with a series of doses ranging from 0.1 ml to 1 ml.
2. Inoculate pairs of guinea pigs subcutaneously with 1 ml of each of the specific sera. Twenty-four hours later, inoculate intramuscularly each of these guinea pigs and two control guinea pigs with a lethal amount of culture. Protection afforded by the serum is specific.

IDENTIFICATION OF *CL. CHAUVOEI*, *CL. SEPTICUM*, *CL. NOVYI*, AND *CL. SORDELLII* BY MEANS OF SPECIFIC FLUORESCEIN-LABELLED ANTISERA

Labelled antisera for these four clostridia are available commercially.† The directions given by the producer are as follows:

1. Smear a portion of the suspected lesion onto a microscope slide and leave for a few moments until reasonably air dry.
2. Fix by immersing in reagent grade anhydrous acetone for ten minutes.
3. Place one drop of the appropriate labelled antiserum on the smear and spread evenly. Other smears may be stained with other labelled antisera if indicated.
4. Leave in a moist chamber for thirty minutes at room temperature. A large Petri dish containing moistened filter paper is quite adequate. .
5. Rapidly wash off gross excess of uncombined reagents with buffered saline (0.15 M sodium phosphate in physiological saline, pH 7.1) and finally leave for a total of at least ten minutes in several changes of buffered saline.
6. Blot gently with clean absorbent paper.

---

*Burroughs Wellcome Co., Research Triangle Park, North Carolina.

†Burroughs Wellcome Co., Research, Triangle Park, North Carolina.

7. Mount in buffered glycerin consisting of nine parts of glycerin to one part of buffered saline.
8. The slide is then examined with a conventional fluorescence microscope employing the filters routinely used in bacterial fluorescence microscopy. For further information, see Chapter 36. Organisms that fluoresce apple green are *Cl. septicum, Cl. chauvoei, Cl. septicum, Cl. novyi,* or *Cl. sordellii* depending upon which antiserum was used (see Plate 1). The rhodamine FA stain used earlier and shown in Plate 1 is no longer used for *Cl. chauvoei.*

CLOSTRIDIUM PERFRINGENS: ENTEROTOXEMIAS

Pathogenicity

TYPE A.:An infrequent enterotoxemia of lambs, sheep, and cattle. Gas gangrene and food poisoning in man.

TYPE B. Lamb dysentery, a severe enterotoxemia of the newborn.

TYPE C: Hemorrhagic enterotoxemia of calves, lambs, and young pigs. Struck (acute intoxication) in mature sheep.

TYPE D: Pulpy kidney disease or overeating toxemia of sheep and less frequently of goats and calves.

TYPE E: Reputed to cause enterotoxemia (dysentery) of lambs and calves.

The principal toxins and other soluble factors of the different types of *Cl. perfringens* considered of importance have been summarized by Brooks et al. (6).

Laboratory Diagnosis of Enterotoxemia

SPECIMEN. Fresh contents from the small intestine.

DIRECT EXAMINATION. Gram's stain smears from the mucosa of the small intestine, and take note of the character and number of Gram-positive rods. There are few bacteria in the healthy jejunum and ileum.

TEST OF INTESTINAL CONTENTS OR CULTURES FOR TOXICITY. The intestinal contents should be collected as soon after death as possible. If the contents are overly mucinous, they should be diluted with an equal amount of sterile physiological saline. Shake well, then centrifuge to separate out the bacteria and particulate material. Draw off the supernatant. Filtration is often difficult and can be omitted. Penicillin and streptomycin can be added to suppress bacteria. Young cultures in cooked meat medium are handled similarly. Trypsinization as described below should be carried out with cultures but is not usually required for intestinal contents.

Each of three mice is inoculated intravenously in the tail vein with 0.3 ml (0.4 ml if contents have been diluted) of the supernatant. Two to three times as much toxin is required to kill mice by the intraperitoneal route. If toxin is present in significant amounts, deaths usually take place within ten hours. Deaths occurring within five minutes are usually attributed to shock. Identification of the toxin type is carried out as described below if the contents have been found toxic.

Mice are easily restrained for intravenous inoculation by pulling the tail through a hole in the metal top of a cylindrical mailing carton. The mouse is allowed to enter the carton, and the lid is screwed on. Suitable mouse holders are available commercially.

IDENTIFICATION OF TYPES OF *CL. PERFRINGENS*

The types can be determined by toxin-antitoxin neutralization tests performed either intradermally in depilated albino guinea pigs or intravenously in mice. The mouse test is less informative and more liable to nonspecific effects.

As the neutralization procedure is rather involved, the laboratory may wish to submit a culture of the *Cl. perfringens*, if available, to a reference laboratory for type identification.

Neutralization Tests in the Skin of Guinea Pigs

For this test, 0.5 ml of suspected toxin is mixed with 0.2 ml of nutrient broth and 0.1 ml of antiserum to give a total volume of 0.8 ml. When more than one antiserum is required in the mixture, the amount of nutrient broth is reduced so that the total volume is still 0.8 ml. The composition of the mixtures required is shown below.*

Similar mixtures are prepared using trypsinised filtrate, i.e. filtrate treated with 1% trypsin powder (Difco 1:250), for one hour at 37°C. The trypsin activates epsilon-protoxin and destroys beta-toxin. The mixtures are left at room temperature for thirty minutes, and 0.2 ml is injected intradermally into guinea pigs. It is convenient to inject mixtures one to four on one flank and mixtures five to eight on the other.

Reactions are read at twenty-four and forty-eight hours. If the guinea pig dies, excess epsilon may be present and the test should be repeated with filtrate diluted one in five with broth.

Serum Neutralization Tests in Mice

Mixtures are prepared as for the skin tests in guinea pigs, and 0.3 ml quantities are injected intravenously into pairs of mice. The results are read up to three days, and the interpretation is similar to that given above except that death or survival are used as indicators of the presence or absence of the

---

*Supplied by Burroughs Wellcome Co., Research, Triangle Park, North Carolina.

	Mixtures	*A*	*B*	*C*	*D*	*E*
				Reactions With Cl. Welchii Type		
Filtrate (untreated)	1. Broth	+	+	+	±	+
	2. Type A serum (anti-alpha)	—	+	+	*—,±	±
	3. Types A + C sera (anti-alpha + anti-beta)	—	*—,±	—	*—,±	±
	4. Types A + C + D sera (anti-alpha + anti-beta + anti-epsilon)	—	—	—	—	±
Filtrate (trypsinised)	5. Broth	±	+	†—,±	+	+
	6. Type A serum (anti-alpha)	—	+	†—,±	+	+
	7. Types A + D sera (anti-alpha + anti-epsilon)	—	†—,±	†—,±	—	+
	8. Types A + C + D sera (anti-alpha + anti-beta + anti-epsilon)	—	—	—	—	+
	‡9. Type A + Type E (anti-alpha + anti-iota)	—	+	†—,±	+	—
Diagnostic toxins identified		α	β+ε	β	ε	ι

*Nearly always —, as epsilon in protoxin form
†Nearly always —, as beta destroyed by trypsin
‡Usually omitted as iota seldom present

— = no reaction
+ = necrotic lesion

diagnostic toxins. If toxin is present in appreciable amounts, deaths usually take place within ten hours.

The tests in mice may also be carried out according to the following schedule:

Test Fluid	*Type Serum*	*Mouse Dose*	*No. of Mice*	*Group*
0.9 ml	0.3 ml A	0.4 ml	2	Serum
0.9 ml	0.3 ml B	0.4 ml	2	Serum
0.9 ml	0.3 ml C	0.4 ml	2	Serum
0.9 ml	0.3 ml D	0.4 ml	2	Serum
0.9 ml	0.3 ml E	0.4 ml	2	Serum
0.9 ml	None	0.4 ml	2	Control

The tests are read over a three-day period and interpreted as follows:

Type A antitoxin neutralizes only the homologous toxin.

Type B antitoxin neutralizes the toxins of types A, B, C, and D.

Type C antitoxin neutralizes the toxins of types A and C.

Type D antitoxin neutralizes the toxins of types A and D.

Type E antitoxin neutralizes the toxins of types A and E.

Readers are referred to Sterne and Batty (7) for additional information on the toxins of *Cl. perfringens* and other pathogenic clostridia.

Isolation and Identification of *Cl. perfringens*

The procedure outlined in the left side of Table 19-I is carried out except that inoculations are made from small bowel contents rather than from a tissue suspension.

Tests for the toxigenicity of strains and the identification of the culture type can be carried out according to instructions provided above.

Some strains of *Cl. perfringens* type A produce an enterotoxin while sporulating. It has been suggested that this toxin, a cause of human food poisoning, may be responsible for diarrhea in animals, particularly horses and rabbits. Procedures for the identification and production of this enterotoxin have been described by Dowell et al. (8).

CLOSTRIDIUM BOTULINUM

Organisms of this species occur widely in the soil.

PATHOGENICITY. Food poisoning in man and animals. Potent toxins are produced as a result of the growth of *Cl. botulinum* in various human and animal foodstuffs. Intoxication results from consumption of materials containing toxin. Wound botulism has been described in man.

The distribution of types of *Cl. botulinum* in relation to some cases and outbreaks of botulism in different animal species is summarized below:

Type	*Principal Hosts*	*Toxin Source*
A	Man, chickens ("limberneck") and mink	Canned vegetables and fruits, meat, and fish
B	Man, horses, cattle and mink	Prepared meat, especially pork
C	Wild birds (aquatic), chickens ("limberneck")	Rotting vegetation, toxin concentrated in fly larvae
	Cattle, horses	Forage poisoning
	Mink	Meat products
D	Cattle (lamziekte)	Carrion, bones
E	Man	Uncooked fish and other marine products
F	Man	Liver paste

Diagnosis of Botulism

Recovery of *Cl. botulinum* from food is not sufficient to establish a diagnosis of botulism. Food can be contaminated without toxin being produced. A diagnosis is based upon the demonstration of a sufficient amount of toxin.

Isolation of Cl. botulinum

Great care should be exercised in working with food or cultures that may contain botulinus toxin. To culture *Cl. botulinum* from the suspected food, several samples are suspended in a small amount of physiological saline and heated at 65–80°C for thirty minutes to eliminate non-sporeforming bacteria. Blood agar plates are inoculated and incubated anaerobically. Identification is carried out with the aid of Table 19-II. If the organism is *Cl. botulinum*, tests are conducted for toxigenicity. Filtrates are prepared from five- to ten-day cultures (30°C) in cooked meat medium. Animal inoculation tests are carried out as described below, the culture filtrate being substituted for the filtrate prepared from food or other materials.

Demonstration and Identification of Toxin

The following may be examined for the presence of toxin: samples of food or anything else that might have been eaten; serum and urine from affected animals; and cecal contents and feces from an affected animal.

Although the demonstration of botulinus toxin is within the capability of most laboratories, it may not be feasible to carry out the rather involved neutralization procedures described below. If the latter is the case, the help of a reference laboratory should be sought.

Food or feed is macerated in physiological saline and left to soak overnight if time permits. The suspension is then centrifuged until a clear supernatant is obtained. The supernatant may be sterilized by Seitz or other filtration, but this is not usually necessary. If a membrane filter is used, a porosity of 0.45 µm will remove bacteria. Without filtration, one must eliminate bacterial infection as a possible cause of deaths. If there is sufficient supernatant, treat nine parts of the test fluid (pH 5.5–6.6) with one part of trypsin solution (1% trypsin; Difco 1:250). The mixture is incubated at 37°C for forty-five minutes. Cecal fluid is handled in a similar fashion.

Adult guinea pigs are each given 2 ml intraperitoneally. At the same time, a control group is given 2 ml of the heated extract (ten minutes at 100°C) intraperitoneally. Guinea pigs in the antitoxin groups receive a mixture of unheated extract and antitoxins of known type. Two or three guinea pigs are usually used for each group. Antitoxin is available from several sources.* It is usually administered in a ratio of one volume of antitoxin to four volumes of test fluid.

---

*Biologics Reagents Section, Laboratory Branch, U.S. Public Health Service, Communicable Disease Center, Atlanta, Georgia. Pasteur Institute, Paris, France. Types A & B: Burroughs Wellcome Co., Research Triangle Park, North Carolina.

Mice may be used instead of guinea pigs. The intraperitoneal dose for the former is 1 ml. The presence within five days of flaccid paralysis followed by death in unprotected animals suggests the presence of botulinal toxin. Feces from animals suspected of having botulism may be examined for toxin as described above.

Proof that the food produced the intoxication can be obtained by feeding normal animals of the same species with the actual food.

Botulism in wild ducks and other animals has been diagnosed by the inoculation of their serum into mice. One milliliter of serum is inoculated intraperitoneally or subcutaneously into each mouse. One group of mice is given serum only, and if toxicity is demonstrated, serum is administered to groups simultaneously inoculated with the different antitoxins. The effectiveness of this test is dependent upon a high level of toxin in the affected animal.

REFERENCES

1. Peckham, M. C.: *Avian Dis, 3*:471, 1959.
2. Lowe, B. R., Fox, J. G., and Bartlett, J. G.: *Am J Vet Res, 41*:1277.
3. Berkhoff, G. A., Campbell, S. G., Naylor, H. B., and Smith, L. DS.: *Avian Dis, 18*:186, 1974.
4. Berkhoff, G. A., Campbell, S. G., Naylor, H. B., and Smith, L.DS.: *Avian Dis, 18*:196, 1974.
5. Holdeman, L. V., Cato, E. P., and Moore, W. C. (Eds.): *Anaerobe Laboratory Manual*, 4th ed. Blacksburg, Va., Virginia Polytechnic Institute and State University, 1977.
6. Brooks, M. E., Stern, M., and Warrack, G. H.: *J Pathol Bacteriol, 74*:185, 1957.
7. Sterne, M., and Batty, I.: *Pathogenic Clostridia.* London, Butterworth Scientific Publications, 1975.
8. Dowell, Jr., V. R., et al.: *Rev Latino-Am Microbiol, 17*:137, 1975.

Chapter 20

LISTERIA AND ERYSIPELOTHRIX

S PECIES OF THESE GENERA are similar in a number of respects. The organisms are slender, Gram-positive, non-sporeforming, facultatively anaerobic, fermentative rods. *Erysipelothrix rhusiopathiae* and *Listeria monocytogenes* are the important pathogenic species; however, three additional species of *Listeria*, of little disease significance, have been described recently. Both *E. rhusiopathiae* and *L. monocytogenes* are capable of survival in the soil and existing in the animal in commensal and carrier states.

It has been proposed that strains of *L. monocytogenes* recovered from nature and animal feces that are nonhemolytic and nonpathogenic be called *L. innocua* (1).

LISTERIA MONOCYTOGENES

PATHOGENICITY. Infections occur in a wide range of animals. Two forms of the disease are seen, the neural and the visceral. The neural form, characterized by microabscesses in the brain stem and meningitis, occurs most commonly in cattle, sheep, and goats and occasionally in the horse, dog, and humans.

The visceral form of the disease, with liver necrosis, is seen in rabbits, guinea pigs, chinchillas, swine, ferrets, raccoons, calves, lambs, and other species. An epizootic form occurs in chickens and turkeys, characterized by necrotic foci of the liver and pericardium.

The organism is considered a cause of abortion in cows and ewes, and in such abortions, it can be recovered from the aborted fetuses. On occasion, *L. monocytogenes* is shed in the milk of infected cows, goats, and ewes.

In humans, the organism is involved in meningoencephalitis, genital infections, and valvular endocarditis.

ISOLATION PROCEDURES. Blood agar and blood agar containing 0.05% potassium tellurite (inhibits many Gram-negative species) are used if listeriosis is suspected. In the visceral form, material is seeded directly onto the solid medium. In the neural form, the medulla and a portion of the cord are cut into small pieces by means of sterile scissors and placed in broth. It is advisable to sample various parts of the medulla. Prepare a 10–20% suspension in a Ten Broeck grinder, with mortar and pestle, or with a Blendor®. After inoculation of media, the suspension is stored in the refrigerator for possible future examinations. These are especially indicated if histopathological sections of the brain suggest listeria infection. If listeria are not recovered initially, the suspension should be examined at the end of the first, third, sixth, and twelfth weeks. The organism can be removed more readily, i.e. with less storage time, from the ovine brain than from the bovine brain.

CULTURAL CHARACTERISTICS. Listeria grow as small, round colonies with a narrow zone of beta-hemolysis. Both smooth and rough colonies are seen. Stained smears reveal small Gram-positive rods (Fig. 20-1). Primary cultures are frequently pleomorphic and can be mistaken for diphtheroids (corynebacteria), diplococci, and streptococci. The organisms are easily decolorized and as a consequence can be mistaken for Gram-negative rods.

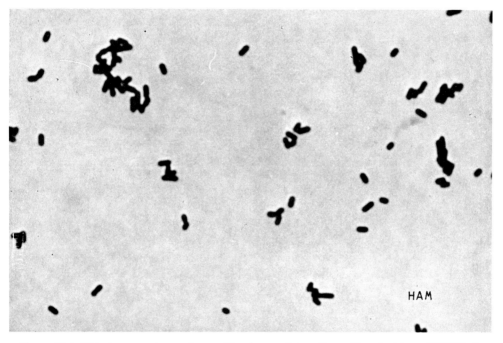

Figure 20-1. *Listeria monocytogenes* from a blood agar plate culture. Gram's stain, ×2250 (H. A. McAllister).

IDENTIFICATION. Some important features that differentiate *L. monocytogenes* from *E. rhusiopathiae* are listed in Table 20-I. Additional characteristics that distinguish *L. monocytogenes* are the following:
- Fermentation: glucose, maltose, levulose, salicin, and trehalose.
- Irregular fermentation: arabinose, sorbitol, sucrose, galactose, and glycerol.
- Not fermented: mannitol, dulcitol, inositol, inulin, and raffinose.
- The oxidase test is negative.
- Nitrate is not reduced.

- Catalase is produced.
- MR and VP tests are usually positive.
- Esculin is hydrolyzed.
- Indole is not produced.
- Gelatin is not liquefied.
- Urease is not produced and starch is not hydrolyzed.

TABLE 20-I

DIFFERENTIATION OF *LISTERIA* AND *ERYSIPELOTHRIX*

	Motility	Beta hemoly-sis	Cata-lase	Nitrate Reduc-tion	Guinea Pig Inocula-tion*	Anton Test	Acid	
							Glucose	Mannitol
L. monocytogenes	+(25°C)	+	+	—	Death, 3-4 days	+	+	—
E. rhusiopathiae	—	—	—	—	Resistant	—	+	—
L. dentrificans	+	—	+	+	Resistant	—	+	—
L. grayi	+	—	+	—	Resistant	—	+	+
L. murrayi	+	—	+	+	Resistant	—	+	+

*0.5 ml of broth culture subcutaneously

If *C. equi* is streaked at right angles to a *L. monocytogenes* streak in a CAMP-fashion, the latter produces a substance that causes *C. equi* to be hemolytic. This effect is not seen with *E. rhusiopathiae*.

A polyvalent fluorescein-labelled globulin is available for the identification of *L. monocytogenes* in smears from cultures. There are reports of the successful employment of the fluorescent antibody staining procedure in the identification of listeria in smears and in fixed sections of tissues (2).

ANIMAL INOCULATIONS. Mice, rabbits, and guinea pigs are susceptible to experimental infection if a considerable number of organisms are used. Several mice are inoculated subcutaneously with graded doses (maximum 0.5 ml) of a twenty-four-hour broth culture. Deaths usually occur three to ten days postinoculation, and necropsy reveals many necrotic foci in the spleen and liver.

If a small amount of a broth culture is placed into the conjunctiva of the rabbit or guinea pig, or swabbed on the everted lid, a severe keratoconjunctivitis develops within twenty-four to thirty-six hours, followed by opacity of the cornea. This is called the Anton Test. It is not necessary for identification.

Four serotypes and several subtypes of *L. monocytogenes* have been recognized for some time. Recently, more than a dozen serotypes or serovars have been identified based upon different O and H antigens (3). Serologic examinations are carried out by some reference laboratories. Typing sera are available commercially.

OTHER SPECIES OF *LISTERIA*

Bergey's Manual describes three additional species of *Listeria: L. denitrificans, L. grayi, and L. murrayi*. These three are not beta-hemolytic, nor are they pathogenic for mice except for possibly *L. denitrificans*. The Anton test in rabbits is negative with the three species, and their potential for causing disease is probably very low. For identification, see Table 20-I.

Their sources according to *Bergey's Manual* are as follows:

L. denitrificans: Cooked blood of beef; natural habitat not known.

L. grayi: Feces of chinchillas.

L. murrayi: Leaves of corn; habitat probably vegetation and soil.

ERYSIPELOTHRIX RHUSIOPATHIAE
Synonym: *Erysipelothrix insidiosa*

PATHOGENICITY. Swine: septicemic form; chronic form with arthritis, endocarditis, or a skin manifestation called diamond skin disease. Sheep and cattle: chronic polyarthritis. Infections in turkeys, chickens, geese, pheasants, pigeons, and other avian species. The acute, septicemic disease occurs most commonly in turkeys. There are sporadic infections in cattle, horses, dogs, and other species. There are several reports of valvular endocarditis in dogs in which the organism has been recovered in blood cultures. Humans: localized infections called "erysipeloid."

ISOLATION PROCEDURES. In suspected erysipelas infection, materials are inoculated onto blood agar. Blood agar containing sodium azide (see Appendix B) may be employed to depress contaminants.

Although *E. rhusiopathiae* grows aerobically, primary growth is accelerated if plates are placed in the candle jar or 10% carbon dioxide.

CULTURAL CHARACTERISTICS. Smooth colonies are small and round; rough colonies are larger and have irregular borders. The latter are more common from chronic infections. Growth is sparse after twenty-four hours incubation but readily apparent after forty-eight hours. A zone of alpha-hemolysis is usually seen around young colonies, followed by a clearing of the zone on further incubation (beta-hemolysis). Smears from smooth colonies disclose slender Gram-positive rods indistinguishable from those of *L. monocytogenes* (Fig. 20-2). Rough colonies yield highly pleomorphic and filamentous forms.

IDENTIFICATION. This is based upon features listed in Table 20-I and upon the characteristics listed below:

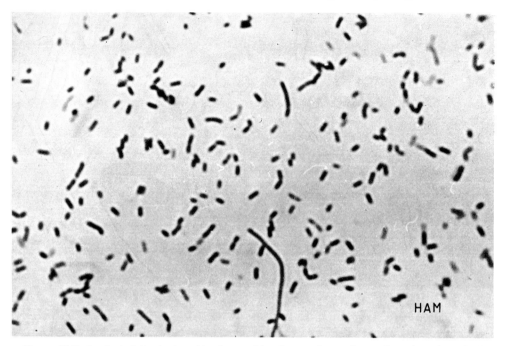

Figure 20-2. *Erysipelothrix rhusiopathiae* from a blood agar plate culture. Gram's stain, ×2440 (H. A. McAllister).

- Fermentation reactions are variable: Lactose, glucose, levulose, and dextrin are frequently fermented; some acid production may be noted in sucrose, arabinose, dulcitol, mannitol, xylose, galactose, and fructose; litmus milk may be acidified.
- Esculin is not hydrolyzed.
- Hydrogen sulfide production and nitrate reduction are variable; oxidase, catalase, and indole are not produced; and gelatin is not liquefied.
- "Test tube brush" growth along stab line in gelatin incubated at 20°C for three to five days.

Most strains of *E. rhusiopathiae* affect triple sugar iron agar and Kliger's iron agar in a characteristic manner. Growth is sparse, with yellowing of slant and butt and the production of hydrogen sulfide along the stab line.

Fluorescein-labelled antibody for the identification of this organism has not been available commercially.

ANIMAL INOCULATION. Mice and pigeons are susceptible, dying within four days after intraperitoneal inoculation of 0.1–0.5 ml of broth culture.

Mice infected with 0.1 ml of broth culture can be protected by the inoculation of 0.3 ml of commercial antierysipelas serum. This protection test can be used to confirm identification.

REFERENCES

1. Seeliger, H. P. R., and Schoofs, M.: Serological analysis of nonhemolyzing strains of *Listeria* sp. *Abstracts VIIth International Symposium on Problems of Listeriosis*. Varna, Bulgaria, September 23–27, 1977.
2. Cherry, W. B., and Moody, M. D.: *Bact Rev*, 29:222, 1965.
3. Seeliger, H. P. R.: In Woodbine, M. (Ed.): *Problems of Listeriosis*. Leicester, England, Leicester University Press, 1975, p. 27.

SUPPLEMENTARY REFERENCES

Listeria Monocytogenes

Buchanan, R. E., and Gibbons, N. E. (Eds.): *Bergey's Manual of Determinative Bacteriology*, 8th ed. *Baltimore*, Williams & Wilkins, 1974, p. 593.

Paterson, J. S.: Diseases due to bacteria. In Stableforth, A. W., and Galloway, I. A. (Eds.): *Infectious Diseases of Animals*. London, Butterworth Scientific Publications, 1959, vol. 1.

Wood, D.: In Shapton, D. A., and Gould, G. W. (Eds.): *Isolation Methods for Microbiologists* (Technical Series No. 3). New York, Academic Press, 1969, p. 63.

Welshimer, H. J.: In Starr, M. P., et al. (Eds.): *The Prokaryotes*. New York, Springer-Verlag, 1981, Vol. II, Chapter 132.

Albritton, W. L., Wiggins, G. R., DeWitt, W. E., and Fellet, J. C.: In Lennette, E. H. (Ed-in-Chief): *Manual of Clinical Microbiology*, 3rd ed. Washington, D.C., American Society for Microbiology, 1980, Chapter 11.

Erysipelothrix Rhusiopathiae

Gledhill, A. W.: Diseases due to bacteria. In Stableforth, A. W., and Galloway, I. A. (Eds.): *Infectious Diseases of Animals*. London, Butterworth Scientific Publications, 1969, vol. 2.

Wilson, G. S., and Miles, A. A.: *Topley and Wilson's Principles of Bacteriology and Immunity*, 5th ed. Baltimore, Williams & Wilkins 1964, vol. 1 and 2.

Ewald, F. W.: In Starr, M. P., et al. (Eds.): *The Prokaryotes*. New York, Springer-Verlag, 1981, Vol II, Chapter 133.

Chapter 21

CORYNEBACTERIUM

$\mathbf{M}$EMBERS of this genus are Gram-positive, small, pleomorphic, non-sporeforming rods. All are nonmotile (except some plant pathogens), usually aerobic, facultatively anaerobic, and fermentative. The group contains a miscellany of species, some of which will ultimately be assigned to other genera. The names given in the "Approved Lists of Bacterial Names" will be used. New names have been proposed, based on extensive taxonomic studies, for the two important animal pathogens *C. equi* and *C. suis: Rhodococcus equi* (1) for the former and *Eubacterium suis* (2) for the latter.

These small, pleomorphic rods display varying degrees of metachromatic staining; *C. diphtheriae* is outstanding in this regard. Many corynebacteria or diphtheroids occur as commensals on the skin and mucous membranes of animals. They are usually of little or no pathogenic significance when recovered from clinical specimens. These miscellaneous diphtheroids are more frequently recovered from tissues taken from animals some hours after death. Ordinarily they are not identified as to species.

PATHOGENICITY

Corynebacterium pyogenes. Swine, cattle, sheep, and goats: suppurative pneumonia, wound and surgical infections, polyarthritis, suppurative mastitis, abortions, and umbilical infections. There are reports of isolations from sinusitis in the horse, abscesses in the head region of chickens, and rare infections in humans.

C. renale. Pyelonephritis in cows and kidney abscesses in swine.

C. equi. Suppurative bronchopneumonia in foals and young horses characterized by abscessation; metritis in mares; abortion in mares; abscesses of the submaxillary and other lymph nodes of swine and cattle. Infrequent infections occur in other animals, and there are several reports of human infections.

C. pseudotuberculosis (*C. ovis*). Caseous lymphadenitis in sheep; occasional cause of abortion in ewes; arthritis and bursitis in lambs; ulcerative lymphangitis in the horse and horsekind; "chest" abscesses in horses with or without habronemiasis. Infrequent infections have been reported as occurring in camels, goats, cattle, deer, laboratory mice, and other animal species (3).

C. suis. Cystitis and pyelonephritis in pigs; occurs in the semen of the boar (4).

C. kutscheri (*C. murium*). This organism causes a disease of mice and rats characterized by caseopurulent foci in the lungs, lymph nodes, and less frequently the liver, kidneys, and subcutis (5).

OTHER CORYNEBACTERIA. These are listed after Identification below. The isolation procedures are essentially the same for all of the corynebacteria except *C. suis*.

ISOLATION PROCEDURES

Gram-stained smears of pus will indicate the presence of corynebacteria, although they can be readily confused with streptococci, staphylococci, and listeria. Identification requires isolation.

Material is streaked on blood agar and incubated aerobically at 37°C. If *C. suis* is suspected, plates should be incubated anaerobically. After inoculation of plates, swabs are placed in tubes of brain-heart infusion semisolid medium, which are also incubated at 37°C.

GUINEA PIG INOCULATION. Intravenous inoculation of *C. pseudotuberculosis* produces abscesses in the lungs and liver, with death in four to ten days. Orchitis is produced in the male guinea pig as a result of intraabdominal inoculation of organisms.

CULTURAL AND COLONIAL CHARACTERISTICS

The cultural characteristics of the most important species are described below. The microscopic morphologic differences among the different species are not significant. Gram-stained smears from colonies disclose small, Gram-positive, pleomorphic rods that may form "palisades" and "chinese letters" (Figs. 21-1, 21-2).

C. pyogenes. Growth is sparse after twenty-four hours incubation. In forty-eight hours, small, beta-hemolytic, streplike colonies appear.

A large colony type of *C. pyogenes* has been described (6). Distinguishing features are colonies 1–2 mm in diameter, no gelatinase, weak catalase production, and, in contrast to the regular *C. pyogenes*, xylose and starch are not fermented while inositol is. In many respects, this so-called variant of *C. pyogenes* resembles *C. haemolyticum*. All isolations have been reported from human beings.

C. renale. In twenty-four hours, small dewdroplike colonies appear. These become ivory colored and opaque in forty-eight hours. In broth, a fine powdery sediment is noted on the side and bottom of the tube.

C. equi. After twenty-four hours incubation, small, moist, creamy white colonies appear. As they age, they enlarge, become pink, then turn to salmon-pink. Growth is distinguished by its abundance. A heavy turbidity with slight sediment is produced in broth.

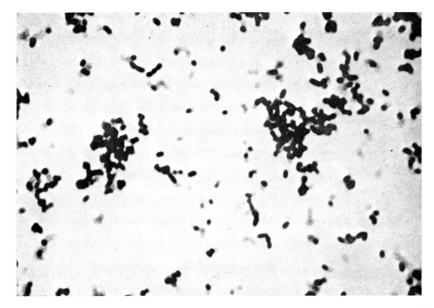

Figure 21-1. *Corynebacterium pyogenes* from a blood agar culture. Gram's stain, × 1900.

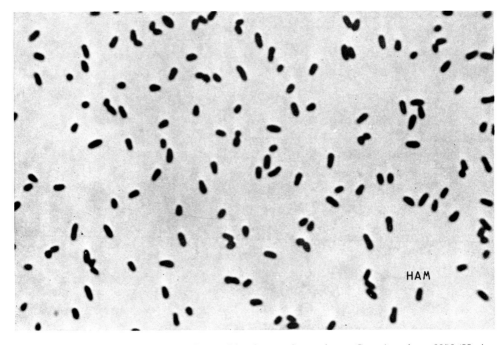

Figure 21-2. *Corynebacterium equi* from a blood agar plate culture. Gram's stain, ×2250 (H. A. McAllister).

C. pseudotuberculosis. Growth is sparse initially. Small, translucent colonies enlarge in three to four days to take on a cream to orange color, becoming opaque, dry, and crumbly. They are beta-hemolytic. A coarse, granular sediment is produced in broth.

C. suis. Blood agar is incubated anaerobically for up to three days. Small, translucent, round, and granular colonies appear that after seven days are flat, dry, and opaque, with crenated edges.

C. kutscheri. Yellow white glistening colonies 1–4 mm in diameter are apparent after two days incubation on blood agar.

IDENTIFICATION

The various corynebacteria are definitively identified according to the criteria presented in Table 21-I. *C. pyogenes* and *C. suis* are catalase negative. By far the greater number of isolates of *C. pyogenes* digest and peptonize litmus milk.

Biberstein et al. (7) have described two types of *C. pseudotuberculosis.* The one that was nitrate-negative was usually recovered from sheep and goats while the nitrate-positive variety was usually isolated from cattle and horses.

OTHER SPECIES

For the identification of these less important species, consult Table 21-I.

C. bovis. This organism is recovered frequently from cow's milk. It is usually considered nonpathogenic but is thought by some to be capable of producing occasional udder infections. It is a slender, nonmotile rod that grows well on enriched media. Colonies on agar are circular, gray, slightly raised, and dry. Colonies are more frequently found in the fatty areas on plates streaked with milk because of their requirement for oleic acid (8). Other differential features are listed in Table 21-I.

C. diphtheriae. It is an important pathogen of humans. Infections in animals have been reported but do not appear to have been well substantiated.

C. minutissimum (Unofficial). This organism causes erythrasma, an infection of the stratum corneum of humans. It has been associated with an acute, moist inflammation of the interdigital space, "scald," in young lambs, with scabs on the brisket and docking wounds, and on the skin of the udder of normal cattle (9).

The lesions in which this organism is found are distinctive in that they have a coral red fluorescence in ultraviolet light. It is claimed that Mueller-Hinton agar is especially suitable for the demonstration of the fluorescence (10).

C. ulcerans. An occasional cause of mastitis in cows (11).

C. metritis (Unofficial). This diphtheroid was considered the cause of systemic infections in rabbits (12).

TABLE 21-I

DIFFERENTIATION OF THE CORYNEBACTERIA

	Catalase	Hemolysis	Urease	Nitrate Reduction	Gelatin Liq.	Glucose	Lactose	Maltose	Sucrose	Trehalose	Starch	Arginine Dihydrolase
C. pyogenes	-	+	-	-	+	+	+	+	v	v	v	:
C. renale	+	v	+	-	-	+	v	v	-	-	v	+
C. equi	+	-	v	+	-	-	-	-	-	-	-	-
C. pseudo-tuberculosis	+	v	v	v	v	+	v	+	v	-	v	+
C. suis (anaerobic)	-	-	+	-	-	v	-	+	(-)	-	+	:
C. bovis	+	-	-	-	-	-	-	-	-	-	-	-
C. kutscheri (C. murium)	+	-	v	+	-	+	-	+	+	+	+	-
C. ulcerans	+	v	+	-	-	+	-	+	v	+	+	-
C. haemolyticum	+	+	-	-	+	+	+	+	v	v	v	:
C. xerosis	+	-	-	+	-	+	-	+	+	-	-	-
C. hofmannii	+	-	+	+	-	-	-	-	-	-	-	-
C. diphtheriae	+	v	-	+	-	+	-	+	-	-	v	-

A diphtheroid similar to the one described by Ray (13) and considered to resemble *C. pseudodiphtheriticum* has been isolated in pure culture on a number of occasions from various tissues of cattle, sheep, and swine.

C. haemolyticum. Roberts (14) reported the isolation of this species, which occurs in humans, from a case of ovine pneumonia.

Fraser (15) has described CAMP-like tests to aid in the identification of *C. equi, C. pyogenes, C. renale, C. pseudotuberculosis,* and *C. haemolyticum.*

REFERENCES

1. Goodfellow, M., and Alderson, G.: *J. Gen Microbiol, 100*:99, 1977.
2. Wegienek, J., and Reddy, C. A.: *Int J Syst Bact, 32*:218, 1982.
3. Benham, C. L., Seaman, A., and Woodbine, M.: *Vet Bull, 32*:645, 1926.
4. Soltys, M. A.: *J Pathol Bacteriol, 81*:441, 1961.
5. Giddens, W. E. Keahey, K. K., Carter, G. R., and Whitehair, C. K.: *Pathol Vet, 5*:227, 1968.
6. Barksdale, W. L. K., Li, K., Cummins, C. S., and Harris, H.: *J Gen Microbiol, 16*:749, 1957.
7. Biberstein, E. L., Knight, H. D., and Jang, S.: *Vet Rec, 89*:691, 1971.
8. Newbould, F. H. S.: *Can J Microbiol, 11*:602, 1965.
9. Pepin, G. A., and Littlejohn, A. I.: *Vet Rec, 76*:507, 1964.
10. Cohen, S. N., and Nickolai, D.: *Appl Microbiol, 17*:479, 1969.
11. Higgs, T. M., Smith, A., Cleverly, L. M., and Neave, F. K.: *Vet Rec, 81*:34, 1967.
12. Lesbouryries, G.: *Bull Acad Vet Fr, 15*:38, 1942.
13. Ray, J. D.: *Vet Med, 23*:490, 1928.
14. Roberts, R. J.: *Vet Rec, 84*:490, 1969.
15. Fraser, G.: *J Pathol Bacterol, 88*:43, 1964.

SUPPLEMENTARY REFERENCES

Buchanan, R. E., and Gibbons, N. E. (Eds): *Bergey's Manual of Determinative Bacteriology,* 8th ed. Baltimore, Williams & Wilkins, 1974, Part 17, p. 599.

Buxton, A., and Fraser, G.: *Animal Microbiology,* London, Blackwell Scientific Publications, 1977, Vol. 1, Chapter 16.

Chapter 22

ACTINOMYCES, NOCARDIA, STREPTOMYCES, AND DERMATOPHILUS

J. F. PRESCOTT

The order Actinomycetales contains Gram-positive bacteria that form hyphae and that show true branching; in some families, the filaments may develop into mycelia. The filaments may also fragment into diploid, coccoid, diphtheroidal, and coccobacillary forms. Morphology is not generally useful as an identifying characteristic because of its variability and because of mimicry by other organisms. Apart from the Mycobacteriaceae, there are four other families of pathogens of veterinary interest within the order. They are described within this chapter.

ACTINOMYCETACEAE

There are three genera in the family Actinomycetaceae: *Actinomyces, Arachnia,* and *Bifidobacteria.* The latter two organisms have been described in Chapter 15 and are uncommon pathogens in animals. All genera (and *Eubacteria*) are capable of causing chronic infections of soft and hard tissues, often with the formation of sulphur granules. The presence of sulphur granules, while striking, is not diagnostic of *Actinomyces* infections, but also occurs in *Nocardia* or *Streptomyces* infections, and with the other genera mentioned. It is important to recognize the presence of sulphur granules in such infections, since the organisms may be hard to demonstrate outside these bodies (Fig. 22-1). It is generally taught that *Actinomyces* are not moderately acid-fast, while *Nocardia* are; exceptions to this rule are not uncommon.

Two species of the genus *Actinomyces* commonly found as pathogens in lower animals are *A. bovis* and *A. viscosus,* both commensals of the mouth of the animal species in which they cause disease. *Actinomyces viscosus* is involved in periodontal disease in humans and animals. It differs from other *Actinomyces* in growing well aerobically and in being catalase positive; for these reasons it had often been misidentified as a *Nocardia.*

Actinomyces infections are often associated with the presence of foreign bodies, whereas *Nocardia* infections are seen in immunocompromised hosts or as iatrogenic infections.

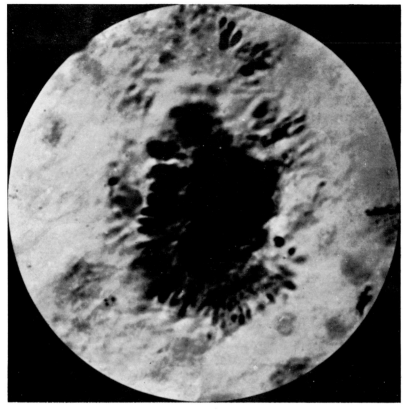

Figure 22-1. *Actinomyces bovis* granule showing the characteristic "ray fungi." ×140. From Nowak, *Documentaria Microbiologica.* Courtesy of Gustav Fischer.

Actinomyces bovis

PATHOGENICITY. This organism is the cause of actinomycosis (lumpy jaw), a chronic suppurative osteomyelitis of the mandible of young cattle. Such infections also occasionally occur in swine, horses, dogs, and humans. On occasion, soft tissues are involved; rarely, these are generalized infections. In horses, the organism was at one time common in fistulous withers and poll evil, associated synergistically with *Brucella abortus* infection or with damage due to ill-fitting harnesses in draft animals. In lumpy jaw, other bacteria may be present such as *Actinobacillus actinomycetemcomitans* and a variety of non-sporeforming anaerobic bacteria.

DIRECT EXAMINATION. The best specimen for direct examination in a case of suspected lumpy jaw is the sulphur granule. This should be crushed and a Gram-stained smear made to demonstrate the delicate, branching Gram-positive filaments (Fig. 22-2).

ISOLATION PROCEDURES. Aerobic and anaerobic cultures are made, and

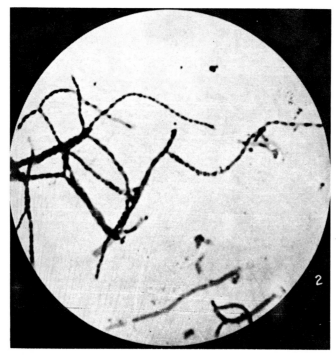

Figure 22-2. *Actinomyces bovis* from agar culture showing beaded filaments. × 2000. From Nowak, *Documenta Microbiologica*. Courtesy of Gustav Fischer.

the plates incubated for two to five days. A plain Sabouraud agar plate should also be inoculated and incubated at room temperature and 37°C. *Actinomyces* species grow very poorly, if at all, on Sabouraud's agar, a clear distinction from the excellent growth seen with *Nocardia* and *Streptomyces*. Liquid thioglycollate media (Chapter 15) must also be inoculated with sulphur granules. Details on anaerobic culture methods are given in Chapter 15. The aerobic plates should be incubated in an atmosphere of 10% CO_2.

CULTURAL CHARACTERISTICS. *Actinomyces bovis* (*naeslundii, odontolyticus,* and *israelii*) grow well anaerobically, but grow poorly, if at all, in an air with 10% CO_2. They do not grow aerobically. Colonies are white, rough, and nodular, 2–3 mm in diameter; smooth and rough forms are common. The colonies adhere tenaciously to solid media and are removed with difficulty. Gram-stained smears from growth on solid or in fluid media show Gram-positive, slightly branched filaments or short hyphae. On subculture, the organisms may become diphtheroidal or coccobacillary. Growth usually occurs within two to four days, but plates should be incubated for seven days and thioglycollate broth for two weeks.

IDENTIFICATION. Presumptive identification can be made on the basis of finding sulphur granules consisting of Gram-positive branching filaments associated with a suppurative, proliferative osteomyelitis. Recovery of an organism with the colonial features of *A. bovis* is usually considered definitive. Proper identification requires that the organism possess the biochemical properties shown in Table 22-I. Fluorescent-antibody identification is most useful but is available only in certain laboratories (1).

Table 22-I

Identification of Actinomyces species

Species	Catalase	Urease	Gelatin hydrolysis	Nitrate reduction	Fermentation of[‡]: Mannitol	Lactose	Sucrose	Salicin	Glycerol	Xylose	Arabinose	Raffinose	Aero-tolerance	Products[‡] by GLC from PYG
A. bovis	−	−	−	−	−	+	+	−	−	−	−	−	M or An	AFLS
A. israelii	−	−	−	$+^-$	$+^-$	$+^-$	+	$+^-$	−	$+^-$	$-^+$	$+^-$	M or An	aFLS
A. odontolyticus[*]	−	−	−	+	−	$+^-$	+	−	−	V	$-^+$	−	M or An	AFIS
A. naeslundii	−	+	−	$+^-$	−	+	+	V	−	−	−	+	F	aFLS
A. viscosus[†]	+	+	V	$+^-$	−	+	+	$+^-$	V	−	−	+	F	AFLS

Key:
+ = positive
− = negative
$+^-$ = most strains positive
$-^+$ = most strains negative
M = microaerophilic
An = anaerobic
F = facultative
[*] Colonies usually red after 4–5 days at room temperature.
[†] Grows best aerobically; 5% CO_2 improves growth.
[‡] See Chapter 15 for details.

Actinomyces Viscosus

PATHOGENICITY. Two forms of canine actinomycosis have been seen (2). The less common is the localized granulomatous abscess involving mainly the skin and subcutis. The other form involves principally the thoracic or abdominal cavities. Pyothorax with granulomatous lesions of thoracic tissues and accumulation of pleural and pericardial fluid containing soft, gray white granules is characteristic of this deep form; it is a serious infection that requires long-term treatment (3). The organism can be found rarely in similar infections in cats and other species.

DIRECT EXAMINATION. Gram-stained and modified acid-fast stained smears are made of purulent material and, whenever possible, from sulphur granules. In the absence of sulphur granules, it is hard to demonstrate the non–acid-fast, Gram-positive filaments. Diphtheroidal forms may sometimes predominate. The sulphur granules in canine actinomycosis are usually soft.

ISOLATION PROCEDURES. Material, particularly sulphur granules, are inoculated onto blood agar plates, and incubated in air with 5–10% CO_2. Discernible colonies are seen after two to four days incubation at 37°C. The organism will grow anaerobically but grows better in air. It is a facultative anaerobe.

CULTURAL CHARACTERISTICS. Two colony forms are seen: One is a smooth, entire, convex, glistening, mucoid or soft colony composed of diphtheroidal forms, and the other consists of rough, irregular, heaped, granular, and slightly dry colonies that yield branching filaments. Both forms are usually seen, but one may predominate.

IDENTIFICATION. The organism is distinguished from other *Actinomyces* in that it grows well aerobically; it is readily distinguished from *Nocardia* or *Streptomyces* by its failure to grow on Sabouraud-dextrose agar. Its distinguishing biochemical properties are shown in Table 22-I.

Other Actinomyces Species

The distinguishing characteristics of other well-recognized *Actinomyces* species are shown in Table 22-I; these are uncommon isolates in lower animals. An organism resembling an actinomycete and designated *A. suis* has been isolated from granules in udder actinomycosis in swine (4) but its significance is unknown. There is evidence that *Corynebacterium pyogenes* should be reclassified as *Actinomyces pyogenes* (5).

NOCARDIA, ACTINOMADURA, AND *STREPTOMYCES*

These species are obligate aerobic saprophytes, found in soil: A few species are rare opportunistic pathogens, causing disease where predisposing factors such as immunity or normal body defense mechanisms are grossly impaired. Colonies vary from heaped, waxy, and variably pigmented to dense, white mycelial and moldlike (6). All generally grow in three to five days at 37°C, and are catalase positive. Of the three genera, *Nocardia* are most commonly encountered, but *Actinomadura* and *Streptomyces* may be cultured from mycetomas in animals in tropical areas. Care should be taken to distinguish these genera from rapidly growing *Mycobacterium* species; the latter differ by the absence of aerial hyphae and by the type of lipid present in the cells (6).

NOCARDIA

The pathogenic *Nocardia* are often acid-fast or partially so, Gram-positive, branching filamentous rods that break up into bacillary or coccoid forms. They are strictly aerobic, and catalase positive. The species of importance is *N. asteroides*. Details on the identification of *Nocardia* are available (6).

Nocardia Asteroides

PATHOGENICITY. Nocardiosis is usually a chronic, progressive disease characterized by suppurating, granulomatous lesions. In cattle, the organism produces acute and chronic mastitis with granulomatous lesions and draining sinus tracts (7).

In the dog, cat, and other animals, localized subcutaneous lesions and/or lymph node involvement are seen. Generalized nocardiosis characterized by pneumonia and the accumulation of large quantities of red-colored fluid in the thoracic and/or abdominal cavity occurs not infrequently in the dog. The fluid in these cavities is serosanguinous and infrequently contains small (< mm) sulphurlike granules. Care must be taken to distinguish any "Nocardia" isolated from *A. viscosus*.

DIRECT EXAMINATION. Granules, if present, are examined as described earlier. Gram-stained smears of pus or crushed granules reveal Gram-positive branching filaments, with or without clubs (Fig. 22-3). The acid-fast stain often shows the retention of some carbol fuchsin, but this is not a reliable characteristic of *Nocardia*. Pus containing the characteristic elements is inoculated onto several blood slants or plates and Sabouraud dextrose agar without inhibitors. Incubate at room temperature and at 37°C for up to a week.

CULTURAL CHARACTERISTICS. Growth is evident in four to five days, and colonies are irregularly folded, raised, and smooth or granular. The color varies from white through yellow to deep orange. Gram-positive, partially acid-fast mycelial filaments, which break up into bacillary forms, are evident under oil immersion. The presence of mycelial elements distinguishes *Nocardia* from saprophytic and atypical mycobacteria. The mycelial forms of the nocardia can be readily seen in slide cultures on Sabouraud dextrose agar. The species grows well at 45°C.

IDENTIFICATION. In animals, a highly presumptive identification of *N. asteroides* infection is based on pathology, demonstration of typical organisms, and colonial, cultural, and morphological characteristics. The species can be distinguished from less common *Nocardia* species, *Streptomyces*, and *Actinomadura* on the properties shown in Table 22-II. Four serotypes of *N. asteroides* have been described in animal isolates (8).

ANTIBIOTIC SUSCEPTIBILITY. Until recently there was no satisfactory method for accurately determining the antimicrobial susceptibility of *Nocardia*.

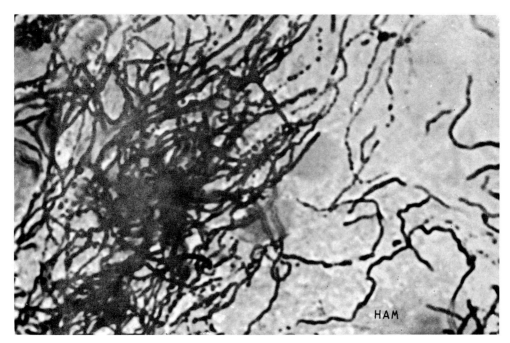

HAM

Figure 22-3. *Nocardia asteroides* in smear of pus. Gram's stain, × 2250 (H. A. McAllister).

Table 22-II

Identification of selected aerobic Actinomycetes (11)

Test	*Actinomadura madurae*	*Nocardia asteroides*	*N. brasiliensis*	*N. caviae*	*N. dassonvillei*	*Streptomyces griseus*	*S. somaliensis*
Acid-fastness	−	V	V	V	−	−	−
Decomposition of:							
Casein	+	−	+	−	+	+	+
Hypoxanthine	+	−	+	+	+	+	−
Tyrosine	+	−	+	−	+	+	+
Urea	−	+	+	+	V	+	−
Xanthine	−	−	−	+	+	+	−
Resistance to lysozyme	−	+	+	+	−	−	−
Diaminopinielic acid*	m	m	m	m	m	l	l

+ = positive reaction
− = negative reaction
V = variable reaction.
* Isomer (meso, laevo) of diaminopinielic acid in hydrolysates.

A recent study of *in vitro* susceptibility has shown that all *N. asteroides* tested were sensitive to sulfisoxazole, trimethoprim-sulfamethazole, doxycycline, and minocycline; most were sensitive to ampicillin; about half were resistant to tetracycline (9).

Other Nocardia

Nocardia species other than *N. asteroides* are uncommon opportunistic pathogens but have been recovered on occasion (10). *Nocardia farcinica*, the cause of bovine farcy in tropical cattle, is thought to be a complex made up of either *N. asteroides* or *Mycobacterium* species (6). The original isolates made in the late nineteenth century by Nocard have long been lost.

STREPTOMYCES AND *ACTINOMADURA*

These are very rare pathogens of animals and are most likely to be found as contaminating bacteria on agar plates. Like *Nocardia*, they grow well on Sabouraud's dextrose agar; details on their differentiation are shown in Table 22-II and given elsewhere (6, 11).

DERMATOPHILUS

The only pathogenic species is *Dermatophilus congolensis*. The one species contained in this genus is an obligate parasite or pathogen of a number of animal species and of humans. It differs from the other organisms discussed in this chapter by its production of motile zoospores and by the peculiar way in which the hyphae or filaments segment.

Dermatophilus Congolensis

The diseases listed below are caused by organisms that are sufficiently alike that they have been given one species name, viz. *D. congolensis*.

PATHOGENICITY. The disease caused by this organism is called rain-scald, streptothricosis, or dermatophilosis. In sheep, the infection is sometimes referred to as mycotic dermatitis. Dermatophilosis is a skin infection of cattle, horses, and goats affecting small areas and characterized by the formation of crusts and a tendency to spread over large areas of the body. Removal of scabs leaves moist depressed areas. Infections have been reported rarely in humans, the cat, and the dog. In sheep, three forms of the disease are described: (1) dermatitis of the wool-covered areas referred to as "lumpy wool," (2) dermatitis of the hairy parts of the face and also of the scrotum, (3) dermatitis of the lower leg and foot referred as to "strawberry foot rot." In temperate climates the disease is a mild, often inapparent infection (12); in tropical countries it may be a cause of severe morbidity and mortality.

DIRECT EXAMINATION. Stained smears are made from the scabs in all of the infections. Prior to preparing the smears, it is essential to soften the scabs in distilled water. Short lengths of narrow, branching, and divided hyphae are seen, as well as numerous Gram-positive cocci (Fig. 22-4). Smears can also be made from the serous exudate after removal of scabs. In Giemsa-stained smears (methyl alcohol fixed), the hyphae and cocci stain deep purple while the epithelial cells are light blue and the nuclei of leukocytes are dark blue. If the organism can be demonstrated in smears, it can usually be cultured. It is generally sufficient for a diagnosis to show the unique appearance, in Giemsa-stained smears, of the thickened, branching filaments dividing both longitudinally and transversely.

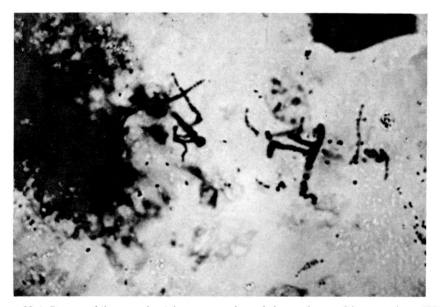

Figure 22-4. *Dermatophilus congolensis* in a smear of a scab from a horse. Giemsa stain, × 1900.

ISOLATION PROCEDURES. The organism grows well on blood agar. Plates are inoculated from clean serous exudate or from the lower aspect of moistened scabs. The organism is not fastidious and grows well on unenriched media such as tryptose agar. The organism also grows well in various broth media. There is no growth on Sabouraud dextrose agar.

CULTURAL CHARACTERISTICS. Pinpoint colonies surrounded by small zones of beta-hemolysis are evident after twenty-four hours incubation at 37°C. After incubation for three to four days, colonies are considerably larger (Fig. 22-5). They may be wrinkled or smooth, convex, and varying in color from grayish white to bright orange.

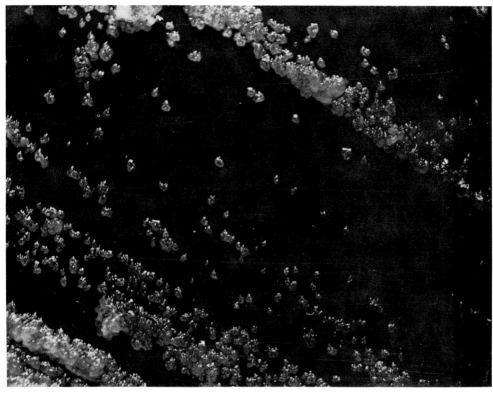

Figure 22-5. Young colonies of *Dermatophilus congolensis* on blood agar. ×8.4. From M. A. Gordon, *J Bacteriol,* 88:509, 1964.

IDENTIFICATION. The characteristic segmenting appearance seen in tissue is often not seen in Gram-stained smears from cultures; these may at times show cocci only, but they usually show Gram-positive branched filamentous organisms (Fig. 22-6). Motile zoospores can be shown after growth in tryptose broth. Definitive identification is made on the unique appearance in tissue, or on the following tests: catalase (+); urease (+); glucose, fructose, maltose—all (+); indole (−); gelatin (+); sucrose, salicin, xylose—all (−).

ANTIBIOTIC SUSCEPTIBILITY. Many drugs inhibit the growth of *D. congolensis in vitro.* These include penicillin, tetracyclines, aminoglycosides, erythromycin, and chloramphenicol. Penicillin-streptomycin is a favored combination.

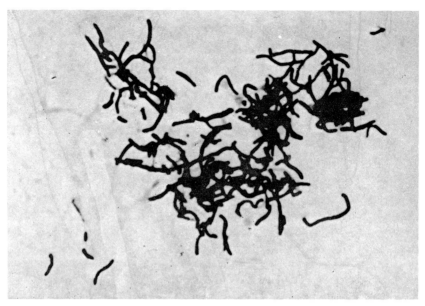

Figure 22-6. *Dermatophilus congolensis* from a blood agar plate culture. Gram's stain, × 1900.

REFERENCES

1. Sonnenwirth, A. C., and Dowell, V. R.: In Lennette, E. H., Balows, A., Spaulding, E. H., and Truant, J. P. (Eds.): *Manual of Clinical Microbiology*, 3rd ed. Washington, D.C., American Society for Microbiology, 1980, Chapter 40.
2. Davenport, A. A., Carter, G. R., and Schirmer, R. G.: *Vet Med Small Anim Clin, 69*:1442, 1974.
3. Hardie, E. M., and Barsanti, J. A.: *J Am Vet Med Assoc 180*:537, 1982.
4. Grasser, R.: *Zblatt Bakt Infekt Hyg Abs 1 Orig, 188*:251, 1963.
5. Collins, M. D., Jones, D., Kroppenstedt, R. M., and Schliefer, K. H.: *J Gen Microbiol, 182*:335, 1982.
6. Gordon, M. A.: In Lennette, E. H., et al. (Eds.): *Manual of Clinical Microbiology*, 3rd ed. Washington, D.C., American Society for Microbiology, 1980, Chapter 15.
7. Bushnell, R. B., Pier, A. C., Fichtner, R. E. et al.: *Proc 22nd Ann Meet Am Assoc Vet Lab Diagn*, 1979, pp. 1–12.
8. Pier, A. C., and Fichtner, R. E.: *J Clin Microbiol, 13*:548, 1981.
9. Carroll, G. F., Brown, J. M., and Haley, L. D. *Am J Clin Path, 68*:279, 1977.
10. Deem, D. A., Harrington, D. D.: *Cornell Vet, 70*:321, 1980.
11. Brown, J. M.: Differentiation and characterization of the clinically important aerobic Actinomycetes. CDC Laboratory Update, CDC 79-56.
12. Lloyd, D. H.: *Vet Rec, 109*:426, 1981.

Chapter 23

MYCOBACTERIUM

CHARLES O. THOEN

THE GENUS *Mycobacterium* includes acid-fast bacteria that are known pathogens, along with numerous saprophytes. The latter are ubiquitous in the environment and occur in soil and water; some are present in the gastrointestinal tract of humans and animals. It is the responsibility of the laboratory to isolate and identify mycobacteria from tissues collected at necropsy or at slaughter. It is necessary to differentiate those organisms that cause contagious or infectious disease from the saprophytes (1, 27). The latter may produce disease in certain situations and are considered opportunists. For example, some otherwise saprophytic mycobacteria may cause infection when introduced into wounds or in humans or animals receiving immunosuppressive therapy. Since the saprophytes have staining properties similar to the pathogenic types, the veterinary microbiologist must be prepared to differentiate them by bacteriologic procedures.

Tubercle bacilli are the most important clinically significant mycobacteria. *Mycobacterium bovis, M. avium,* and *M. tuberculosis* have been recognized for more than eighty years and continue to cause disease in animals resulting in serious economic losses. Remarkable progress has been achieved toward eliminating bovine tuberculosis in cattle in the United States, yet outbreaks of disease still occur (8, 26). The importance of tuberculosis in exotic captive animals has been emphasized by the widespread occurrence of disease in various species maintained in zoos, animal parks, and primate colonies (42).

MICROSCOPIC EXAMINATION

Preliminary examinations of tissues suspected of being tuberculous should include the preparation of suitably stained smears. The demonstration of acid-fast bacilli by microscopic examination of lesions is important because cultures of material with positive smears will usually grow mycobacteria. In contrast, a negative smear only means that no acid-fast bacilli are seen; it does not mean that no acid-fast bacilli are present. The significance of microscopic findings on smears of specimens in which tuberculosis is suspected should be discussed with veterinarians and epidemiologists responsible for conducting follow-up investigations on exposed animals.

An identifiable smear can be made on a new slide from scrapings of the cut surface of tissue. The smear should be air dried and fixed by flaming for one to two seconds. The Kinyoun modification of the Ziehl-Neelsen stain is recommended because no heat is required (34). The stained slides are observed with an ordinary light microscope for the presence of acid-fast bacilli, which appear as red coccoidal or bacillary cells 1–3 microns in length occurring singly or in clumps. An alternate method utilizing fluorescein dyes such as auramine and rhodamine has been described (24). These slides must be observed using an ultraviolet light with a microscope equipped with special filters. Acid-fast bacilli are seen as short, yellow-staining rods.

ISOLATION

Appropriately stained smears provide information on the presence of acid-fast bacilli; however, it is important to remember that a specific diagnosis of tuberculosis requires the isolation and identification of the organism from exudates, body discharges, or from lesions. The procedures for isolating mycobacteria are different from the methods used for culturing other bacteria because specimens must be treated to kill contaminants. Acid-fast bacteria are relatively resistant to alkali and to acids that have been used for this purpose. Most pathogenic mycobacteria of clinical significance grow very slowly; therefore, isolation and final identification may require several weeks. The mycobacteria that cause disease in animals may also infect humans; therefore, laboratories where mycobacteriologic examinations are conducted should be equipped with biologic safety cabinets with separate air exhaust filters. Moreover, routine tuberculin skin tests and/or radiologic examinations should be available for tuberculin-negative employees working in the laboratory and for animal caretakers exposed to infected animals.

Specimens for bacteriologic examination should be collected aseptically and immediately submitted to the laboratory for processing. In instances where specimens must be transported by mail, the specimens should be placed in sodium borate (saturated solution) to minimize the growth of bacterial contaminants (2, 45).

A flow diagram of a procedure for isolating mycobacteria is shown in Figure 23-1. When excessive contamination is present, it may be necessary to increase the time interval for treatment with 2% sodium hydroxide solution from ten to twenty or thirty minutes to obtain suitable cultures (41). The treated tissue suspensions are used to inoculate various kinds of culture mediums (17, 43). Inoculated mediums should be incubated at 37°C in a slanted position so that the inoculum covers the surface. The cultures should be placed in a tray at about a 45 degree angle for one week and then

Decontaminate tissue in 1:1,000 hypochlorite, 4 to 16 hr

↓

Trim off fat and grind tissue in a sterile mortar using
broth with phenol red indicator

↓

Treat with equal volume of 2% NaOH for 10 min*

↓

Neutralize with 6N HCl

↓

Centrifuge at 1,450 RCF for 20 min

↓

Decant supernatant

↓

Inoculate 8 tubes of culture media with cotton-tip applicator†

↓

Observe slants at weekly intervals for 8 wk

↓

Stain smears of growth with carbol fuchsin

↓

Test to identify acid-fast cultures

---

*Urine, milk, and body fluids may be treated with equal
volume of 2% NaOH and neutralized with HCl; media may
be inoculated with supernatant.
†One slant each of Lowenstein-Jensen, Lowenstein-Jensen
with glycerol, Egg Yolk Agar (EYA), EYA with glycerol
and malachite green, EYA with glycerol, malachite green,
and mycobactin (2 mg/liter), Middlebrook 7H10 with
pyruvate (4.1 g/liter), and two slants of Stonebrinks medium
with pyruvate (5 g/1.2 liter).

Figure 23-1. Procedure for isolating mycobacteria. From *Veterinary Microbiology*, 8th ed., 1978.
Courtesy of Iowa State University Press.

transferred to a vertical position. The cultures should be observed for
the appearance of growth at weekly intervals for eight weeks. When *Mycobacterium paratuberculosis* is suspected, the incubation period should be increased to twenty weeks. When growth is detected, smears should be made
of colonies and stained by the Ziehl-Neelsen method to determine their
acid-fastness. A magnifying lens is useful in examining medium slants
for the presence of young colonies. Colony morphology is an important
characteristic of certain mycobacteria; however, it is not always reliable.
Although variations occur, *M. bovis* colonies usually appear mammilate
with a raised center, *M. tuberculosis* colonies have a crumbled-bread appearance, and *M. avium* colonies are often doughnut shaped. It should be emphasized that colonies of some saprophytes may closely resemble these
pathogens; therefore, additional laboratory tests are necessary for their
identification.

IDENTIFICATION

The specific identification of acid-fast bacilli involves the use of biochemical and drug susceptibility tests. Characteristics for identifying certain pathogenic mycobacteria are shown in Table 23-I. The details for supplemental procedures used in identifying mycobacteria will not be included here because they have been described in readily available microbiology textbooks (24). When problems are encountered in conducting *in vitro* tests for identification, it may be necessary to obtain assistance from a reference laboratory that routinely conducts the tests.

Seroagglutination tests are of value in identifying strains of *M. avium* (25, 29), which includes certain slowly growing mycobacteria previously called *M. avium — M. intracellulare* (Battey bacilli). Currently, twenty-five serotypes of *M. avium* are recognized (18, 50). Serologic tests may be conducted using antisera prepared in rabbits or mice for representative strains (25, 31).

Animal pathogenicity tests may also be employed in identifying mycobacteria; however, these procedures are not routinely used because they are expensive and time-consuming. Animal tests are useful in the identification of unusual strains that have aberrant drug or biochemical characteristics. In instances where mixed cultures present problems, guinea pigs, rabbits, and/ or chickens may be inoculated (Table 23-II). Animal inoculation may also be of value where small numbers of organisms are present or where excessive contamination is not controlled by the decontamination procedure.

CLINICALLY SIGNIFICANT MYCOBACTERIA

Some reliable characteristics are included for each of the species.

Mycobacterium bovis (the bovine tubercle bacillus). This fastidious organism causes tuberculosis in cattle, swine, and cats. It has also been associated with outbreaks of tuberculosis in nonhuman primates and certain exotic hoofed animals (42). Growth appears after three to eight weeks of incubation at 37°C. The colonies appear almost colorless and have raised centers with irregular edges. Growth in Proskauer and Beck liquid medium with 5% serum (P & B) is granular in appearance, and strains form short cords. The organism fails to produce niacin or reduce nitrate and is inhibited by thiophen-2-carboxylic acid hydrazide (TCH) and by 5% glycerol.

Mycobacterium tuberculosis (the human tubercle bacillus). This organism causes pulmonary disease in humans, monkeys, baboons, and certain hoofed animals; it has also been isolated from dogs and parrots (35). Buff-colored, raised, rough colonies usually appear after two to four weeks of incubation at 37°C. Growth in P & B is flocculent, and most strains form serpentine cords. The organism produces niacin, reduces nitrates, and growth is not inhibited by TCH. Most strains of *M. tuberculosis* are inhibited by para-

TABLE 23-I

IN VITRO TESTS FOR IDENTIFYING
SOME CLINICALLY SIGNIFICANT MYCOBACTERIA

Mycobacterium	Growth Rate			Niacin	Nitrate Reduction	Cord Formation	Thiophen-2-carboxylic acid-hydrazide	NaCl Tolerance	Tween Hydrolysis	Chromogenicity	Arylsulfatase at 3 days	Glycerol Inhibition
	43°C	37°C	31°C									
M. tuberculosis		M		+	+	+	+	-	-	-	-	+
M. bovis		M		-	-	+	-	-	-	-	-	-
M. avium*	M	M	M	-	-	-	+	-	-	-	-	+
M. kansasii		M	S	-	+	V	+	-	+	+	-	+
M. marinum			S	-	-	-	+	-	+	+	-	+
M. scrofulaceum		M	S	-	-	-	+	-	-	+	-	+
M. xenopi	S	S		-	-	-	+	-	-	+	V	+
M. fortuitum		R	R	-	+	V	+	+	-	-	+	+
M. chelonei		R	R	V	-	-	-	-	-	-	+	+

- = negative — absence or inhibition; + = positive — production or growth; V = variable; R = rapid (1–6 days); M = moderate (6–14 days); S = slow (more than 14 days)

* Includes strains previously identified as *M. intracellulare* (12). *M. avium* serotypes 1 and 2 grow best at 43°C; some strains of serotypes 3 through 25 grow best at 22° to 30°C.

TABLE 23-II

PATHOGENICITY OF *MYCOBACTERIUM BOVIS, M. TUBERCULOSIS,* AND
M. AVIUM IN LABORATORY ANIMALS

Animal	Organism		
	M. bovis	M. tuberculosis	M. avium
Guinea Pig	+	+	—
Rabbit	+	—	+*
Chicken	—	—	+

*Yersin-type reaction (1).

aminosalcylic acid (PAS), isonicotinic acid hydrazide (INH), streptomycin, and ethambutol; however, drug-resistant strains have been isolated from human patients (47).

Mycobacterium avium — intracellulare complex. Currently, there are twenty-five different serotypes, including some organisms previously called "Battey bacilli" (49). *Mycobacterium avium* is the most common cause of tuberculosis in swine (32). However, the serotypes isolated from pigs in different geographical areas of the United States vary (38). Serotypes 1, 2, and 3 commonly isolated from birds are usually pathogenic for chickens, whereas serotypes 4 to 20, which have been isolated from humans and other animals, fail to produce progressive disease in chickens (13, 33, 39). Colonies usually appear in two to three weeks incubation at 37°C; they are raised, rounded, smooth, buff or slightly yellow color. Growth in P & B is turbid; cords are not formed. The organisms fail to produce niacin, to reduce nitrates, or to hydrolyze Tween 80.

Mycobacterium kansasii. The organism has been isolated from pulmonary and extrapulmonary lesions in man (12); it has been isolated from lymph nodes of cattle, swine, and certain exotic animals (34). The importance of this organism in inducing tuberculin-skin reactivity to mammalian tuberculin has been discussed (11). Growth appears in two to four weeks of incubation at 37°C. The colonies are colorless to buff color when incubated in the dark but are photochromogenic, i.e. the colonies develop yellow color when exposed to light. The microorganism does not produce niacin, but it does reduce nitrates and hydrolyze Tween 80. Some strains form cords and produce carotene crystals.

Mycobacterium marinum. The organism causes swimming pool granuloma in humans; isolations have been made from cold-blooded animals (24). Growth appears in three to six weeks of incubation at 30°C. The organism, like *M. kansasii,* is photochromogenic but may be differentiated by its failure to reduce nitrate. Tween 80 is hydrolyzed. Cords and carotene crystals are not formed, and niacin is not produced.

Mycobacterium scrofulaceum. The organism has been isolated from cervical lymph nodes of children and from lymph nodes of pigs (32, 47). Growth appears in two to three weeks of incubation at 37°C as raised, rounded, yellow to orange colonies in the dark or in light. The organism does not produce niacin, reduce nitrates, or hydrolyze Tween 80.

Mycobacterium xenopi. A few reports are available on isolations of this organism from animals (34). Yellowish, rounded, smooth colonies usually appear in three to five weeks incubation at 42–45°C. The organism does not produce niacin, reduce nitrates, or hydrolyze Tween 80. On cornmeal agar, the colonies have a characteristic matted appearance with radiating filaments that may be observed with low-power magnification (×7) after two weeks incubation.

Mycobacterium fortuitum. This organism causes pulmonary disease in humans. The microorganism is often resistant to chemotherapeutic agents used for treating tuberculosis (24). Growth appears in two to six days of incubation at 25–37°C as raised, rough, buff-colored colonies. Some strains form loose cords. The cultures fail to produce niacin but do reduce nitrates and produce arylsulfatase. Growth is observed in 5% sodium chloride.

Mycobacterium chelonei. This mycobacterium has been isolated from injection abscesses and from patients with valvular endocarditis (24); it has also been cultured from lesions in animals (34). It is similar to *M. fortuitum,* but it fails to reduce nitrates. No growth is observed in 5% sodium chloride.

Mycobacterium lepraemurium. The rat leprosy bacillus has not been cultivated on solid medium used routinely for cultivating other mycobacteria. However, growth of this organism has been observed on liquid medium enriched with cytochrome-*c* and α-ketoglutarate (22). This organism has been suggested as the etiologic agent of a leprosylike disease in cats (16); efforts to culture the organism on routine media used for mycobacteria have been unsuccessful.

Mycobacterium leprae. The leprosy bacillus was first observed by Hansen in Norway; therefore, infection with this organism is sometimes called Hansen's disease. *Mycobacterium leprae* has recently been associated with a naturally occurring disease in armadillos (46). Definitive information is not available on the growth of *M. leprae* in cell-free culture medium; however, the organism will multiply in the footpads of mice.

Mycobacterium paratuberculosis. The Johne's bacillus was first identified in 1895 by Johne and Frothingham while studying chronic dysentery in cattle. The organism has been isolated from cattle, sheep, goats, swine, and exotic animals in the United States and Canada (7, 10, 14, 30, 32, 36, 42, 48). *M. paratuberculosis* is mycobactin dependent and grows very slowly at 37°C incubation (5, 19, 20). Growth usually appears at four to ten weeks. The colonies are colorless to white and translucent. Because a prolonged interval is required to isolate this mycobacterium, a presumptive diagnosis may be obtained by preparing appropriately stained smears of biopsies of mucosa of the rectum (34). On microscopic examinations, acid-fast bacilli are readily observed in clumps within macrophages.

Cultures for *M. paratuberculosis* may be made on mycobactin-enriched medium and medium without mycobactin after the specimens (intestinal mucosa or feces) are treated with benzalkonium chloride (45). A portion of intestine 3–4 cm anterior to the ileocecal valve and 3–4 cm posterior with associated lymph nodes should be collected on necropsy, washed to remove fecal material, and submitted to the laboratory frozen. The tissues may be ground in 3% trypsin and the sediment decontaminated in 0.1% Zephiran® for sixteen to twenty hours. Fecal specimens (about 1 oz.) collected from the rectum of animals suspected of having Johne's disease should be placed in appropriately identified ointment containers. No refrigeration is required. The sample should be transported to the laboratory and processed within twenty-four to thirty-six hours to minimize contamination. An aliquot of the feces is suspended in sterile water and processed with 0.3% Zephiran for sixteen to twenty hours. Herrold's egg yolk agar slants are inoculated with the treated suspensions (three tubes with mycobactin and one without mycobactin). Inoculated mediums should be incubated at 37°C and examined at periodic intervals for twenty weeks. Cultures are identified by growth rate, mycobactin dependency, and by seroagglutination tests (6). It should be noted that when large numbers of fecal samples are collected at one time, it is possible to store specimens at −70°C prior to processing in the laboratory.

Measures for controlling and eliminating Johne's disease in cattle have been based on the diagnosis and elimination of infected cattle shedding the organism in the feces (3, 15). The success of controlling Johne's disease has been limited by the lack of a rapid diagnostic test. Further investigations are needed to determine the practical value of lymphocyte immunostimulation tests in detecting animals infected with *M. paratuberculosis* (21). Lymphocyte immunostimulation tests conducted on blood of calves inoculated with *M. paratuberculosis* revealed that infected calves could be detected 120 days after inoculation (9). The production of refined *M. paratuberculosis* antigens may improve the specificity of the cell responses (4, 44).

The importance of developing a rapid, simplified diagnostic test has stimulated an interest in enzyme-linked immunosorbent assays (ELISA) for detection of *M. paratuberculosis* infected cattle (37, 40). Production of immunoglobulin class or subclass specific conjugates for use in ELISA could improve the sensitivity and specificity of ELISA for identification of diseased animals (28). The importance of Johne's vaccine containing killed *M. paratuberculosis* suspended in oil on the development and persistance of humoral and cellular responses should be investigated.

REFERENCES

1. Feldman, W. H.: *Avian Tuberculosis Infections*. Baltimore, Williams & Wilkins, 1938.
2. Feldman, W. H., and Karlson, A. G.: *J Am Vet Med Assoc*, 96:141–149, 1940.
3. Gilmour, N. J. L.: *Vet Record*, 99:433, 1976.
4. Gunnarsson, E., and Fodstad, F. H.: *Acta Vet Scand*, 20:200–215, 1979.
5. Francis, J., Macturk, H. M., Madinaveita, J., and Snow, G. A.: *Biochem J*, 55:596–607, 1953.
6. Jarnagin, J. L., Champion, M. L., and Thoen, C. O.: *Clin Microbiol*, 2(3):268–269, 1975.
7. Jessup, D. A., Abbas, B., Behymer, D., and Gogan, P.: *J Am Vet Med Assoc*, 179(11): 1252–1254, 1981.
8. Johnson, D. C., Rogers, A. N., Andrews, J. F., Downard, J. A., and Thoen, O.: *J Am Vet Med Assoc*, 167:833–837, 1975.
9. Johnson, D. W., Muscoplat, C. C., Larsen, A. B., and Thoen, C. O.: *Am J Vet Res*, 38(12):2023–2025, 1977.
10. Julian, R. J.: *Can Vet J*, 16(2):33–43, 1975.
11. Karlson, A. G.: *Adv Vet Sci*, 7:143, 1975.
12. Karlson, A. G.: *Surg Clin North Am*, 53(4):905–912, 1973.
13. Karlson, A. G. and Thoen, C. O.: Tuberculosis. In Hitchner S. B., Domermuth, C. H., Purchase, H. G., and Williams, J. E. (Eds.): *Isolation and Identification of Avian Pathogens*, 2nd ed. Endwell, N. Y., Creative Printing Co., Am Assoc Avian Pathologists, 1980, pp. 36–39.
14. Kopecky, K. E.: *J Am Vet Med Assoc*, 162:787–788, 1973.
15. Larsen, A. B.: *J Am Vet Med Assoc*, 161:1539–1541, 1972.
16. Leiker, D. L., and Poelma, F. G.: *Int J Lepr*, 42:312–315, 1974.
17. Matthews, P. R. J., McDiarmid, J. A., Collins, P. and Brown, A.: *J Med Microbiol*, 2:53–57, 1977.
18. Meissner, G., et al.: *J Gen Microbiol*, 83:207–235, 1974.
19. Merkal, R. S., McCullough, W. G. and Takayama, K.: *Bull Institute Pasteur*, 79:251–259, 1981.
20. Merkal, R. S., and Curran, B. J.: *Appl Microbiol*, 28:276–279, 1973.
21. Muscoplat, C. C., Thoen, C. O., Chen, A. W., and Johnson, D. W.: *Am J Vet Res*, 36:(4):395–398, 1975.
22. Nakamura, M., Itoh, T., and Yoshii, Z.: *Proc Soc Exp Biol Med*, 148:183–186, 1975.
23. Richards, W. D., and Thoen, C. O.: *Clin Microbiol*, 6(4):392–395, 1977.
24. Runyon, E. H., Karlson, A. G., Kubica, C. P., and Wayne, L. G.: *Mycobacterium*. In Lennette, E. H., Spaulding, E. H., and Truant, J. P. (Eds.): *Manual of Clinical Microbiology*, 2nd ed. Washington, D.C., American Society for Microbiology, 1974, pp. 148–174.
25. Schaefer, W. B.: Serological identification of atypical mycobacteria. In Norris, J. R., and Bergan, T. (Eds.): *Methods in Microbiology*. New York, Academic Press, 1980, Vol. 13, pp. 323–344.
26. Spencer, P. L.: *Proceedings 79th Annual Meeting US Animal Health Association*. Portland, Oregon, 1975, pp. 305–306.
27. Thoen, C. O.: Factors associated with the pathogenicity of mycobacteria. In Schliessinger, D. (Ed.): *Microbiology*. Washington, D.C., American Society for Microbiology, 1979, pp. 162–167.
28. Thoen, C. O., Bruner, J. A., Luchsinger, D. W. and Pietz, D. E.: *Am J Vet Res* (In Press), 1983.
29. Thoen, C. O., Jarnagin, J. L., and Champion, M. L.: *Clin Microbiol*, 1:469–471, 1975.
30. Thoen, C. O. and Johnson, D. W.: Johne's disease (Paratuberculosis). In Davis, J. W., Karstad, L. H., and Trainer, D. O.: *Infectious Diseases of Wild Mammals*, 2nd ed. Ames, Iowa St U Pr, 1981, pp. 275–279.

31. Thoen, C. O., and Karlson, A. G.: *Appl Microbiol*, 20:847–848, 1970.
32. Thoen, C. O., and Karlson, A. C.: Tuberculosis. In Leman, A. D., Glock, R. D., Mengeling, W. L., Penny, R. H. C., Scholl, E. and Straw, B. (Eds.): *Diseases of Swine*, 5th ed. Ames, Iowa St U Pr, 1981, pp. 508–516.
33. Thoen, C. O., and Karlson, A. C.: Tuberculosis. In Hofstad, M. S., Calnek, R. W., Helmboldt, C. F., Reid, W. M., and Yoder, H. W. (Eds.): *Diseases of Poultry*, 7th ed. Ames, Iowa St U Pr, 1977, Chapter 6, pp. 209–224.
34. Thoen, C. O., and Karlson, A. G.: The genus *Mycobacterium*. In Packer, R. A., Mare, C. J., and Merchant, I. A. (Eds.): *Veterinary Microbiology*, 8th ed. Ames, Iowa St U Pr, 1983.
35. Thoen, C. O., Karlson, A. G., and Himes, E. M.: *Mycobacterium tuberculosis* complex infections in animals. In Kubica, G. P. and Wayne, L. G. (Eds.): *The Mycobacteria: A Sourcebook*. New York, Marcel Dekker, Inc., 1983.
36. Thoen, C. O., and Muscoplat, C. C.: *J Am Vet Med Assoc*, 174(8):838–840, 1979.
37. Thoen, C. O., Armbrust, A. L., and Hopkins, M. P.: *Am J Vet Res*, 40:1096–1099, 1979.
38. Thoen, C. O., Jarnagin, J. L., and Richards, W. D.: *Am J Vet Res*, 36(9):1383–1386, 1975.
39. Thoen, C. O., Karlson, A. G., and Himes, E. M.: *Rev Infect Diseases*, 3(5):960–972, 1981.
40. Thoen, C. O., Mills, K., and Hopkins, M. P.: *Am J Vet Res*, 40:833–835, 1980.
41. Thoen, C. O., Richards, W. D., and Jarnagin, J. L.: *Appl Microbiol*, 27:448–451, 1974.
42. Thoen, C. O., Richards, W. D., and Jarnagin, J. L.: *J Am Vet Med Assoc*, 170:(9):987–990, 1977.
43. Thoen, C. O., Himes, E. M., Jarnagin, J. L., and Harrington, R.: *Clin Microbiol*, 9(2):194–196, 19−9.
44. Thoen, C. O., Jarnagin, J. L., Muscoplat, C. C., Cram, L. S., Johnson, D. W., and Harrington, R.: *Comp Immunol Microb Infect Dis*, 3:355–361, 1980.
45. *U.S.D.A. Laboratory Methods in Veterinary Mycobacteriology*, rev. ed. Ames, Iowa, Veterinary Services Laboratories, APHIS, 1974.
46. Walsh, G. P., Storrs, E. E., Burchfield, H. P., Cottrell, E. H., Vindine, M. F., and Binford, C. H.: *J Reticuloendothel Soc*, 18:347–351, 1976.
47. Washington, J. A. II: *Laboratory Procedures in Clinical Microbiology*. Boston, Little, 1974.
48. Williams, E. S., Spraker, T. R., Schoonveld, G. S.: *J Wildlife Diseases*, 15:221–227, 1979.
49. Wolinsky, E.: *Am Rev Resp Dis*, 119:107–159, 1979.
50. Wolinsky, E., and Schaefer, W. B.: *Int J Syst Bacteriol*, 23:182–183, 1973.

Chapter 24

MISCELLANEOUS PATHOGENIC AND NONPATHOGENIC BACTERIA

I NCLUDED in this chapter are a number of bacteria that cause infrequent infections in the lower animals. Also listed, with only a brief mention of their occurrence and principal characteristics, are some species and groups that may be recovered occasionally from clinical materials. They are not generally considered to be pathogenic.

CHROMOBACTERIUM

Bergey's Manual lists two species of this genus, *Chromobacterium violaceum* and *C. lividum*. Both occur in soil and water, but only the former has been incriminated in disease. *C. violaceum* is mesophilic and ferments sugars, while *C. lividum*, which produces leaden or dark blue pigment, is psychrophilic, and oxidizes sugars. Both species are motile and catalase positive. The former species is a facultative anaerobe, while the latter only grows aerobically. Readers should consult *Bergey's Manual* for the identification of *C. lividum*.

CHROMOBACTERIUM VIOLACEUM

PATHOGENICITY. A cause of suppurative pneumonia of swine and cattle. Serious infections also occur in humans.

ISOLATION AND CULTIVATION. This species grows well on ordinary unenriched media, including MacConkey's agar, but it is usually isolated on blood agar. On this medium, colonies are black, smooth, and shiny, attaining a size of about 3 mm in diameter after incubation at 37°C for twenty-four hours. A black to purple pigment is produced when the organism is grown in broth.

IDENTIFICATION. Principal characteristics are the following (1):
- Gram-negative rods of moderate size.
- Motile and non-sporeforming.
- Indole is negative.
- Nitrates are reduced.
- Litmus milk is peptonized.
- Hydrogen sulfide is not produced.
- MR and VP tests are negative.

- Fermentation with acid but no gas: glucose, trehalose, sucrose, levulose, and mannose.
- Not fermented: maltose, inositol, salacin, raffinose, xylose, dulcitol, arabinose, galactose, rhamnose, mannitol, sorbitol, lactose, esculin, inulin, and glycerol

FLAVOBACTERIUM

This genus includes many free-living species. On appropriate media, they produce colonies with a bright yellow pigment. There do not appear to be reports of members of this genus causing disease in animals; however, in view of their occurrence in human clinical materials, it seems likely that they will be recovered on occasion from pathologic materials of animal origin. The flavobacteria that have been recovered from human clinical materials include *F. meningosepticum*, *F.* species Group 11 B, *F. odoratum* (formerly M-4F), and *F. breve* (2). Workers are referred to King's tables as revised by Weaver et al. (3) and the discussion by Rubin et al. (2) for the identification of these organisms.

Some of the characteristics of *F. meningosepticum* are as follows:
- Colonies possess a yellow pigment.
- Nonmotile.
- Catalase positive.
- Oxidase positive.
- Indole is produced.
- Sugars are fermented slowly.
- Hydrogen sulfide is produced (lead acetate paper strips).
- TSI slant: alkaline.
- Gelatin is liquefied and litmus milk peptonized.

NOGUCHIA

This genus was described in the 7th edition of *Bergey's Manual* but was omitted in the 8th edition. The author has not been able to locate cultures representing species of this genus. The three species of the genus are found in the conjunctiva of human and animals affected with follicular conjunctivitis. They are small, slender, capsulated, Gram-negative, motile rods. The three species, along with their corresponding hosts, are *N. granulosis* (human), *N. simiae* (monkey), and *N. cuniculi* (rabbit).

ISOLATION AND CULTIVATION. If organisms of this genus are suspected, material is inoculated onto blood agar and incubated at the optimum temperature, which is in the range of 15–30°C. Organisms grow less well at 30°C. In forty-eight hours, minute, nonhemolytic, shiny, round, somewhat raised, and slightly grayish colonies appear. Good growth can also be obtained in leptospira media. Enrichment of media is necessary.

IDENTIFICATION. Identification of organisms of this genus is based upon the cultural and morphologic characteristics referred to above, motility in the temperature range of 30–37°C, and certain biochemical characteristics. Litmus milk is unchanged; gelatin is not liquefied; nitrate is not reduced to nitrite; and indole is not formed.

N. cuniculi differs from the other two species in not fermenting carbohydrates. *N. granulosis* is most readily differentiated from *N. simiae* by means of an agglutination test in which specific antisera are used.

The fermentative reactions listed in *Bergey's Manual* (4) are presented below. These organisms are not listed in the current *Bergey's Manual*.

N. simiae. Acid but no gas from glucose, fructose, mannose, galactose, xylose, arabinose, and rhamnose. A small amount of acid is reported from dextrin. Raffinose, salacin, dulcitol, maltose, trehalose, sorbitol, inositol, and amygdalin are not fermented.

N. granulosis. Acid but no gas from glucose, fructose, mannose, sucrose, galactose, maltose, salacin, xylose, mannitol, dextrin, arabinose, amygdalin, and lactose. A small amount of acid is produced from raffinose, inulin, rhamnose, and trehalose. No acid is produced from sorbitol, dulcitol, and inositol.

STREPTOBACILLUS MONILIFORMIS

This extremely pleomorphic, aerobic, Gram-negative rod produces L forms spontaneously, especially on initial isolation.

PATHOGENICITY. This organism is a normal inhabitant of the throat and nasopharynx of rats, both wild and laboratory varieties. It is associated with mycoplasmas in pneumonic lesions in rats and is considered to be the cause of a disease of mice characterized by diverse manifestations, including septicemia, septic arthritis, hepatitis, and lymphadenitis. Outbreaks of *S. moniliformis* infection attributed to rat bites have been reported in turkeys (5). Clinical signs and lesions were associated with the joints. This species has also been recovered from guinea pigs with cervical abscesses (6).

Infections in humans result from rat bites and are characterized by septicemia and polyarthritis.

ISOLATION PROCEDURES. Blood or other clinical materials can be inoculated directly into thioglycollate broth or BHI semisolid containing 10–20% horse serum. Other material is inoculated onto serum or blood agar. Plates should be incubated in air containing 5% CO_2. Mohamed et al. (5) incubated their plates in a candle jar.

CULTURAL AND MORPHOLOGIC CHARACTERISTICS. Growth in thioglycollate broth or BHI semisolid is characterized by the formation of small, fluffy colonies referred to as "fluffballs." Colonies are removed with a Pasteur pipette, and smears are stained by the Giemsa or Gram procedure.

On solid media incubated aerobically, small discrete colonies develop within two to three days. They are glistening, smooth, colorless to grayish, and irregularly round. Minute L-type colonies may be seen under low power or with a dissecting microscope. Giemsa-stained smears of these minute colonies reveal a variety of forms resembling somewhat those of the pleuropneumonialike organisms. They are best examined *in situ* by Dienes' staining procedure (see Appendix A). Gram-stained smears from the large colonies on solid media and from the "fluffballs" disclose a remarkable variety of morphological elements including small, slender rods and abundant curved filaments with numerous moniliform (necklacelike) swellings.

IDENTIFICATION. This is generally based upon cultural and morphologic characteristics, source, and associated disease if present.

Two of the turkey isolates of Mohamed et al. (5) when incubated in carbohydrate broths enriched with 15% horse serum fermented maltose, dulcitol, sorbitol, lactose, arabinose, sucrose, and dextrose; trehalose and mannitol were not fermented, and the results with salacin were variable. For additional information on the identification of *S. moniliformis*, readers are referred to the discussion by Rogosa (7).

BACILLUS PILIFORMIS
(Synonym: *Actinobacillus piliformis*)

This interesting organism, which in some respects resembles the Gram-negative anaerobes and *Streptobacillus moniliformis*, has not been satisfactorily cultivated in artificial media.

PATHOGENICITY. Natural Tyzzer's disease has been observed in mice, foals, dogs, foxes, rats, rabbits, gerbils, and monkeys. The epizootic form is a major cause of losses among laboratory mice, particularly those subjected to various stresses such as thymectomy, irradiation, and cortisone treatment. The disease is frequently enzootic and present in a latent or subclinical state. The principal lesion seen in the acute disease consists of diffusely distributed, pale gray necrotic foci in the liver.

DIRECT EXAMINATION. Smears are made from the necrotic foci and stained with Giemsa stain. Long, slender organisms staining bluish purple are seen in the cytoplasm of hepatic cells. Very long, thin, and tortuous filaments are sometimes seen, as well as shorter bacillary forms and occasional forms with subterminal swellings (Fig. 24-1). The latter resemble the moniliform structures that are characteristic of *Streptobacillus moniliformis*. Bacillary forms are frequently tapered, and beading is often noted. The motility of *B. piliformis* can be seen in phase-contrast microscopy.

ISOLATION PROCEDURES. Methods for the reliable cultivation of *B. piliformis* in artificial media have not yet been developed. The organism has been cultivated successfully in the yolk sac of embryonated chicken eggs (8). The

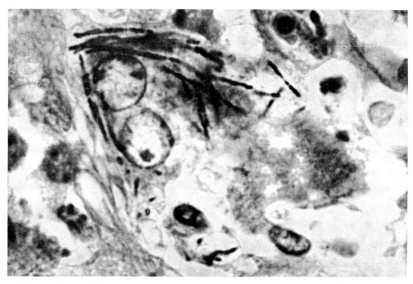

Figure 24-1. *Bacillus piliformis* in hepatic cells. Gomori methenamine-silver nitrate and HE method, ×2000. Courtesy of Doctor K. Fujiwara.

following procedure has been employed successfully (9): A 10% suspension of affected liver is prepared in brain-heart infusion broth with the aid of a Ten Broeck grinder. After centrifuging in a clinical centrifuge at 2000 rpm for ten minutes, 0.2 ml of supernatant is injected into the yolk sac of each of six eggs. The embryos usually die between the fourth and seventh day postinoculation.

Giemsa-stained smears from the embryonic liver and yolk sac disclose the highly pleomorphic and characteristic forms of *B. piliformis*. The viability of organisms can be preserved for a number of months if the yolk material is frozen and stored at −70°F.

IDENTIFICATION. A diagnosis of Tyzzer's disease is based upon the demonstration of the typical organisms with the varied pleomorphic forms in Giemsa-stained smears from the necrotic foci in the liver. Cultivation in the yolk sac of the embryonated egg is confirmatory.

The indirect immunofluorescence technique has been used for the identification of *B. piliformis* (10).

NEISSERIA

Species of this genus are found occasionally in animals as commensals on mucous membranes, particularly of the nasopharynx, and conjunctiva. They are aerobic, Gram-negative cocci that split sugars oxidatively or not at all. They are catalase and oxidase positive and taxonomically closely related to *Moraxella* spp.

Several species of animal origin have been described, but none is known for certain to have a primary role in the causation of disease.

The organism previously named *Neisseria catarrhalis* is now called *Branhamella catarrhalis*. This organism has been isolated from the pneumonic lungs of calves, but its significance is not yet fully known. The characteristics of the strains recovered by Hunter and Harbourne (11) were as follows:

- Diplococcus.
- Small, circular, convex, grayish white, nonhemolytic colonies are produced on blood agar.
- Indole negative.
- Catalase and oxidase positive.
- No oxidation or fermentation of twelve of the frequently used carbohydrates.

Three additional species from animals have been described. They are listed below, and workers are referred to the original references for information on their identification. The identification of various *Neisseria* and *Branhamella* spp. are dealt with in detail by Morello and Bohnhoff (12).

N. caviae (13) and *N. animalis* (14) have been recovered from the nasopharyngeal region of guinea pigs. These and *N. ovis* are not official species.

N. ovis was recovered from sheep with keratoconjunctivitis (15, 16). It is not known for certain to be the primary cause of this disease.

MISCELLANEOUS USUALLY NONPATHOGENIC BACTERIA

Listed here are several genera of bacteria that occur occasionally in clinical materials as contaminants. Other genera in this category have been referred to elsewhere in the manual. Information on the definitive identification of these nonpathogenic bacteria is found in the publications listed under the Supplementary References.

Kurthia

OCCURRENCE. Manure and putrefying organic matter.

CHARACTERISTICS. Gram-positive rods; non-sporeforming; motile, facultatively anaerobic; catalase positive; do not attack sugars.

Lactobacilli

OCCURRENCE. Associated with dairy products; normal flora of the mouth, vagina, and intestinal tract.

CHARACTERISTICS. Long, slender, Gram-positive rods; usually motile; non-sporeforming; anaerobic and facultatively anaerobic; catalase negative; ferments sugars; and grow best at a pH near 6. Strains can be confused with nonsporulating *Cl. perfringens*.

Erwinia spp.

This group includes Gram-negative organisms that are plant pathogens. They are included in the *Enterobacteriaceae*. Some have been isolated from human clinical materials, and on occasion, infections have been attributed to them. Workers are referred to the recent report by von Graevenitz (17) for a discussion of their taxonomy, identification, and significance in human infections. It seems likely that they will be encountered occasionally from animal specimens.

Alkalescens-Dispar Group (A–D Group)

This group includes the organisms earlier referred to as *Bacterium alkalescens* and *B. dispar*. They are nonmotile and are generally considered to be anaerogenic biotypes of *Escherichia*. They can be confused with shigellas. *B. alkalescens* does not ferment lactose, and *B. dispar* is a delayed lactose fermenter. For additional information, readers are referred to Cowan (18).

EF-4

Organisms designated EF-4 have been isolated from human, cat, and dog bite wounds. Their occurrence as part of the normal oral flora of the dog has been reported (19). There is as yet no evidence that they cause disease in either cats or dogs.

Some important characteristics of this bacterium are as follows:
- Small Gram-negative rods or coccobacilli.
- Colonies on blood agar 1–2 mm in diameter, circular, entire, convex, and mucoid.
- Yellow to tan pigment frequently produced.
- Nonhemolytic; oxidase and catalase positive.
- Nonmotile; indole and urease negative; and poor or slow growth on MacConkey's agar.

For definitive identification, readers should consult the tables prepared by Weaver et al. (3) or Rubin et al. (2).

Simonsiella

This Gram-negative bacterium occurs as part of the normal oral flora of man and animals. Saphir and Carter (19) identified an organism recovered from the gingival flora of dogs as *Caryophanon*. Isolations were made on blood agar; colonies were 1–3 mm in diameter, smooth, convex, and translucent. The identification of *Caryophanon* was questioned by Nyby et al. (20). In view of the fact that *Simonsiella* occurs rather commonly in the oral cavity of dogs and that it could be mistaken morphologically for *Caryophanon*, it seems likely that the identification of the latter organism by Saphir and

Carter (19) was mistaken. The author has seen organisms on a number of occasions resembling those in Figure 24-2 from the oral cavities of dogs. Workers are referred to Skerman (21) and *Bergey's Manual* for detailed descriptions of these apparently harmless but interesting organisms.

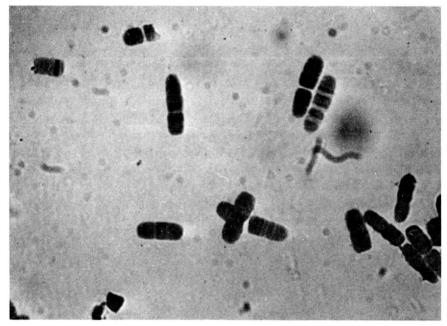

Figure 24-2. Gram-stained smear of *Simonsiella* (originally identified as *Caryophanon*) from the canine gingiva. × ca. 1600. From D. A. Saphir and G. R. Carter, *J Clin Microbiol, 3*:344, 1976.

Prototheca

Species of this genus are microscopic, colorless, achlorophyllic algae of the family Chlorellaceae. They occur widely in nature and are occasionally recovered from clinical specimens. There are several reports of human infections and severe bovine mastitis (22) due to these algae. Small colonies resembling *Cryptococcus* are produced on Sabouraud agar (25°C) and blood agar (37°C) in twenty-four hours. They do not have a capsule and are hyaline and globose in form, with width and length as great as 13–16 µm. As many as eight or more characteristic endospores are produced by internal segmentation (Fig. 24-3). Five species of *Prototheca* have been identified with fluorescent antibody reagents (23).

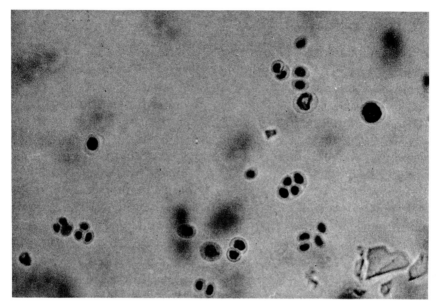

Figure 24-3. *Prototheca* from milk of a cow with mastitis. Note endospores. Lactophenol cotton blue, × 330 (Paul C. Watkins).

REFERENCES

1. Sippel, W. L., Medina, G., and Atwood, M. B.: *J Am Vet Med Assoc, 124*:467, 1954.
2. Rubin, S. J., et al.: In Lennette, E. H. (Ed.-in-Chief): *Manual of Clinical Microbiology*, 3rd ed. Washington, D.C., American Society for Microbiology, 1980, Chapter 21.
3. Weaver, R. E., Tatum, H. W., and Hollis, D. G.: *The Identification of Unusual Pathogenic Gram-negative Bacteria* (Elizabeth O. King), Preliminary Revision, 1972. Published as training materials by the Center for Disease Control, Public Health Service, Department of Health, Education and Welfare, Center for Disease Control, Atlanta, Georgia.
4. Breed, R. S., Murray, E. G. D., and Smith, N. R.: *Bergey's Manual of Determinative Bacteriology*, 7th ed. Baltimore, Williams & Wilkins, 1957.
5. Mohamed, Y. S., Moorhead, P. D., and Bohl, E. H.: *Avian Dis, 13*:379, 1969.
6. Fleming, M. P.: *Vet Rec, 99*:256, 1976.
7. Rugosa, M.: In Lennette, E. H. (Ed.-in-chief): *Manual of Clinical Microbiology*, 3rd ed. Washington, D.C., American Society of Microbiology, 1980, Chapter 28.
8. Craigie, J.: *Proc R Soc B, 165*:35, 1966.
9. Carter, G. R., Whitenack, D. L., and Julius, L. A.: *Lab Anim Care, 19*:648, 1969.
10. Savage, N. L., and Lewis, D. H.: *Am J Vet Res, 33*:1007, 1972.
11. Hunter, D., and Harbourne, J. F.: *Vet Rec, 76*:738, 1964.
12. Morello, J. A., and Bohnhoff, M.: In Lennette, E. H. (Ed.-in-chief): *Manual of Clinical Microbiology*, 3rd ed. Washington, D.C., American Society for Microbiology, 1980, Chapter 9.
13. Pelczar, M. J., Jr.: *J Bacteriol, 65*:744, 1953.
14. Berger, U.: *Z Hyg Infektkr, 147*:158, 1960.

15. Spradbrow, P. B., and Smith, I. D.: *Aust Vet J, 43*:40, 1967.
16. Lindquist, K.: *J Infect Dis, 106*:162, 1960.
17. von Graevenitz, A: *Ann N Y Acad Sci, 174*:436, 1970.
18. Cowan, S. T.: *Manual for the Identification of Medical Bacteria*, 2nd ed. Cambridge, Cambridge U Pr, 1974.
19. Saphir, D. A., and Carter, G. R.: *J Clin Microbiol, 3*:344, 1976.
20. Nyby, M. D., Gregory, D. A., Kuhn, D. A., and Pangborn, J.: *J Clin Microbiol, 6*:87, 1977.
21. Skerman, V. B. D.: *A Guide to the Identification of the Genera of Bacteria*, 2nd ed. Baltimore, Williams & Wilkins, 1967.
22. Frank, N., Ferguson, L. C., Cross, R. F., and Redman, D. R.: *Am J Vet Res, 30*:1785, 1969.
23. Sudman, M. S., and Kaplan, W.: *Appl Microbiol, 25*:981, 1973.

SUPPLEMENTARY REFERENCES

Buchanan, R. E., and Gibbons, N. E. (Eds.): *Bergey's Manual of Determinative Bacteriology*, 8th ed. Baltimore, Williams & Wilkins, 1974.

Cowan, S. T.: *Manual for the Identification of Medical Bacteria*, 2nd ed. Cambridge, Cambridge U Pr, 1974.

Sippel, W. L., Medina, G., and Atwood, M. B.: *J Am Vet Med Assoc, 124*:467, 1954.

Skerman, V. B. D.: *A Guide to the Identification of the Genera of Bacteria*, 2nd ed. Baltimore, Williams & Wilkins, 1967.

Weaver, R. E., Tatum, H. W., and Hollis, D. G.: *The Identification of Unusual Pathogenic Gram-negative Bacteria* (Elizabeth O. King), Preliminary Revision, 1972. Published as training materials by the Center for Disease Control, Public Health Service, Department of Health, Education and Welfare, Center for Disease Control, Atlanta, Georgia.

Appropriate chapters: In Lennette, E. H. (Ed.-in-chief): *Manual of Clinical Microbiology*, 3rd ed. Washington, D.C., American Society for Microbiology, 1980.

Chromobacterium

Sneath, P. H. A.: In Gibbs, B. M., and Skinner, F. A. (Eds.): *Identification Methods for Microbiologists* (Technical Series No. 1, Part A). New York, Academic Press, 1966, p. 15.

Bacillus Piliformis

Fries, A. S.: *Tyzzer's Disease and the Importance of Inapparent Infection in Biomedical Research.* Copenhagen, The Royal Veterinary and Agricultural College, 1981.

Chapter 25

RICKETTSIAE AND CHLAMYDIAE

JOHANNES STORZ

Rickettsiae and chlamydiae are host-cell-dependent, obligate intracellular, prokaryotic organisms pathogenic for man and animals. These infectious agents are currently classified in the orders Rickettsiales and Chlamydiales. Laboratory diagnostic methods for detecting these infections in animals as well as approaches to cultivate and identify these agents are presented.

ORDER RICKETTSIALES

This order contains the following families: Rickettsiaceae, Hemobartonellaceae, and Anaplasmataceae.

Family Rickettsiaceae

The majority of the members of this family are rod-shaped, coccoid, or pleomorphic. Spherical forms have a diameter of 200–700 nm, while the rods measure 300–600 nm in width and 800–2000 nm in length. They are Gram-negative, possess typical bacterial cell walls containing muramic acid, and multiply by binary fission only inside eukaryotic cells. They parasitize the gut cells of arthropods, which transmit the infection to animals. Capillary endothelial cells of infected animals are parasitized. The organisms grow to high titers in the endothelial cells lining the yolk sac of developing chicken embryos (3). Some also are adapted to growth in insect or animal cell cultures (20). These agents retain basic fuchsin when stained by the method of Gimenez, a modification of the procedure of Macchiavello (4). Growth is inhibited significantly by chlortetracycline, oxytetracycline, chloramphenicol, and erythromycin but to a lesser extent by penicillin and streptomycin. These agents are rapidly inactivated at 56°C, and they are rather unstable outside host cells when separated from host component, except in diluent containing proteins or in Bovarnick's buffer consisting of phosphate-buffered sucrose and glutamate (1,2,6). Included in the Rickettsiaceae are the genera *Rickettsia, Cowdria, Ehrlichia, Neorickettsia,* and *Coxiella* (6,8,17,18).

239

Rickettsia rickettsii

PATHOGENICITY. This infectious agent is maintained in animals and is transmitted to humans by ticks (*Dermatocentor, Amblyomma, Haemaphysalis*), causing Rocky Mountain spotted fever. It multiplies in the cells of the small peripheral blood vessels and induces thrombosis and extravasation. Dogs are susceptible to natural infection and develop mild illness with fever, loss of appetite, and lassitude.

DETECTION. The organisms can be demonstrated in ticks by immuno-fluorescence. Isolation from blood and tissues of infected animals is rarely accomplished. Antibodies against *R. rickettsii* are detected by the Weil-Felix reaction, by indirect immunofluorescence with cell culture- or yolk sac-propagated antigen, or by indirect hemagglutination (5).

Cowdria ruminantium

PATHOGENICITY. This organism causes heartwater, an acute septicemic disease of cattle, sheep, and goats in Africa. It is transmitted by ticks of the genus *Ablyomma*.

DIRECT EXAMINATION. Smears are made from the hippocampus, the cerebral cortex, the spinal cord, and the intima of large veins. It is important to have fresh tissues. They are fixed in methanol and stained with Giemsa stain. The blue- and purple-staining rickettsiae are found in large numbers in the endothelial cells of the intima of capillaries and veins (9).

ANIMAL INOCULATION. Inoculate blood or minced spleen into heartwater-susceptible and bluetongue virus-immune sheep. The incubation period of heartwater is about eleven days.

The organism of heartwater will survive for weeks in mice inoculated intraperitoneally with infectious material. Mice can be inoculated from field cases and the organism maintained viable in this way if difficulties are experienced in delivering clinical materials to the laboratory (9).

Ehrlichia canis

PATHOGENICITY. *E. canis* produces an infection in dogs usually character-ized by a long course with recurrent fever and hemorrhagic hepatoencephalitis. Cases terminating fatally are frequently complicated with babesiasis or hemobartonellosis. *E. canis* is usually transmitted by the brown dog tick *Rhipicephalus sanguineous*. There are reports of its occurring in Africa and countries of the Mediterranean basin, Asia, the West Indies, and United States. Canine ehrlichiosis is also known as tropical canine pancytopenia (8).

CYTOLOGIC EXAMINATION. The optimum time for the demonstration of rickettsiae in leukocytic and mononuclear cells of Giemsa-stained blood smears is approximately postinfection day 13. The morulae may be difficult to find in the peripheral blood, and it is advisable at necropsy to prepare impression smears from organs, especially the lung (Fig. 25-1). Concurrent infections with the erythrocytic parasites *Babesia canis* and *Haemobartonella canis* are sometimes seen. The organism has been propagated in canine monocyte cultures (Fig. 25-2), and fluorescent antibody staining of blood and impression smears has been employed as a diagnostic aid (13).

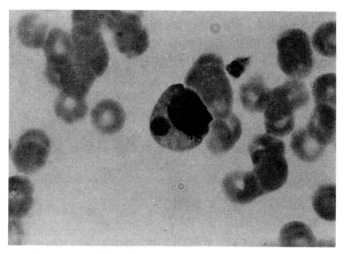

Figure 25-1. Lymphocyte containing inclusion of *Ehrlichia canis* in cytoplasm. × 1800. From a dog twelve days after inoculation. (Courtesy of Dr. K. A. Gossett.)

Ehrlichia bovis

This parasite of cattle was discovered in North Africa. It is transmitted by ticks of the genus *Hyalomma* and produces a mild febrile disease of cattle. Organisms are massed in monocytic cells and neutrophiles.

Ehrlichia ovina

This organism has been described as the cause of an infection of sheep in Algeria. It is found in the cytoplasm of monocytic cells, and the vector is considered to be *Rhipicephalus bursa*. Ehrlichiae infections were observed also in horses, and the agent is referred to as *E. equi*.

Neorickettsia helminthoeca and the Elokomin Fluke Fever (EFF) Agent

PATHOGENICITY. This organism is the cause of the salmon poisoning disease complex of dogs. It occurs when dogs consume raw salmon and trout that carry the rickettsia-infected metacercariae of the intestinal fluke *Nanophyetus salmincola*.

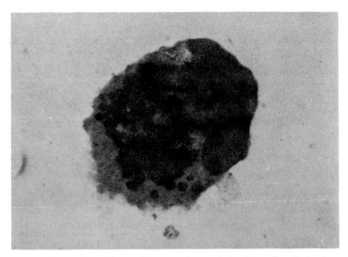

Figure 25-2. Cultured canine peritoneal macrophage eight days after infection with *Ehrlichia canis*. × 1000. (Courtesy of Dr. E. H. Stephenson)

The incubation period is usually five to seven days. The disease, which is characterized by high fever and severe diarrhea, can terminate fatally in a week to ten days.

PARASITOLOGICAL AND CYTOLOGICAL EXAMINATION. Laboratory diagnosis is based on demonstration of the characteristic fluke eggs in the feces of affected dogs. The rickettsiae are found in cells of smears of fluid aspirated from mandibular lymph nodes. Various lymph nodes may contain white foci of necrosis, and smears from these disclose small (about 0.3 µm), coccoid or coccobacillary intracytoplasmic bodies (6, 17). They stain purple with Giemsa and are red or blue after staining with the methods of Gimenez or Macchiavello (4).

The elementary bodies of a similar rickettsia, the EFF agent, are also seen in smears of lymph nodes and resemble *N. helminthoeca*. This organism, along with *N. helminthoeca*, occurs in the canine disease, but the EFF agent alone can infect dogs, foxes, coyotes, racoons, and ferrets.

SEROLOGICAL PROCEDURES. Complement fixation, serum neutralization, and animal protection tests have been used for diagnosis on an experimental basis.

Coxiella burnetii

PATHOGENICITY. *C. burnetii* causes mild or inapparent enzootic infections in cattle, sheep, goats, rats, and other wild and domesticated animals. The disease is of little practical importance in animals, but it is important because humans may be infected from animals. Sources of human infection are cow's milk and infectious dust derived from organisms shed from animals.

Large numbers of organisms may be shed with the placenta and in fetal fluids. Abortions in sheep are also associated with this infection. The disease caused by *C. burnetii* is called Q fever. In humans, it is an acute systemic infection usually accompanied by interstitial pneumonia (18).

LABORATORY DIAGNOSIS. Workers are referred to the manual of procedures of the American Public Health Association for detailed diagnostic methods.

C. burnetii can be readily isolated and cultivated in the yolk sac of five- to seven-day-old chicken embryos. Heparinized blood is recommended for inoculation. Embryos usually die in four to five days, and organisms are demonstrable in yolk-sac smears stained by Gimenez or Macchiavello's staining procedures.

Guinea pigs and hamsters are inoculated intraperitoneally for isolation from clinical materials that may contain bacterial contaminants.

Family Hemobartonellaceae

PATHOGENICITY. *Hemobartonella felis* causes an acute or chronic disease of cats. The causative organisms are found on the surfaces of erythrocytes and occasionally free in the plasma. The number of red blood cells affected varies with the severity of the infection. Infected cats may form antibodies to their own erythrocytes, resulting in hemolytic anemia. The modes of natural transmission have not been established. A significant proportion of the cat population may have clinically inapparent infection (19).

CYTOLOGICAL EXAMINATION. The organisms are found in various numbers on the surface of erythrocytes of peripheral blood or bone marrow and occasionally free in the plasma. *H. felis* consists of small, coccoid, rodlike or ring-shaped organisms with diameters of 0.2–1.0 μ for coccoid forms and a length of 3.0 μ for the rod forms.

Family Anaplasmataceae

PATHOGENICITY. This family includes *Anaplasma marginale*, which parasitizes erythrocytes of cattle or related ruminants and induces anemia. The host range of this organism is limited to ruminants and does not include laboratory animals. The relatively nonvirulent *A. centrale* occurs in Africa with *A. marginale* and can be differentiated by the more central location in the erythrocytes. Numerous species of ticks (*Boophilus, Rhipicephalus, Dermatocenter, Hyalomma,* and *Ixodes*) transmit the infection experimentally (11).

DETECTION. The organism can be found as marginal bodies of erythrocytes of infected cattle in Giemsa-stained blood smears. Up to 50 percent of the erythrocytes may be parasitized. Prolonged cases with extensive red blood cell destruction may have few anaplasma bodies in the blood. Complement fixation, direct and indirect immunofluorescence, and capillary tube agglutination are used to detect subclinical forms of anaplasmosis.

ORDER CHLAMYDIALES

A monogeneric family, Chlamydiaceae, is described; two species are recognized in the genus *Chlamydia. Chlamydia trachomatis* and *C. psittaci* are differentiated from each other by stable characteristics. Chlamydiae associated with trachoma, inclusion conjunctivitis, lymphogranuloma venereum, and other genital tract infections of man belong to the species *C. trachomatis*, which includes three biotypes. Chlamydial agents that cause infections in birds and various domestic and wild mammals belong to the species *C. psittaci*, which comprises a large collection of chlamydiae that differ antigenically and culturally (14, 15).

Chlamydiae are obligate intracellular parasites that multiply in the cytoplasm of animal cells and form membrane-bound cytoplasmic inclusions (Figs. 25-3, 25-4). During multiplication, they go through a complicated developmental cycle. The morphology of the organism changes sequentially from small, infectious elementary bodies to larger, reticulate bodies that reach a diameter of 1000 nm, divide by binary fission, and reorganize into highly stable, infectious elementary bodies with a diameter of 200–250 nm. The reticulate bodies are extremely fragile extracellularly and are not infectious. All chlamydial agents share group- or genus-specific antigens that are heat stable and consist of a lipoprotein-carbohydrate complex. These organisms cannot generate high-energy compounds (15). Chlamydiae are sensitive to antibiotics such as chloramphenicol, tetracycline, cycloserine, and penicillin. Chlamydial strains of the species *C. trachomatis* are inhibited by sulfonamides.

PATHOGENICITY. Chlamydiae of the species *C. psittaci* are infectious in a wide range of animals, as well as humans, through heterologous chains of transmission. Epithelial cells of mucous membranes are infected, but infections may generalize after penetration of mucous membranes and endothelial and other cells become infected. Chlamydial infections of animals may lead to the following disease syndromes: intestinal infection and diarrhea, pneumonia, placental infections and abortion, genital infections, mastitis, polyarthritis-polyserositis, encephalitis, and conjunctivitis. Avian chlamydial infections may lead to pneumonia and airsacculitis, pericarditis, conjunctivitis, encephalitis, as well as intestinal infections and diarrhea. Clinically inapparent intestinal infections usually lead to prolonged fecal shedding of chlamydiae (14).

Intestinal Infection

The frequently overlooked intestinal chlamydial infections may cause primary diarrhea in young animals, initiate pathogenetic events in other

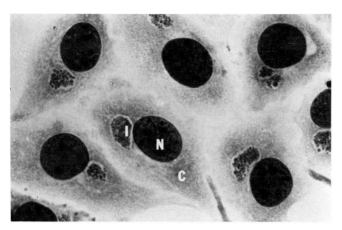

Figure 25-3. Mouse L cells infected with a bovine chlamydial strain. Notice early inclusion (I) in cytoplasm (C) of L cell. Nucleus (N) is prominent. × 900, Giemsa stain.

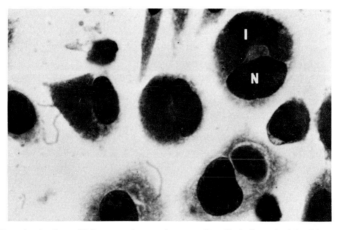

Figure 25-4. Late inclusions (I) in cytoplasm of mouse L cells infected with chlamydial strain of feline pneumonitis. N = nucleus. × 900, Giemsa stain.

chlamydial diseases, and play a role in perpetuating and spreading this infection. Chlamydiae were isolated from feces or diarrhea fluid of calves, sheep, goats, swine, and dogs, as well as domestic poultry and many bird species (14).

DETECTION. Isolation of chlamydiae in cell cultures or developing chicken embryos is required. The isolation methods are described in Appendix E.

Respiratory Infections and Pneumonia

Chlamydial pneumonia is found worldwide in laboratory mice, cats, calves, foals, lambs, goats, swine, and rabbits. This infection is expressed as rhinitis and conjunctivitis, particularly in cats; tracheitis; and interstitial broncho-pneumonia (14).

DETECTION. Isolation of chlamydiae from nasal secretions or specimens from affected lungs in cell cultures or developing chicken embryos or by intranasal inoculation of three-week-old mice free of chlamydial infections should be attempted. Evaluation of Giemsa-stained exfoliative cytological preparations of scraping from the conjunctivae or respiratory tissues of affected animals gives reliable and fast diagnostic results. Demonstration of circulating or secretory antibodies in the indirect microimmunofluorescent (IMIF) or the complement fixation (CF) tests gives evidence of this infection.

Placental and Fetal Chlamydial Infections and Abortions

Placental and fetal infections with chlamydiae, with ensuing abortion or birth of weak offspring, are now recognized as a significant cause of reproductive failure in sheep, goats, cattle, and other domestic animals. During a chlamydemic phase in pregnant subjects, the placental junction is breached, and infection is established in chorionic epithelial cells, which develop large cytoplasmic inclusions. Local spread involves cotyledons and pericotyledonary tissue with focal necrosis. Fetuses become also infected (14).

DETECTION. Exfoliative cytological examination of affected placental tissue after Gimenez staining is a reliable tool to detect chlamydial elementary bodies singly or in clusters (Fig. 25-5). Fluorescent antibody (FA) techniques make this approach even more specific. Enzyme immune assays (EIA) to detect chlamydial antigens should be employed for objective spectrophotometric evaluation. Isolation of chlamydiae from placental and fetal tissues by cell culture or chicken embryo techniques was accomplished by many workers, but this approach is time-consuming and not sensitive enough, because chlamydial infectivity in specimens from aborted placentas and fetuses may be inactivated by the time samples reach the laboratory. An antibody rise following abortion links this infection causatively. The CF and IMIF tests, as well as EIA, are reliable for detecting antibodies (14).

Urogenital Infection and Seminal Transmission

Chlamydial agents were isolated from semen and genital tissues of rams and bulls, as well as guinea pig boars affected with breeding disorders, in different parts of the world. Poor semen quality, orchitis, and seminal vesiculitis are associated with this infection.

DETECTION. Isolation of chlamydiae from semen or genital tissues is currently required to diagnose this infection. The presence of secretory antibodies has not been evaluated sufficiently for diagnostic purposes (14).

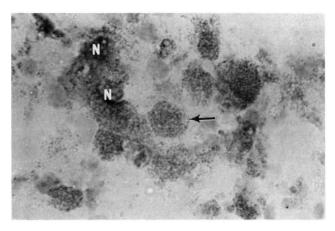

Figure 25-5. Cryostat section of a chlamydia-infected cotyledon from a goat. The arrow points towards one of several large clusters of elementary bodies. Nucleus (N) of infected cells with disintegrating cytoplasm. X 1800, Gimenez stain.

Polyarthritis-Polyserositis and Sporadic Bovine Encephalitis

Some chlamydiae induce systemic infection and have a propensity for synovial tissue of lambs, calves and young cattle, swine, and foals. This infection leads then to polyarthritis of lambs and to polyarthritis with fibrinous serositis and pericarditis in calves. Young cattle with this infection may develop encephalitis in sporadic cases. The chlamydial strains causing this type of infection in cattle and sheep differ biologically and antigenically from chlamydiae associated with abortions and are distinguished as immuno-type 2.

DETECTION. Isolation of chlamydiae from affected joints is required for a specific diagnosis and differentiation from other arthropathogenic infection. Exfoliative cytologic examination of synovial fluids and affected serous membrane also reveal this infection through the presence of elementary bodies. The serological tests mentioned detect rising antibody responses to this infection during the acute phase of the disease (14).

Conjunctivitis and Ocular Disease

Conjunctivitis and keratoconjunctivitis are caused by strains of *C. psittaci* in cats, lambs, calves, piglets, and guinea pigs. Ocular involvement has long been recognized as a clinical sign of chlamydiosis in pigeons, ducks, and geese. Chlamydiae infect conjunctival cells, leading to exfoliation, hyperemia, conjunctival inflammation with pannus formation, and keratitis as well as mucopurulent ocular discharge (14).

DETECTION. Exfoliative cytological examination of conjunctival scrapings from subjects with conjunctivitis gives a clear indication of this infection if conjunctival and monocytic cells contain chlamydial inclusions after Giemsa or FA staining (Fig. 25-6). Elementary bodies can be detected in the background of such smears. Chlamydiae can be isolated by the methods described from conjunctival scrapings that contain follicular contents. Secretory antibodies in lacrimal fluids detected by the IMIF test have diagnostic significance in inclusion conjunctivitis of guinea pigs.

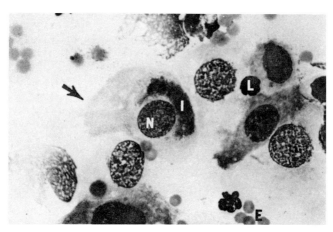

Figure 25-6. Exfoliative cytological preparation of conjunctiva from chlamydia-infected lamb. The arrow points towards an infected conjunctival cell with late chlamydial inclusion (I) and nucleus (N). L = leukocyte, E = erythrocyte. × 690, Giemsa stain.

Avian Chlamydiosis

Avian chlamydiosis is an infectious disease of domesticated and wild birds caused by *C. psittaci*. The disease induced by *C. psittaci* in human beings and psittacine birds was previously called *psittacosis* or "parrot fever," whereas the infection involving nonpsittacine birds and humans was called *ornithosis*. Birds are the most frequent source of human infection, but *C. psittaci* infection can also be transmitted to humans by cats and other mammals. *C. psittaci* is often harbored in the intestinal tracts, spleen, and kidneys of birds appearing clinically normal; but these birds shed the agent in the feces over long time periods. Avian chlamydiosis is a systemic chlamydial infection leading to pneumonia and airsacculitis, pericarditis, enteritis with diarrhea, and conjunctivitis (7, 14).

The disease in humans begins with inhalation of infectious dust and leads to pharyngitis and pneumonia with chlamydemia and systemic involvement. Coughing, headache, fever, and malaise are prominent clinical signs.

SPECIAL CAUTION. Chlamydia-infected live and dead birds shed large quantities of chlamydiae and may generate dangerous aerosols through infectious dust. Great care should be exercised in handling birds suspected of having chlamydiosis. Dead birds should be dipped in effective disinfectant before necropsies are done (7).

DETECTION. Exfoliative cytological examination of impression smears of affected serosal surfaces, exudates of body cavities, or lung, liver, and spleen reveals chlamydial elementary bodies free or within cytoplasmic inclusions of infected cells. Isolation of the agent can be accomplished by the methods described. The CF test can be used for most avian sera, but the indirect CF test is used for sera from chickens and turkeys to detect chlamydial antibodies in infected flocks (7).

Exfoliative Cytological and Histological Techniques

Exfoliative cytology can be a powerful tool for diagnosing chlamydial infections of birds with airsacculitis and pericarditis and of mammals with conjunctivitis, polyarthritis, placentitis, pneumonia, or peritonitis (14). Samples of conjunctiva are collected by scraping the affected surface with an edged tool to collect conjunctival and inflammatory cells. Synovial or peritoneal fluids are applied directly onto slides, or the cells in these fluids are harvested, washed, and applied to slides. These cytological preparations may be fixed with methanol, but cells are best preserved by Bouin or Zenker fixing solution. Specimens of this type are stained by the Giemsa or Gimenez method. Cytoplasmic chlamydial inclusions in infected cells are the diagnostic indicators (Fig. 25-6). The types of inflammatory and other cells present and the occurrence of single or connected groups of conjunctival cells are also diagnostic considerations. Chlamydial inclusions must not be confused with pigment granules or artifacts. These staining methods also reveal chlamydial infections in cultured cells by differentiating between inclusions and other cytological features. Chlamydial elementary bodies released from infected cells are not markedly differentiated by the Giemsa stain.

The Gimenez (4) or the Machiavello procedure is the staining method for detecting chlamydial elementary bodies in yolk-sac impression smears, in touch preparations of peritoneal surfaces, spleen, and liver of infected birds and mice, and in exfoliative ctyological analysis of chlamydia-infected placentas (Fig. 25-5). Elementary bodies, single or in aggregates, stand out as bright red dots. Microscopic demonstration of elementary bodies in impression smears of infected placentas provides firm proof of infection. Bacteria having diverse properties are also stained and readily seen. Cryostat sections of placental tissues from the margins of lesions fixed with paraformaldehyde and Gimenez stained may give more information, since elementary bodies in heavily infected trophoblastic cells stand out as red chlamydial colonies.

Histological examination of specimens from intestinal, placental, lung, or joint infections may locate chlamydia-infected cells, but success of this technique depends on using sections with a thickness of 4 µ or less. Tissues embedded in plastic give best results. Both Giemsa and Gimenez staining reveal inclusions in infected cells of these tissue preparations. Washing sections in ammonia water before Gimenez staining enhances the contrasting red of the elementary bodies in inclusions.

Fluorescent antibody techniques bring immunological specificity to all histological and cytological diagnostic procedures. The indirect FA method is more flexible, once established, for a given system. The immunofluorescence preparation may be restained by Giemsa or Gimenez methods to relocate inclusions.

Isolation and Identification of Chlamydiae from Clinical Specimens

Cell culture methods for isolating chlamydiae from diagnostic samples are effective. The advantages of cell culture methods are the high sensitivity when enhancing methods are used, the short time period required until results can be evaluated—days instead of weeks—and the reduced danger of loss of the culture because of bacterial contamination (12). Chlamydia-infected cells can be detected even in cell cultures contaminated with bacteria.

The infectivity-enhancing method of choice is centrifugation of the inoculum onto the cell monolayer. This procedure should be carried out at 37°C. Cycloheximide (2 µg/ml) is useful for most strains of *C. psittaci* isolated from animals; however, strains associated with polyarthritis, encephalitis, and conjunctivitis of ruminants generate aberrant forms in the developmental cycle and produce fewer infectious particles in the presence of cycloheximide.

To trace infection of cultured cells in isolation procedures, the cells grown on cover slips are fixed in Bouin's fluid or methanol forty to sixty hours after inoculation, stained by the Giemsa or Gimenez method, and examined microscopically for chlamydial inclusions in the cytoplasm. Parallel cultures are kept for subpassages. If subpassages are made, as depicted in Figure 25-7, these cultures are treated with sound to disrupt infected cells to liberate chlamydial elementary bodies. At least three subpassages at two- to three-day intervals should be made before a sample is considered free of chlamydiae.

The chicken embryo technique for isolating chlamydiae from animals has been quite successful (10, 14). This method may remain useful for laboratories not equipped to use the more sophisticated cell culture methods. Sometimes, several blind passages requiring weeks have to be made before results can be assessed. The seven-day-old chicken embryos inoculated by the yolk-sac route must be candled daily. Embryos that die within three days are tested for sterility, then discarded; the average time of death of all dying chicken embryos is recorded. The pathological changes in yolk-sac membranes and chicken embryos are evaluated. Chlamydial infection is verified by the presence of elementary bodies in Gimenez-stained impressions of washed

ISOLATION PROCEDURE

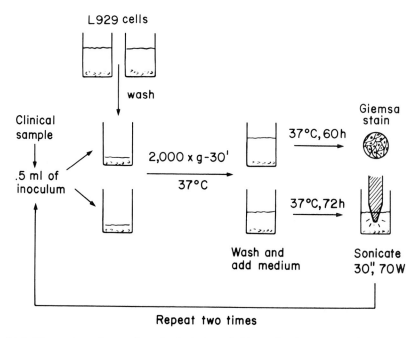

Figure 25-7. Illustration of procedure for isolation of chlamydiae by the use of cell cultures.

pieces of infected yolk-sac membranes. Boiled suspensions of chlamydia-infected yolk-sac membranes contain antigens that react with chlamydial antibodies in the CF test. In isolation procedures, chicken embryos should be tested for bacterial, mycoplasmal, or viral infections.

Some other indicator hosts used to cultivate rickettsial and chlamydial infections are summarized in Table 25-I. Intranasal or intraperitoneal inoculation of three-week-old mice yields chlamydial isolates from avian chlamydiosis. Chlamydiae from cattle, sheep, and goats usually do not readily multiply in mouse tissues during primary isolation steps. This host has an advantage for isolating chlamydiae from diarrhea or fecal specimens of birds because low levels of bacterial contamination are eliminated by the mouse. The naturally occurring pneumonic chlamydial infection of mice may confuse isolation results unless mice are free of this infection.

Similarly, guinea pigs may have naturally occurring chlamydial infections, but they have been used to isolate chlamydiae from fecal specimens of cattle and sheep. Elimination of contaminating bacteria after intraperitoneal inoculation of guinea pigs can be considered an advantage of this method; however, isolation of chlamydiae in cultured cells or chicken embryos after careful bacterial decontamination is better.

Table 25-I

Preferred Systems for the Isolation and Cultivation of Some
Rickettsial and Chlamydial Agents

Agent	Host System	Route of Inoculation
Rickettsia rickettsii	Cell cultures	Induce plaques
	Chicken embryo	Yolk sac
	Guinea pig	Intraperitoneal
Cowdria ruminantium	Mice	Intraperitoneal
Ehrlichia canis	Dog monocyte cultures	—
Neorickettsia helminthoeca	Dog monocyte cultures	—
Coxiella burnetii	Cell cultures	Induce plaques
Hemobartonella felis	Cats	Oral or parenteral
Anaplasma marginale	Rabbit bone marrow cells	—
	Deer	Parenteral
	Sheep	Parenteral
Chlamydia psittaci	Cell cultures (L cells)	—
	Chicken embryos	Yolk sac
	Mice	Intranasal, intraperitoneal
	Guinea pig	Intraperitoneal

An important step in isolating chlamydiae from heavily contaminated specimens such as feces, placental tissues, or semen is elimination of contaminating bacteria. Streptomycin (500 µg/ml), vancomycin (75 µg/ml), gentamicin (50 µg/ml), and mycostatin (500 units/ml) suppress bacterial contamination in cell cultures or chicken embryos used for chlamydial propagation. These antibiotics do not interfere with chlamydial multiplication. Differential centrifugation at 2000 G for thirty minutes at 4°C is most effective. The aim is to leave chlamydiae in suspension but to centrifuge out all larger, contaminating bacteria. Repeated cycles of differential centrifugation may be required with transfering only the top components of the supernatant fluid because of the motile bacteria in fecal specimens. This decontamination scheme is illustrated in Figure 25-8.

Serological Procedures to Detect Chlamydial Antibodies

The serological test most widely used to detect chlamydial antibodies in animal and human infections is the CF test. Active systemic chlamydial infections induce titers that have diagnostic significance if fourfold or higher antibody rises develop that are related to a disease episode. A highly antigenic strain should be selected as a substrate for antigen, which should be prepared, if possible, from purified chlamydial suspensions. Since guinea pigs have

PROCESSING OF SAMPLES

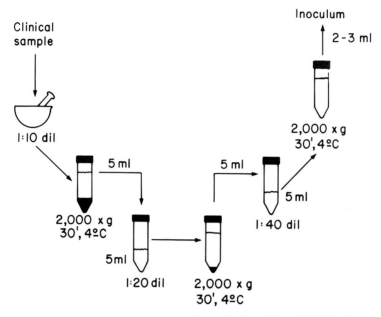

Figure 25-8. Processing of contaminated samples for chlamydial isolation.

naturally occurring chlamydial infections, guinea pig complement must be checked for freedom from chlamydial antibodies (14).

The indirect CF test or a modification of the direct test must be used with sera from turkeys and chickens because their sera do not fix guinea pig complement. Similarly, supplementation of guinea pig complement with fresh bovine serum free of chlamydial antibodies enhances the sensitivity of the CF test for bovine antibodies.

Another test system for group-specific chlamydial antibodies, which promises to gain importance, is the enzyme-linked immunosorbent assay (ELISA). It is more rapid and sensitive than the CF test and can be read objectively by automatic spectrophotometry. Antibodies of the various immunoglobulin classes can be detected simultaneously. A shortcoming is that separate antigammaglobulin conjugates are needed for every animal species.

The IMIF test has become a standard serological technique for detecting antibodies against *C. trachomatis* in human sera or secretions and is also used to type isolates of *C. trachomatis* (16). For this purpose type-specific antibodies are produced in mice by intravenous inoculation of test strains. Formalinized antigens can be stored for long periods without loss of type specificity and sensitivity. Unfortunately, this test has not been perfected to differentiate among members of the genus *C. psittaci*.

REFERENCES

1. Brezina, R., E. S. Murray, M. L. Tarrizzo, and K. Bögel. 1973. Rickettsiae and rickettsial diseases. *Bull WHO, 49*:433–442.

2. Bovarnick, M. R., J. C. Miller, and J. C. Snyder. 1950. The influence of certain salts, amino acids, sugars and proteins on the stability of rickettsiae. *J Bacteriol, 59*:509–522.

3. Cox, H. R. 1938. Use of yolk sac of developing chick embryo as the medium for growing rickettsiae of Rocky Mountain spotted fever and typhus groups. *Public Health Rep, 53*:2241–2247.

4. Giménez, D. F. 1964. Staining rickettsiae in yolk-sac cultures. *Stain Technol, 39*:135–140.

5. Keenan, K. P., W. C. Buhles, Jr., D. L. Husxoll, R. G. Williams, P. K. Hildebrandt, J. M. Campbell, and E. H. Stephenson. 1977. Pathogenesis of infection with *Rickettsia rickettsii* in the dog: A disease model for Rocky Mountain spotted fever. *J Infect Dis, 135*:911–917.

6. Moulder, J. W. 1974. Order I. Rickettsiales. In *Bergey's Manual of Determinative Bacteriology*, 8th ed. (R. E. Buchanan and N. E. Gibbons, editors), pp. 882–883. Baltimore, The Williams and Wilkins Company.

7. Page, L. A. 1978. Avian chlamydiosis (Ornithosis). In *Disease of Poultry*, 7th ed. (M. S. Hofstand, editor), pp. 337–366. Ames, Iowa State University Press.

8. Philip, C. B. 1974. Tribe II. Ehrlichieae. In *Bergey's Manual of Determinative Bacteriology*, 8th ed. (R. E. Buchanan and N. E. Gibbons, editors), pp. 893–897. Baltimore, The Williams and Wilkins Company.

9. Pienaar, T. G. 1970. Electronmicroscopy of *Cowdria ruminantium* in the endothelial cells of the vertebrate host. *Onderstepoort J Vet Res, 37*:67–78.

10. Rake, G., C. M. McKee, and M. F. Shaffer. 1940. Agent of lymphogranuloma venereum in the yolk sac of the developing chick embryo. *Proc Soc Exp Biol Med, 43*:332–335.

11. Ristic, M. 1980. Anaplasmosis. *Bovine Med Surg, 1*:324–348. (Am. Vet. Publications, Inc., Santa Barbara, Calif.)

12. Spears, P., and J. Storz. 1979. *Chlamydia psittaci*: Growth characteristics and enumeration of serotypes 1 and 2 in cultured cells. *J Infect Dis, 140*:959–967.

13. Stephenson, E. H., and J. V. Osterman. 1977. Canine peritoneal macrophages: Cultivation and infection with *Ehrlichia canis*. *Am J Vet Res, 38*:1815–1819.

14. Storz, J. 1971. *Chlamydia and Chlamydia Induced Diseases*. Springfield, Illinois, Charles C Thomas, Publ, 358 pp.

15. Storz, J., and P. Spears. 1977. Chlamydiales: Properties, developmental cycle and effect on eukaryotic host cells. *Curr Top Microbiol Immunol, 76*:165–212.

16. Wang, S. P. 1971. A micro-immunofluorescence method. Study of antibody response to TRIC organisms in mice. In *Trachoma and Related Disorders Caused by Chlamydial Agents* (R. L. Nichols, editor), pp. 273–288. Amsterdam, Excerpta Medica.

17. Weiss, E., and J. W. Moulder. 1974. Genus I. Rickettsia. In *Bergey's Manual of Determinative Bacteriology*, 8th ed. (R. E. Buchanan and N. E. Gibbons, editors), pp. 883–890. Baltimore, Maryland, The Williams and Wilkins Company.

18. Weiss, E., and J. W. Moulder. 1974. Genus III. Coxiella. In *Bergey's Manual of Determinative Bacteriology*, 8th ed. (R. E. Buchanan and N. E. Gibbons, editors), pp. 891–893. Baltimore, The Williams and Wilkins Company.

19. Weinman, D. 1974. Family II. Bartonellaceae. In *Bergey's Manual of Determinative Bacteriology*, 8th ed. (R. E. Buchanan and N. E. Gibbons, editors), pp. 903–906. Baltimore, The Williams and Wilkins Company.

20. Wisseman, C. L., E. A. Edlinger, A. D. Waddell, and M. R. Jones. 1976. Infection cycle of *Rickettsia rickettsii* in chicken embryo and L-929 cells in culture. *Infect Immun, 14*(8):1052–1064.

Chapter 26

AVIAN MYCOPLASMAS®

HARRY W. YODER, JR.

AVIAN MYCOPLASMAS are primarily associated with respiratory diseases of chickens and turkeys, although synovitis is sometimes significant. They are the smallest free-living bacterial organisms, having no rigid cell wall but possessing a limiting membrane that is antigenic. Their reproductive cycle is complex but variable, frequently with elementary bodies often arising from a budding-type process. The mycoplasmas are fastidious organisms that require a protein-base medium enriched with various sterols.

Freundt (1) published the characteristics of the most important avian mycoplasmas in the Eighth Edition of *Bergey's Manual of Determinative Bacteriology*. The family Mycoplasmataceae was placed under the order Mycoplasmatales. It contains the genus *Mycoplasma*, which includes numerous species of significance in animals and man. The family Acholeplasmataceae contains the common saprophyte *Acholeplasma laidlawii*, which does not require added sterols for growth and readily multiplies at room temperature.

The possible relationship of avian *Mycoplasma* to so-called L phase organisms of various bacteria is still controversial. However, most workers consider *Mycoplasma* to be true bacteria, with some possible L forms present in media due to the inhibition of other bacterial vegetative forms by the common addition of bacterial inhibitors (penicillin and thallous acetate) to *Mycoplasma* media.

During recent years approximately twenty serotypes of *Mycoplasma* have been characterized from avian sources (2, 3, 4, 5, 6, 7). Those serotypes representing significant pathogens have been given genus and species designations, as have a few of lesser importance. Thus, it has become necessary to discuss the entire group under the broad designation of avian mycoplasmas, with more complete treatises presented for only the three most significant members: *Mycoplasma gallisepticum, M. meleagridis*, and *M. synoviae*.

A summary concerning the major aspects of *Mycoplasma* associated with poultry is presented in Table 26-I.

*Reprinted in part from *The Laboratory Diagnosis of Mycoplasmosis in Food Animals*, with permission of the American Association of Veterinary Laboratory Diagnosticians.

TABLE 26-I
CHARACTERISTICS OF AVIAN MYCOPLASMAS

Serotype	Genus & Species	HA	Ferments Glucose	Requires Serum	Requires NAD	Arginine Decarboxylation	Pathogenicity Chickens	Pathogenicity Turkeys
A	Mycoplasma gallisepticum	+	+	+	−	−	+	+
B	M. gallinarum	−	−	+	−	+	−	−
CO	(M. pullorum)	−	+	+	−	−	−	−
DP	(M. gallinaceum)	−	+	+	−	−	−	−
EG	M. iners	−	−	+	−	+	−	−
F	(M. gallopavonis)	−	+	+	−	−	−	−
H	M. meleagridis	−*	−	+	−	+	−	+
IJKNQR	(M. iowae)	+	+	+	−	+	−	+
L	(M. columbinasale)	−	−	+	−	+	−	−
S	M. synoviae	+	+	+	+	−	+	+
	Acholeplasma laidlawii	−	+	−	−	−	−	−
	M. anatis	−	+	+	−	−	(+ducks)	−

*Most isolates are HA negative, but may be HA positive.

(Names proposed by the International Organization for Mycoplasmology at their Conference September 10-11, 1980, Custer, S.D.)

Original isolates of serotype K seem to be more closely related to serotype F than I (during recent studies).

MYCOPLASMA GALLISEPTICUM

Pathogenicity

Mycoplasma gallisepticum infection is commonly designated as chronic respiratory disease (CRD) of chickens and infectious sinusitis of turkeys. It has (rarely) caused infection in pheasants and quail. It is characterized by respiratory rales, coughing, nasal discharge, and frequently in turkeys by a sinusitis. The clinical manifestations are usually slow to develop, and the disease has a long course. Air sac disease designates a severe airsacculitis that is the result of *M. gallisepticum* infection complicated by some respiratory virus infection and also usually *Escherichia coli*.

M. gallisepticum can be spread by direct contact, indirect contact with contaminated objects, apparently by air transmission to a limited extent, and via egg transmission. Voluntary control of *M. gallisepticum* in chickens and turkeys is a common procedure throughout most of the world based on selection of serologically negative breeder flocks to avoid egg transmission of the infection to progeny flocks.

Isolation Procedures

Mycoplasmas are fastidious organisms that require a protein-base medium enriched with 10–15% serum or serum factors. No single medium formulation has been accepted as suitable for the various avian mycoplasmas, and mediums for antigen production are frequently further modified. See "Mediums for Growth" in Appendix F.

Since *Mycoplasma* infections tend to be respiratory and involve most birds in a flock, culturing five or ten tracheal swabs is often sufficient. If airsacculitis is present, culturing air sac swabs may be done directly into media, or from a suspension prepared with a mortar and pestle using a small amount of nutrient broth. Sinus exudate may usually be aspirated with a syringe and needle.

Once specimens have been collected and the broth medium inoculated, the tubes should be incubated at 37°C for at least fourteen days before they are discarded as negative. Cultures should be transferred daily whenever evidence of growth (increased turbidity) or change in pH is observed. Always transfer a generous amount with a pipette, such as 1 ml into fresh tubes of 5–10 ml of medium. If there is no evidence of growth by day 4 or 5 of incubation, one drop from a pipette should be transferred to a plate of mycoplasma agar medium streaked with a soft wire loop. Several cultures can be streaked on a single plate. Sometimes agar medium can be streaked directly with swab specimens, but initial broth passage tends to give better results and fewer cultures are lost because of gross bacterial contamination.

Inoculated agar plates should be incubated for three to five days at 37°C in a moist container. Maintenance of a moist atmosphere is very important. The addition of 5% carbon dioxide or the use of a candle jar is sometimes suggested. Some isolates appear to prefer anaerobic conditions for isolation and a few initial passages, but this is rarely provided.

Cultural Characteristics

Colonies are best observed with the aid of approximately ×20 to ×50 magnification through a dissecting microscope employing oblique indirect lighting. Typical colonies are 0.1–1 mm in diameter, with a central, elevated, more dense portion (note colony types as presented in Fig. 26-1). These elevated centers may not be present in initial cultures but tend to be more obvious after two or three passages in media. Colonies may be transferred with a wire loop, but a flame-sterilized scalpel blade is more satisfactory. A small block of agar is most readily excised and transferred into broth or inverted and spread over the surface of a new plate of agar medium. This procedure may be repeated three or four times to select reasonably pure colony cultures. However, there may be *Mycoplasma* cells on the agar surface adjacent to single colonies that are also picked up with the excised agar block.

Observation of coccoid bodies approximately 0.250 μm in diameter in Giemsa-stained colonies or broth culture sediment is suggestive of *Mycoplasma* but rarely proves much.

Cultural and biochemical characteristics of selected avian *Mycoplasma* species are presented in Table 26-I. However, such information is rarely sufficient for serotyping purposes.

Identification

Since avian *Mycoplasma* are classified by serological procedures, no other method actually is a valid substitute. Cultures with somewhat typical *Mycoplasma* colonies that fail to grow on most ordinary bacteriological media can most rapidly be identified by the fluorescent antibody (FA) test applied directly to colonies on an agar surface. High titer FA conjugate prepared from *M. gallisepticum* antiserum produced in rabbits is used (8).

Preparation of agglutinating or hemagglutinating antigens for use against homologous and heterologous prepared antisera is relatively slow and expensive. Isolates may be typed by the agar gel precipitation (AGP) system employing centrifuged broth culture sediment as antigen after it is frozen and thawed ten times (9). AGP typing serum can be prepared by inoculating type cultures into the footpads of young chickens and waiting five or six weeks to collect the serum.

Serological tests to identify antibodies against *M. gallisepticum* in infected poultry may be used as laboratory diagnostic aids.

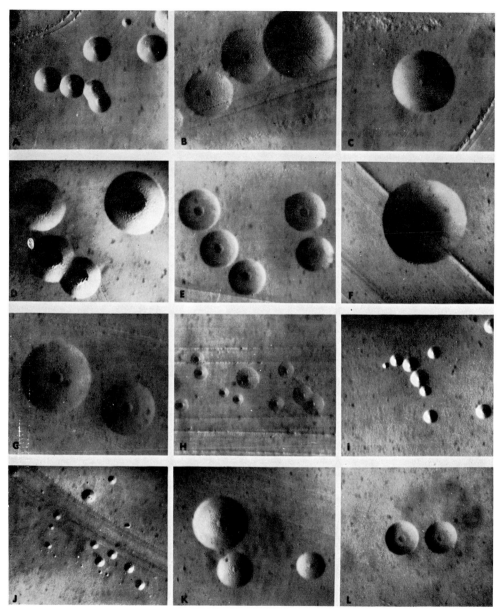

Figure 26-1. Typical colonies of avian mycoplasmas representing stereotypes A through L as indicated. Approximately forty-eight hours incubation (37°C) on turkey serum agar with 5% yeast autolysate, ×28. From H. W. Yoder and M. S. Hofstad, *Avian Dis, 8*:481, 1964.

The rapid serum plate test employs stained antigens prepared specifically for *M. gallisepticum, M. synoviae,* or *M. meleagridis* separately. Approximately 0.02 ml of serum is mixed with 0.03 ml of antigen on a glass plate. Rotate the plate for five seconds, then read the results at one and two minutes. Be sure to have the antigen, serum samples, and plate at room temperature. Several obvious reactors or strong suspicious samples indicate possible reactions that should be confirmed by the hemagglutination inhibition (HI) test. HI titers of 1:80 or greater are considered positive, while titers of 1:40 are strongly suspicious when conducted in well-controlled series employing known positive and negative samples. The tests should be run with 2 HI units for turkey sera and 4 HI units for chicken sera. Some judgment is often needed, and repeat testing in two or three weeks is frequently suggested before a firm status is indicated.

The tube agglutination test is similar to the plate test in being a simple screening procedure for flocks. Serum dilutions of 1:12.5, 1:25, and 1:50 are frequently tested, and reactions at 1:25 or greater are considered positive. They should be confirmed by the HI test.

Experimental Animals

Inoculation of avian tissues or exudates containing *M. gallisepticum* into seven-day-old embryonated chicken eggs via the yolk sac route usually results in embryo deaths within five to seven days. However, one or more yolk passages may be necessary before typical deaths and lesions occur. Dwarfing, generalized edema, liver necrosis, and enlarged spleens are most typical. The frequent presence of other bacterial contaminants makes embryo inoculation studies far less productive than broth culture studies employing high levels of penicillin and thallous acetate.

In general, turkeys appear to be more susceptible than chickens—at least turkeys develop more severe sinusitis, airsacculitis, and tendovaginitis following inoculation of those sites. However, the production of specific antibodies within three to five weeks is readily detected employing standard antigens to test sera of turkeys or chickens inoculated into the tendon sheath areas of the hock and footpad.

MYCOPLASMA SYNOVIAE

M. synoviae infection is commonly designated as infectious synovitis, an acute to chronic infectious disease of chickens and turkeys involving primarily the synovial membranes of joints and tendon sheaths, producing an exudative synovitis or tendonvaginitis. However, during recent years *M. synoviae* has less frequently been associated with synovitis but more frequently associated with airsacculitis in chickens and sometimes in turkeys.

This newer clinical manifestation closely resembles the air sac disease (airsacculitis) historically associated with *M. gallisepticum* infection complicated by some respiratory virus disease. It definitely seems to become more evident during the winter months, and like *M. gallisepticum*, it can be spread by direct contact, indirect contact with contaminated objects, apparently by air transmission to a limited extent, and via egg transmission. Voluntary control of *M. synoviae* in poultry is based on selection of serologically negative breeder flocks to avoid egg transmission of the infection to progeny flocks.

Isolation and Identification

M. synoviae grows best in Frey's medium (see Appendix F) supplemented with 10–15% normal swine serum. The swine serum should be heat inactivated at 56°C for thirty minutes. In addition, *M. synoviae* requires the addition of 0.1% reduced nicotinamide adenine dinucleotide (NAD) to broth and agar media. *M. synoviae* readily ferments glucose, as does *M. gallisepticum*; thus, evidence of growth in liquid medium can be noted by a change in the phenol red indicator from red to yellow as the medium becomes acid. Note other characteristics recorded in Table 26-I.

Colonies on agar surfaces are very similar to those of *M. gallisepticum*. However, they are readily identifiable by use of the fluorescent antibody (FA) system employing specific *M. synoviae* and *M. gallisepticum* conjugates. Broth culture sediment can be prepared as antigen for typing by the AGP system (9).

Broth cultures may be inoculated into the tendon sheath areas of the hock and footpad of young chickens or turkeys to produce specific antibodies, which can readily be identified by standard plate, HI, and AGP tests. This is a very practical method for culture typing if one can tolerate the three to five weeks required for adequate antibody titers to develop.

MYCOPLASMA MELEAGRIDIS

M. meleagridis infection is very common in turkeys throughout the world. It tends to localize in tracheas and cloacal areas, where it may live for many months. It produces airsacculitis in young turkeys and seems to be associated with decreased hatchability, certain skeletal abnormalities, and poor growth performance. *M. meleagridis* can be spread by direct and indirect contact, and apparently by air transmission to a limited extent. However, the major means of transmission is through infected eggs to progeny flocks. Considerable progress has been made to establish breeder flocks free of *M. meleagridis*.

Isolation and Identification

M. meleagridis is rather difficult to cultivate, since it grows rather slowly upon initial isolation and does not ferment glucose nor reduce most tetrazolium salts that might serve as growth indicators. It tends to prefer the surface of agar medium and frequently is aided by the presence of a small amount of broth medium added to the base of agar slants prepared from media best suited for the growth of *M. meleagridis* (note the media descriptions in Appendix F). The inclusion of 10–15% horse serum is especially important.

As noted in Table 26-I, *M. meleagridis* has few cultural characteristics to aid in its identification. Therefore, it is almost essential to employ the FA test procedure to identify cultures grown on agar surfaces.

Mycoplasma of the other serotypes are sometimes encountered during routine culture work, since almost all of them will grow in the usual media formulations. However, serotypes B, EG, and L do not ferment glucose. Thus, evidence of growth is not readily apparent. Subculture onto agar medium is necessary. Serotypes CO, DP, F, IJKNQR, and *M. anatis* as well as *Acholeplasma laidlawii* do ferment glucose. Serotypes representing IJKNQ and R are primarily isolated from turkeys, CO and DP from chickens, L only from pigeons, and *M. anatis* only from ducks; *A. laidlawii* is a true saprophyte that does not require added serum in the medium for growth and can multiply at room temperature. Pigeon isolates are generally nonpathogens.

Specific identification of most of these "other" serotypes is rarely conducted, since specific FA conjugates are mainly available only for *M. gallisepticum, M. synoviae,* and *M. meleagridis*. Serotype M is no longer described separately, since it apparently was identical to serotype B.

REFERENCES

1. Freundt, E. A.: The mycoplasmas. In Buchanan, R. E., and Gibbons, N. E. (Eds.): *Bergey's Manual of Determinative Bacteriology,* 8th ed. Baltimore, Williams & Wilkins, 1974, Part 19.
2. Yamamoto, R., and Adler, H. E.: *J Infect Dis, 102:*243–250, 1958.
3. Kleckner, A. L.: *Am J Vet Res, 21:*274–280, 1960.
4. Yoder, H. W., Jr., and Hofstad, M. S.: *Avian Dis, 8:*481–512, 1964.
5. Dierks, R. E., Newman, J. A., and Pomeroy, B. S.: *Ann NY Acad Sci, 143:*179–189, 1967.
6. Barber, T. L., and Fabricant, J.: *Appl Microbiol, 21:*600–605, 1971.
7. Barber, T. L., and Fabricant, J.: *Avian Dis, 15:*125–138, 1971.
8. Corstvet, R. E., and Sadler, W. W.: *Poult Sci, 43:*1280–1288, 1964.
9. Nonomura, I., and Yoder, H. W., Jr.: Identification of avian mycoplasma isolates by the agar-gel precipitin test. *Avian Dis, 21:*370–381, 1977.

SUPPLEMENTARY REFERENCES

Yoder, H. W., Jr.: Mycoplasmosis. In Hitchner, S. B., Domermuth, C. H., Purchase, H. G., and Williams, J. E. (Eds.): *Isolation and Identification of Avian Pathogens,* 2nd ed. College Station, Texas, Am Assoc Avian Pathologists, Texas A & M University, 1980, Chapter 13.
Yoder, H. W., Jr., Yamamoto, R., and Olson, N. O.: Avian Mycoplasmosis. In Hofstad, M. S., Calnek, B. W., Helmboldt, C. F., Reid, W. M., and Yoder, H. W., Jr.: *Diseases of Poultry,* 7th ed., Ames, Iowa St U Pr, 1978, Chapter 8.

Chapter 27

MYCOPLASMAS OF ANIMALS

OLE H. V. STALHEIM

THE MYCOPLASMAS (members of the class Mollicutes, order Mycoplasmatales) are the smallest free-living organisms. Unlike bacteria, they have no cell wall but are bounded by a membrane. This explains their remarkable pleomorphism. In stained smears, they are seen as ring forms, globules, small coccobacilli, or filaments. Although the "fried egg" colony is the hallmark, a high percentage grow with aberrant colonial morphology when first isolated or grow as tiny colonies (T mycoplasmas or ureaplasmas) visible only under the low power microscope (Fig. 27-1). Because most mycoplasmas require sterol for growth, suggestions were made that they be placed in the animal kingdom, but the current tendency is to place them into a sixth division of microbiology as distinct from bacteria, viruses, fungi, protozoa, and the blue-green algae. The non–sterol-requiring organisms are known as acholeplasmas, whereas the family Mycoplasmataceae contains five genera, including pathogens of humans, animals, plants, and insects, and parasitic or free-living mycoplasmas of such diverse environments as the intestine and rumen of the cow, hot springs, and the waters of abandoned coal mines.

The class Mollicutes (Table 27-I) contains more than forty pathogens of humans and animals including the causative agents of three respiratory diseases formerly thought to be due to viruses: primary atypical (virus) pneumonia of humans (*M. pneumoniae*), mycoplasmal (virus) pneumonia of swine (*M. hyopneumoniae*), and bovine contagious pleuropneumonia (*M. mycoides* subsp. *mycoides*). Mycoplasmas are mostly host-specific but *M. bovis*, for example, which causes a variety of lesions in cattle, has been isolated from the lungs of pneumonic sheep and from human patients with respiratory disease. The mycoplasmas of bovine origin were classified into eight groups by Leach (1). Most isolates from sick cattle can be identified by serologic procedures as members of these groups. Several other species have been recovered from the eyes or genital secretions of cattle but their pathogenicity has not been demonstrated.

The ureaplasmas (see Table 27-I) or T mycoplasmas are distinguished by their ability to split urea into ammonia and carbon dioxide. Those of animal

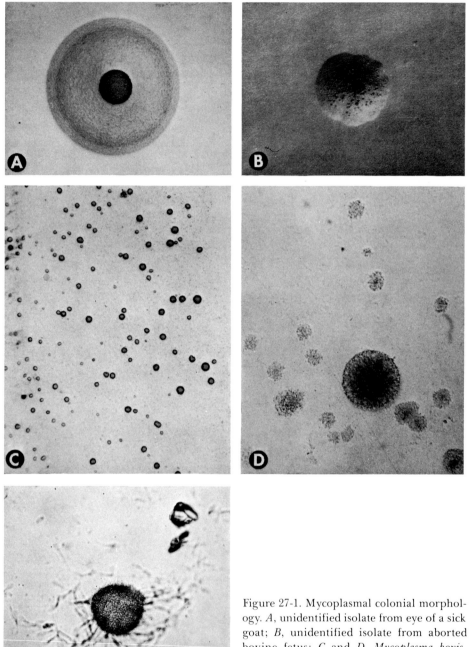

Figure 27-1. Mycoplasmal colonial morphology. *A*, unidentified isolate from eye of a sick goat; *B*, unidentified isolate from aborted bovine fetus; *C* and *D*, *Mycoplasma bovis*, showing small colonies without halos, and granular colonies, recovered from experimentally infected cows (×40); *E*, ureaplasma from bull semen (×200).

TABLE 27-I
TAXONOMY OF CLASS MOLLICUTES

Class: Mollicutes
  Order I: Mycoplasmatales
    Family I: Mycoplasmataceae
      Genus I: *Mycoplasma*
        Species: *M. pneumoniae* and numerous animal pathogens
        1. Sterol required for growth
        2. Sensitive to digitonin
        3. Genome size: 4.5×10^8 daltons
      Genus II: *Ureaplasma*
        Species: *U. urealyticum* of man and animals
        1. Sterol required for growth
        2. Sensitive to digitonin
        3. Genome size: 4.5×10^8 daltons
    Family II: Acholephasmataceae
      Genus I: *Acholeplasma*
        Species: *A. equifetale*
        1. Sterol not required for growth
        2. Insensitive to digitonin
        3. Genome size: 1.0×10^9 daltons
    Family III: Spiroplasmataceae
      Genus I: *Spiroplasma*
        1. Sterol required for growth
        2. Sensitive to digitonin
        3. Genome size: 1.0×10^9 daltons
    Genera of uncertain affiliation:
      Anaeroplasma
        1. Some strains require sterols; some do not
      Thermoplasma
        1. Sterol not required
        3. Genome size: 1.0×10^9 daltons

Reprinted with permission from *Laboratory Diagnosis of Mycoplasmosis in Food Animals.* Courtesy of the American Association of Veterinary Laboratory Diagnosticians, Columbia, Missouri.

origin appear to be antigenically distinct from those associated with nongonococcal urethritis of humans, and they have different nutritional requirements. The role of ureaplasmas in animal diseases appears to be expanding.

At the present time, the acholeplasmas (see Table 27-I) are not known to be pathogenic for animals. Spiroplasmas (motile, helical mycoplasmas) cause diseases of plants. Because they can cause cataracts in suckling mice and rats and microphthalmia in hamsters (2), they are considered potentially pathogenic for animals.

Strictly anaerobic mycoplasmas (anaeroplasmas) exist in the rumen of cattle and sheep (3). They are antigenically distinct from intestinal mycoplasmas (4).

Many established tissue culture cell lines are now known to be contaminated with mycoplasmas. They were ignored for a long time by laboratory workers since they seemed to have no effect on cell cultures. In recent years, it has become evident that contaminating mycoplasmas are capable of altering the activity of cells and their sensitivity to viruses and drugs. Primary cell cultures are much less apt to be contaminated.

CLINICAL MANIFESTATIONS

Clinical observations, laboratory studies, and the results of controlled research have demonstrated causal relationships for several mycoplasmas (Table 27-II). Pathogenic mycoplasmas have a predilection for serous surfaces, i.e. the thoracic, abdominal, and articular cavities of cattle and swine and the air sacs of poultry, where they localize and persist protected from antibody and therapeutic agents by the fibrinous tissue reactions that characterize mycoplasmosis. In swine, they cause pneumonia, arthritis, and serositis; in cattle, sheep, and goats, they cause mastitis (5), arthritis, and pneumonia. The significance of mycoplasmas in diseases of the genital tract is not clear. Under experimental conditions, some species (*M. bovis*) caused severe diseases (seminal vesiculitis of the bull, salpingitis abortion (6) of the cow, for example), whereas other species are considered of potential etiologic importance.

The clinical signs of mycoplasmosis are not distinctive. Mycoplasmal pneumonia cannot be diagnosed by observations of affected animals or by the appearance of lesions at necropsy, except perhaps for the characteristic sequestra of contagious bovine pleuropneumonia.

SWINE. The clinical signs of swine mycoplasmosis include pneumonia (caused by *M. hyopneumoniae*), polyserositis, and arthritis in young pigs (*M. hyorhinis*) and arthritis in older pigs associated with *M. hyosynoviae*. Mycoplasmal pneumonia of swine is a chronic disease with a high morbidity and a low mortality characterized by chronic, nonproductive coughing. With stress and secondary bacterial infections, infected pigs may develop clinical signs of pneumonia. They may die, recover, or be stunted. The economic losses from mortality, poor weight gains, and feed conversion are high (7).

CATTLE. The clinical signs of bovine mycoplasmosis include pneumonia, arthritis, and mastitis; mycoplasmas have also been associated with bovine abortion, conjunctivitis, salpingitis, seminal vesiculitis, and vaginitis. Mycoplasmas (*M. dispar, M. bovis*, and ureaplasmas) caused subclinical infections, and with other bacteria or viruses resulted in severe clinical disease (8).

Mycoplasmal mastitis should be suspected whenever there is an increase in cases of severe purulent mastitis involving more than one quarter (usually all 4) with tannish or brownish secretions and marked loss of milk production. The outbreak may be explosive or insidious; affected cows are not septic and continue to eat. The most common cause was *M. bovis*; treatment was not effective and infected cows shed mycoplasmas indefinitely (9). The same mycoplasma, *M. bovis*, has been recovered from cases of severe, intractable arthritis, often following outbreaks of mastitis or respiratory disease.

A granular vulvo-vaginitis syndrome in Ontario dairy cattle was associated with ureaplasmas. It was characterized by hyperemia, epithelial cysts, and a purulent vulvar discharge. The morbidity was highest in winter months and had a significant effect on fertility (10). Many cows are vaginal carriers of large numbers of ureaplasmas without any clinical signs. Clinically normal bulls may carry virulent strains of ureaplasmas, which are transmitted during coitus and induce the vulvitis syndrome.

SHEEP and GOATS. Mycoplasmas were first isolated from small ruminants in the Middle East with contagious agalactiae. The causal agent, *M. agalactiae*, was isolated and identified from an arthritic goat in California (11).

In 1969, a mycoplasma was isolated from the conjunctival sac of a goat in Connecticut. At first, it was identified as *M. mycoides* subsp. *capri*; further studies revealed it to be *M. mycoides* subsp. *mycoides* (hereinafter referred to as *M. mycoides*). Additional isolations have been made from goats in six states with pneumonia, arthritis, mastitis, and conjunctivitis, but most were from the milk of apparently normal or mastitic goats. In some flocks of goats, *M. mycoides* caused severe morbidity of young animals; in other flocks, the organism persisted in the udder and was shed in the milk but the goats were apparently healthy. Treatment with heavy and repeated doses of antibiotics failed to eradicate the infections. Still another mycoplasma, *M. putrefaciens*, also caused caprine mastitis (11). These organisms have not been isolated from sheep in this country (12). Other mycoplasmas were associated with disease outbreaks of both goats and sheep (Table 27-II). In Kenya, MacOwan and Minette recovered an antigenically distinct strain, F38, from the lungs of goats with contagious caprine pleuropneumonia. It has not been identified in the Americas.

RATS AND MICE. Chronic respiratory disease has been a problem in laboratory rats and mice for a long time (see review by Cassell and Hill). In addition to mortality losses, the presence of intercurrent disease undermines the validity of research performed with pneumonic animals. The causative agent of murine respiratory mycoplasmosis (MRM) is *M. pulmonis*. Rats and mice acquire MRM from their mothers at an early age. The infection is usually subclinical; it may persist for life. Mortality is cumulative over many months; mice are less severely affected than rats. Elimination of *M. pulmonis* from an infected colony of rats or mice is a very difficult undertaking. Spontaneous polyarthritis in laboratory rats is caused by *M. arthritidis*. The disease is usually enzootic with only a few clinical cases. An infection of *M. neurolyticum* in mice is manifested as conjunctivitis. Inoculations into mice

TABLE 27-II

MYCOPLASMAS AND ACHOLEPLASMAS
ASSOCIATED WITH DISEASES OF ANIMALS

Animal Species Affected	Disease	Organisms
Cattle	Pneumonia	*M. mycoides* var. *mycoides*
		M. bovis
		M. dispar
		Ureaplasmas
	Arthritis	*M. bovis*
		M. bovigenitalium
	Mastitis	*M. bovis*
		M. californicum
	Abortion	*M. bovis*
	Vaginitis	Ureaplasmas
		M. bovigenitalium
	Seminal vesiculitis	*M. bovis*
		M. bovigenitalium
	Uncertain	*M. bovirhinis*
		M. alkalescens
		M. arginini
		A. modicum
		A. laidlawii
		M. bovoculi
		M. verecundum
		M. canadense
		M. alvi
Swine	Pneumonia	*M. hyopneumoniae*
	Arthritis	*M. hyorhinis*
		M. hyosynoviae
	Uncertain	*M. flocculare*
		M. sualvi
		A. axanthum
		A. granularum
Sheep and Goats	Pneumonia	*M. ovipneumoniae*
	Pneumonia, arthritis, mastitis	*M. mycoides* subsp. *capri*
		M. mycoides subsp. *mycoides*
		M. agalactiae
		M. putrefaciens
	Conjunctivitis	*M. conjunctivae*
	Arthritis	*M. capricolum*

TABLE 27-II continued

MYCOPLASMAS AND ACHOLEPLASMAS
ASSOCIATED WITH DISEASES OF ANIMALS

Animal Species Affected	Disease	Organisms
	Uncertain	*M. arginini*
		M. oculi
Horses	Uncertain	*A. equigenitalium*
		M. equirhinis
		M. subdolum
		M. felis
		M. arginini
		M. salivarum
		A. equifetale
		A. hippikon
		A. laidlawii
Rats and Mice	Pneumonia	*M. pulmonis*
	Arthritis	*M. arthritidis*
	Rolling disease	*M. neurolyticum*
Guinea pigs	Uncertain	*M. caviae*
Dogs	Pneumonia	*M. cynos*
	Uncertain	*M. spumans*
		M. maculosum
		M. edwardii
		M. molare
		M. canis
		M. opalescens

of infected tissue, cultures of *M. neurolyticum*, or cell-free filtrates cause rolling disease.

DOGS AND CATS. Mycoplasmas were first isolated from dogs with distemper in 1934. They were recovered from the nose, throat, or vagina of 30 percent of apparently healthy dogs in Japan (13) and from the lung, liver, kidney, and spleen of dogs with respiratory and other diseases, but there is no evidence that mycoplasmas are responsible for any specific disease of dogs. The *Mycoplasma* species of canine origin are *M. spumans*, *M. maculosum*, *M. edwardii*, *M. cynos*, *M. molare*, and *M. canis*. The latter species was recovered from a woman with respiratory illness similar to an illness in her pet dog. The woman and her three children had close contact with the dog. All four persons and the dog harbored *M. canis* in their throats, and all had serologic reactions to this mycoplasma (14).

Of the mycoplasmas isolated from dogs (Table 27-II), only *M. cynos* was found consistently in the lungs of dogs with distemper. When the culture was given to puppies, it induced focal pneumonia. The organism persisted in the lungs as long as three weeks after inoculation; clinical signs were not manifested (15).

Cats harbor *M. gatae, M. arginini,* and *M. felis.* The latter species has been associated with outbreaks of acute conjunctivitis in cats (16), but attempts to induce the disease experimentally have failed or yielded equivocal results. Switzer (17) isolated a mycoplasma from the pneumonic lung of a kitten but could not induce the disease experimentally. Eighty-one strains of *M. arginini* were isolated from a series of 555 cats. When experimentally inoculated into young kittens, *M. arginini* colonized the inoculated sites but clinical disease was not induced (18).

Ureaplasmas were isolated from the throat or vagina of twenty-five of thirty-six apparently healthy cats (19). When given by the oral or intra-peritoneal route to normal cats, clinical signs were not induced. However, when given by the vaginal route to three pregnant cats, two gave birth to kittens that died in ten and twelve days, and the third cat aborted ten days postexposure. Ureaplasmas were recovered from the uterus or vagina of all three cats (20).

HORSES. The mycoplasmas most frequently isolated from horses (*M. equirhinis* and *M. felis*) were recovered from clinically normal horses as well as those with respiratory disease. By serologic procedures, specific antibod-ies were detected in serums from two of six horses that harbored *M. equirhinis* (21). The virulence of mycoplasmas for horses has not been demonstrated. Two unidentified mycoplasmas of equine origin cross-reacted with strains of *M. mycoides* of caprine origin (22).

LABORATORY PROCEDURES

The isolation and further cultivation for identification of most mycoplasmas from clinical materials is not difficult if a suitable, well-prepared medium is used. Satisfactory mediums are described in Appendix F. The basic ingredi-ent of each is an infusion of meat that is supplemented with yeast extract, serum, and perhaps deoxyribonucleic acid or nucleotides. The quality of water used in the medium is extremely important. It should equal that used for cell cultures. Some batches of phenol red have been shown to be toxic and have either delayed or inhibited the growth of mycoplasmas. Arginine is inhibitory for *M. bovoculi.* Thallium acetate at a fungistatic concentration is inhibitory to ureaplasma species. Penicillin G and benzyl penicillin may be inhibitory to the growth of *M. dispar.* When either a new bottle or batch of a component is prepared, it should be pretested to determine that it has no inhibitory effect.

Special Techniques

SWINE. The specific-pathogen-free (SPF) program of swine health was developed to eliminate mycoplasmal pneumonia of swine (MPS) (and other diseases); herds are monitored for continued freedom from MPS and are dropped from the program if it is recognized in the herd. The monitoring procedures include serology, clinical and slaughter inspections, and laboratory attempts to recover *M. hyopneumoniae*. Because the latter is expensive, laborious, and often unsuccessful, it is not routine at most laboratories. The Danish isolation method, which claimed success in virtually all attempts, was recently assessed in Indiana; *M. hyopneumoniae* was isolated from seventeen of twenty lungs with uncomplicated lesions of MPS. Samples of lung, including bronchii and advancing edges of atelectic areas, were processed at once (or frozen), homogenized, serially diluted in Friis' medium (to 10^{-4}), and incubated (37°C) for three weeks. One ml of the 10^{-2} dilution was added to 50 ml of medium and passed through a 0.20 μm cellulose filter. Each inoculated membrane was placed in growth medium until the pH changed, rinsed, dried, and examined for colonies of *M. hyopneumoniae* by the indirect immunofluorescence test. The procedure should be useful for the definitive laboratory diagnosis of MPS (23).

European swine harbored a relatively fastidious mycoplasma that grew slowly and formed small colonies (0.5 mm) without halos; it was named *M. flocculare* because of the tendency of some isolates to form small aggregates in liquid medium. Several isolations of *M. flocculare* were recently made from swine in the U.S. by supplementation of Friis' medium with cycloserine (0.5 mg per ml) and *M. hyorhinis* antiserum (5%) to inhibit the faster growing *M. hyorhinis* (24). Neither the incidence nor the role of *M. flocculare* in MPS is known.

RUMINANT ANIMALS. Special mediums are useful for the isolation of *M. dispar* (M-96 medium), *M. arginini* (add 1% L-arginine, adjust pH to 7.0), the ureaplasmas (modified Hayflick's), and the anaeroplasmas. When first isolated on agar medium, some mycoplasmas display aberrant colonial morphology (Fig. 27-1). Colonies of *M. dispar* and *M. ovipneumoniae* lack the central downward growth that contributes the "fried egg" appearance regarded as classic for mycoplasmas.

When samples of ovine lung were cultured for mycoplasmas, the most common strains grew slowly in broth, fermented glucose, formed small colonies without halos, and reacted with *M. ovipneumoniae* antiserum (25). Some laboratory diagnosticians reported that the most common mycoplasma isolated from dairy goats was *M. mycoides* (11). Australian workers compared strains of *M. mycoides* from cattle and goats. Strains of caprine origin grew to greater turbidity in broth and formed larger colonies in agar (diameters of

2.2 mm vs. 1.0 mm), digested casein, liquefied inspissated serum, and survived longer at 45°C. Most strains from goats were the large-colony (LC) type, whereas strains from cattle were the small-colony type (26). Strains of *M. mycoides* from goats in this country were the LC type, with one exception, but only a few strains have been examined.

Use and Source of Serum

Serum from several species of animals may be used as a supplement for the growth of mycoplasmas. These may be obtained from commercial suppliers or may be processed in the laboratory. The growth of mycoplasmas may be influenced by the species of animal used as the serum source and whether or not the serum was inactivated. Serum and serum fractions are generally utilized at concentrations of 10–20%, depending in part on the animal source of the serum and in part on the mycoplasma that may be in the clinical material. In general, 5–15% is used for maintenance mediums and up to 20% for isolation mediums. Horse serum is considered the best for general use because of its ready availability and economy. Pooled serum is probably best, but each lot should be pretested for the purpose intended (isolation or maintenance). Unsatisfactory serum should not be mixed with acceptable serum but should be discarded. Whether or not fresh serums should be inactivated by heat and/or acid treatment is a choice each laboratory diagnostician must make. For acid inactivation, add 1 N HCl until a pH value of 4.2 is attained. Refrigerate the serum overnight, and readjust the pH to 7.1 with 1 N NaOH. Clarify the inactivated serum by centrifugation or filtration. Inactivation may be attained by storage combined with freezing and thawing.

Human serum may be prepared from either human whole blood or human plasma as follows. Add one part of a 2.5% solution of calcium chloride to one part blood or plasma. Remove coagulated material by centrifugation, and adjust the pH to 7.2. Human serum is probably the best supplement for mycoplasmal growth medium. Swine serum may be excellent but varies markedly between animals and batches. The donor animals must be free of mycoplasmas.

Inhibitors

Two inhibitors are frequently added to mycoplasmal growth medium: Thallium acetate at a concentration of 1:2000 to 1:4000 is a fungal inhibitor, and crystalline penicillin G at 250 to 1000 units/ml is a bacterial inhibitor. Amphotericin B (5 µg/ml) is commonly used as a replacement for thallium acetate in ureaplasmal growth medium.

The pH of the medium may significantly alter the isolation of mycoplasmas. Mycoplasma of bovine origin grow well between pH 7.0 and 7.6; others are tolerant of a low pH. Ureaplasmas require a low pH for growth. Some mycoplasmas will survive longer at incubation temperature if the pH of the medium is 8.5.

Plastic tubes and Petri dishes have replaced glassware in many laboratories. These have been reported as the cause of poor mycoplasmal growth. To prevent such problems, quality control tests could be made when a new batch or lot is received. A simple comparative test between old and new plates or tubes is all that is required.

Quality Control of Medium and Components

Qualitative checks for the acceptability of a new component may be conducted by preparing two lots of medium, one with the old and the other with the new component. For solid medium, inoculate duplicate plates of each medium with a measured amount of inoculum of a titrated culture. The mediums are compared on the basis of rate of growth, colony size, and number of colonies that develop. A similar study should be carried out for liquid medium using a tenfold titration method. Duplicate titrations for each test medium should be used for such comparisons.

Pretitrated cultures for inoculum in quality control may be prepared as follows. Grow the test species in broth, and dispense 0.1 ml in each well of a disposable microtitre "U" plate. Seal the plate with cellulose tape, and freeze at $-70°C$ for fourteen to twenty-one days. Titrate the culture by cutting off two or three wells with sterile scissors. Aseptically remove the covering tape, and place each pellet in a tube containing 0.9 ml of either glucose or arginine broth. Make tenfold dilutions from the initial tube to 10^{-9}. Inoculate duplicate plates of solid medium from 10^{-4} to 10^{-8} dilution tubes. Incubate the broth and solid medium at $37.5°C$. After four to five days, count the colonies and determine the average colony count per milliliter. Incubate the broth medium and determine the most probable number of organisms per milliliter. The two figures obtained should be almost equal. If the pretitrated culture is maintained at $-70°C$, there will be little change from the original in the postfreezing titration result over a six-month period (27).

Isolation Techniques

To isolate mycoplasmas from sick animals, swabs containing exudates from the eye, nostril, trachea, and the genitalia are placed in tubes of mycoplasmal broth medium. The inoculated tubes should be chilled to $2-3°C$ and transported to the laboratory as quickly as possible. They should be processed on the day they are collected, if possible, but viability will be maintained for forty-eight to seventy-two hours at $2-3°C$. Beyond this time, storage should be at $-70°C$. After thorough mixing, 0.1 ml is delivered to agar plates and dilutions made into broth mediums.

Cervico-vaginal mucus is aspirated from the vaginal fornix with aseptic precautions, sealed in the collecting rod, and transported to the laboratory. The mucus (0.2 ml) is added to 1.8 ml of broth and mixed thoroughly; 0.1 ml of the mixture is used as the inoculum for agar medium and 0.2 ml for broth medium. Plating the mucus directly on solid medium may increase the recovery rate if only a few mycoplasmas are present.

Preputial material may be collected either on a swab inserted through a sterile tube to the preputial fornix or by flushing the preputial fornix with 10 ml sterile distilled water or broth medium. The fluid is centrifuged (9,000 G for 15 minutes) at 5°C in an anglehead rotor. The sediment is resuspended in 1 ml of broth and inoculated into both liquid and solid mediums.

Semen from bulls may be examined for mycoplasma either as whole fresh semen or after being processed by an artificial insemination unit. Whole semen (0.2 ml) is diluted with 1.8 ml of a medium without inhibitors, mixed thoroughly, and inoculated to solid and liquid mediums. Processed semen is added directly to solid medium (0.1 ml) and liquid medium (0.2 ml). Because processed semen contains microbial inhibitors, direct cultures are often negative in low dilutions but positive in higher dilutions. To certify that semen destined for interstate commerce is free of mycoplasmas, special procedures are being developed.

Fluid materials such as milk, body wastes, fetal fluids, synovial fluids, and fetal stomach contents are inoculated directly into liquid and solid mediums.

Tissues such as lung, liver, spleen, and kidney may be examined in two ways. (a) Cut a block of tissue and streak it directly across the surface of an agar plate. A sterile loop can be used to spread the inoculum. (b) A block of tissue can be suspended in 9 ml of broth medium and homogenized in a Ten Broeck tissue grinder. Both solid and liquid mediums are inoculated. The tissue suspensions should be serially diluted to 10^{-6} or 10^{-7}. Frequently, lower dilutions will be negative for mycoplasmal growth due to a carry-over of inhibitory agents in the tissue, while higher dilutions are positive (27).

For routine attempts to isolate mycoplasmas, inoculate the clinical materials into tubes of usually two liquid mediums (modified Hayflick's with inhibitors and Livingston's modification of Hayflick's medium for ureaplasmas) and on two plates of solidified Hayflick's medium. The addition of semisolid medium (Hayflick's plus 0.15% agar) was reported to increase the rate of isolations. If fastidious organisms such as *M. dispar* are to be isolated, special medium (Gourlay and Leach) must be included.

The inoculated plates are incubated in a humid atmosphere at 37.5°C; one plate should be incubated aerobically, the other in 5% carbon dioxide and 95% nitrogen. After forty-eight and ninety-six hours incubation, the plates should be examined using obliquely transmitted light and a stereoscopic microscope at 25–40 magnifications. If the growth is sparse (1 to 10 colonies/

plate), flood the plate with approximately 1.5 ml of sterile broth and incubate for an additional forty-eight to seventy-two hours or longer. If colonies are not detected after fourteen days, the plates may be discarded as negative. Tubes of liquid medium are checked daily. When slight turbidity is detected (or when the color changes), transfers are made to an appropriate solid medium. Some workers "blind-passage" from broth to broth at intervals of seven, fourteen, and twenty-one days.

Purification of Isolates

Before fresh isolates can be identified, they must be purified by the single colony technique. A plate with discrete colonies is selected for cloning. Individual colonies are removed by cutting out a small block of agar using a sterile scalpel. Transfer the colony to a tube containing 2–3 ml of broth and incubate it for forty-eight hours or longer. Draw all of the culture into a sterile syringe and express through a Swinney filter (0.45 μm pore size). Dilute the filtrate 1:10 and 1:100 in broth, spread 0.05 ml of the dilutions on plates of solid medium, and incubate. Repeat the colony selection and filtration procedure for at least three cycles. If morphologically distinct colonies are present, pick and clone representative colonies for each type. Because not all colonies grow, start with four or five individual colonies of each morphologic type.

Colony cloning of primary isolates may be performed without filtration. Cut and remove an agar block containing a single colony. Invert it on an agar plate, and push it back and forth across the surface of the agar. After incubation (forty-eight to seventy-two hours), pick an isolated colony and repeat the transfer a second time and again for a third time. A single colony picked from the third transfer is considered to be a cloned strain. The only serological test applicable to an uncloned isolate that can be reliably interpreted is the fluorescent antibody (FA) test (27).

Maintenance of Reference Cultures and Isolates

To identify isolates, type cultures and specific hyperimmune antiserums against the type cultures may be obtained from the American Type Culture Collection. Maintenance of reference or type cultures in a pure condition requires constant vigilance. The following suggestions may be useful. Have only one reference culture in use at any one time. If possible, work only in a biohazard or similar type of hood. Sterilize the hood before introducing the culture and container. Chemically sterilize the outside of the culture container. Make all transfers in a sterile manner with a pipette filler; do not put the pipette in your mouth. Transfers of cultures should be performed in a manner that minimizes the formation of aerosols. Because the pathogenicity of mycoplasmas of animal origin for laboratory personnel is unknown, all reasonable precautions should be taken to minimize self-exposure.

Reference cultures of mycoplasma may be stored in the lyophilized condition, or a simple method to preserve reference cultures for up to four years is as follows. Cut blocks from solid medium with almost confluent colonial growth and place in a small vial (two per vial). Add approximately 0.2 ml of liquid medium and incubate overnight (37.5°C). Verify the viability of the cultures by testing random samples of the vials, and freeze the balance of the vials at −70°C. Field isolates may be stored in the same manner (27).

Collaborating Center for Animal Mycoplasmas

The rapid expansion in recent years of the known mycoplasmal infections in animals led to the formation by the International Organization of Mycoplasmologists of several teams of investigators in more than thirty laboratories around the world who collaborate on programs to evaluate reference reagents and to develop new and improved techniques for the isolation and identification of mycoplasmas. A Collaborating Center has been established at the Institute of Medical Microbiology, University of Aarhus, Aarhus, Denmark. Inquiries about the identification of newly isolated mycoplasmas should be addressed to the Director.

BIOCHEMICAL TESTS

Biochemical tests for the characterization of mycoplasmas were recommended by the Subcommittee on the Taxonomy of Mollicutes (28). In many situations, a much less detailed approach will provide significant information useful in identifying mycoplasmas isolated from animals. Before proceeding with biochemical tests, it is necessary to purify the unknown isolate by cloning and to show that it is in fact a mycoplasma as follows.

ABSENCE OF CELL WALL. Electron microscopy is the preferred method to determine the absence of a cell wall, but since this is not practical for most laboratories, a substitute procedure is the examination of a broth culture by phase-contrast or darkfield microscopy to show pleomorphic morphology (small coccoid bodies, ring forms, and fine filaments).

DETECTION OF BACTERIAL L-FORMS. One is technically required to complete five consecutive subcultures on medium without antibacterial agents to test for reversion from an L-form to a bacterium. For most diagnostic purposes, this is excessive. If the culture appears to be a mycoplasma and can be characterized biochemically and serologically as a known species, this is rather conclusive evidence.

COLONIAL MORPHOLOGY. A stereo-dissecting microscope with magnification to ×30 is needed for examining mycoplasmal colonies. Greater magnification may be necessary, especially for the study of ureaplasmal colonies.

Dienes' staining procedure (29) is also a useful tool. Colony morphology is variable, depending upon the culture medium and the conditions of incubation. Some combination of these should, in most instances, grow the typical "fried egg" colony with the central spot growing down into the medium and a peripheral zone of surface growth. Variations are seen, especially with recent isolates, from all "central spot" to all surface growth.

STEROL REQUIREMENT. Dependence on sterol for growth is the criterion for determining the family: Mycoplasmataceae are sterol dependent, and Acholeplasmataceae are sterol independent. Procedures for a direct determination of this dependence have been reported. However, sensitivity to digitonin was shown to closely parallel the sterol requirement and is much easier to determine. The procedure is similar to disc growth inhibition with specific antiserum. The discs are saturated with 0.025 ml of a 1.5% (w/v) ethanolic solution of digitonin and allowed to dry. A plate of an agar medium that supports good growth of the culture is inoculated so as to give a uniform population of colonies. Flooding the surface with diluted broth culture and removing the excess fluid works well. After the inoculum has dried, the digitonin disc is placed on the surface and the culture is incubated. The test may be interpreted when colonies can be detected. In order to be considered sensitive, the zone of growth inhibition around the disc should exceed 5 mm. Mycoplasmas are sensitive, and acholeplasmas are resistant.

Because it is not practical to run a long series of biochemical tests in a busy veterinary diagnostic laboratory, some judgment must be used in the selection of tests that will provide the most information for the time and effort expended (30). Three tests (glucose fermentation, arginine hydrolysis, and urea hydrolysis) are of prime importance. Other tests that have shown some usefulness and are not too difficult to perform are phosphatase activity and film and spot formation.

GLUCOSE FERMENTATION. Two medium controls are essential: (a) Base medium without the test substrate must be inoculated and incubated along with the test; (b) medium with the test substrate must be incubated without inoculation.

TEST MEDIA. Add 10 ml of sterile, heat-inactivated (56°C/30 min) horse serum, 5 ml of yeast extract, 10 ml of 10% (w/v) solution of glucose, and 1 ml of 0.5% (w/v) solution of phenol red to 74 ml of heart infusion broth (Difco). After the pH is adjusted to 7.5, sterilize by filtration, and dispense in 5 ml amounts to screw-capped tubes. Alterations in this medium that may prove helpful with some mycoplasmas are (a) substitution of a different serum supplement, or PPLO serum fraction (Difco) for the horse serum, (b) deletion of the yeast extract, or (c) the addition of bacterial inhibitors. Occasionally it will be necessary to use a more complex base medium. The medium of either Frey or Friis is recommended. Each lot of medium should be checked with cultures known to give positive and negative reactions (30).

YEAST EXTRACT. A suspension of 125 g of Fleischmann's® pure dry yeast, Type 2040 (Standard Brands Inc., New York) in 750 ml of water is placed at 37°C for twenty minutes and then heated to 95°C for five minutes. After cooling, centrifuge at 1000 G for thirty minutes. Dispense the supernatant fluid in 60 ml volumes and autoclave at 115°C for five minutes. This preparation may be stored at −20°C for three months.

INOCULATION. Test cultures should be subcultured to adapt them to the medium. The test and appropriate control media are inoculated with 1 ml of twenty-four-hour broth culture.

INCUBATION. Tests are incubated at 37°C for two weeks. Broth medium exposed to increased atmospheric CO_2 during incubation can show false positive reactions.

INTERPRETATION. Tests are examined daily for one week and every two days for the second week. The pH value of uninoculated control media should not change. To be considered positive, inoculated glucose medium should develop an acid shift that exceeds any acid shift in the inoculated base medium by at least 0.5 pH unit. It is helpful to prepare a set of pH standards in test medium for reference purposes.

ARGININE HYDROLYSIS AND UREA HYDROLYSIS. Use the procedure as for glucose fermentation with the following changes: (a) Replace the glucose in the medium with arginine or urea as the case may be; (b) adjust pH of the medium to 7.0; (c) a positive test is an alkaline shift of 0.5 pH unit.

PHOSPHATASE ACTIVITY. The following medium is recommended: sterile, heat-inactivated horse serum, 20 ml; yeast extract, 5.0 ml; 1% (w/v) sodium phenolphthalein diphosphate solution, 1 ml; and heart infusion agar (Difco), 74.0 ml. Adjust the pH to 7.8. Plates of this medium are inoculated in triplicate with a drop from a twenty-four-hour broth culture. Uninoculated control plates are also incubated at 37°C for three, seven, and fourteen days respectively. One inoculated and one uninoculated plate are examined on each test day by flooding with 5 N NaOH. The appearance of a red color after approximately one-half minute on the inoculated plate indicates phosphatase activity. Uninoculated plates may turn red but at a much slower rate.

FILM AND SPOTS. A plate of agar medium is inoculated with a drop from a twenty-four-hour broth culture and incubated for two weeks at 37°C. The plates should be examined at three– to four-day intervals for the appearance of a film over the heavily inoculated area. Examination under a dissecting microscope will help detect the crinkled appearance of the film and the small black dots in the upper layer of the medium. The biochemical reactions to be expected with mycoplasmas are summarized Table 27-III.

TABLE 27-III

BIOCHEMICAL REACTIONS OF SELECTED ORGANISMS
OF THE ORDER MYCOPLASMATALES

	Digi-torin	Glucose	Arginine	Urea	Phos-phatase	Film & Spots	Serum Digestion
M. mycoides subsp. *mycoides*	S	+	−	−	−	+ or −	−
M. bovigenitalium	S	−	−	−	+	+	−
A. laidlawii	R	+	−	−	−	−	−
M. bovirhinis	S	+	−	−	−	+ or −	+ or −
M. bovis	S	−	−	−	+	+	−
A. modicum	R	+	−	−	−	−	−
M. sp. (group 7)	S	+	−	−	−	−	+
M. alkalescens	S	−	+	−	+	−	−
M. arginini	S	−	+	−	−	−	−
M. dispar	S	+	−	−	−	−	−
M. bovoculi	S	+	−	−	−	−	−
M. canadense	S	−	+	−	+	−	−
M. verecundum	S	−	−	−	ND	+	N
M. mycoides subsp. *capri*	S	+	−	−	−	−	+
M. agalactiae subsp. *agalactiae*	S	−	−	−	+	+	−
M. conjunctivae	S	+	−	−	−	−	−
A. oculi	R	+	−	−	−	−	−
M. ovipneumoniae	S	+	−	−	−	−	−
M. capricolum	S	+	+	−	+	−	+
M. putrefaciens	S	+	−	−	+	−	−
M. hyopneumoniae	S	+	ND	−	ND	ND	ND
M. flocculare	S	+	−	−	−	−	ND
M. hyorhinis	S	+	−	−	+	−	−
M. hyosynoviae	S	−	+	−	−	+	−
A. granularum	R	+	−	−	−	−	−
M. gallisepticum	S	+	−	−	−	−	−
M. meleagridis	S	−	+	−	+	−	ND
M. synoviae	S	+	ND	−	ND	+	ND
M. iners	S	−	+	−	−	+	−
M. gallinarum	S	−	+	−	−	+	−
M. antis	S	+	−	−	+	+	ND
Ureaplasma sp.	S	−	ND	+	ND	−	−

Key:
S = sensitive
R = resistant
+ = positive
− = negative
+ or − = variable
ND = no data.

Reproduced with permission, from Blackburn, B. O.: *Proc 19th Meet Am Assoc Vet Lab Diagnosticians,* 118, 1976.

SEROLOGIC TESTS

GROWTH INHIBITION TESTS. The inhibitory effect of homologous antiserum on growth of mycoplasmas is highly specific. Although several methods of application are possible, probably the most common growth inhibition (GI) test is that of Clyde (31). The identification of mycoplasmas by GI tests requires a battery of high-quality antiserums. They must have a known zone of growth inhibitory effect against the homologous species with little or no effect against the heterologous species. Such antiserums are not generally available from commercial sources.

ANTIBODY PRODUCTION. Concentrated suspensions (100× to 200×) of washed mycoplasmas are homogenized with an equal volume of Freund's complete adjuvant. Homogenization may be accomplished by ultrasonification or with a hypodermic syringe. The latter is filled and emptied several times using an 18 gauge or smaller needle. This usually produces a stable emulsion. A properly prepared emulsion will not separate for several weeks; however, if separation occurs, repeat the process. To minimize abscess formation, 1,000 units of penicillin and 0.1 mg of streptomycin per milliliter may be added.

If rabbits are used to produce antibody, the toepad injection procedure supplemented with intramuscular (IM) inoculation produces a good antiserum. The first injection of 0.1 ml is given in the one toepad and 0.9 ml IM into the same leg. The following week, the same procedures are used with the other leg; a third injection of 1 ml IM is given one week later with two subcutaneous injections of 0.25 ml along the back. A small amount of blood is taken from the marginal ear vein seven to ten days later for testing. Animals that have a low titer or poor spectrum of antibody can be boosted by giving them one or more IV injections of 0.1–0.5 ml of mycoplasma concentrate without adjuvants diluted 1:10 at weekly intervals until the desired level of precipitating antibody is attained. Rabbits may be bled seven to fourteen days after the last injection. It is possible to get up to 50 ml of blood from the marginal ear vein by rubbing the ear with equal parts of toluene-ethanol mixture and making a lateral cut along the vein. The serum should be stored at −70°C.

TEST PROCEDURE.

1. Sterile filter paper discs (6 mm diameter) are saturated with antiserum of known potency and specificity. Commercially available discs of the type used for antibiotic sensitivity testing are recommended. They uniformly absorb 0.025 ml of antiserum and have no tendency to stick together. The discs are separated in a sterile Petri dish and loaded with a microdiluter calibrated to deliver 0.025 ml. Saturated discs may be frozen and stored at −20°C in a screw cap vial until used, or the saturated discs may be dried overnight under ultraviolet light; dried discs are easier to work with.

2. Cultures should be cloned to assure purity. It is advisable to determine growth characteristics in the mediums of choice.

3. The agar medium is inoculated with a broth culture in log phase of growth. Usually the inoculum will need to be diluted 1:10 or 1:100 to obtain a suitable population of colonies. They should be numerous but not too crowded to permit normal development. The entire agar surface should be inoculated. This can be done by flooding the surface and removing the excess fluid or by using a bent glass rod to spread the inoculum. After inoculation, the plates are dried with the lids ajar in a 37°C incubator for one-half to one hour.

4. The dried, inoculated plates must be marked to identify the discs that will be used to test for inhibition of growth. The method used is not important, but an area about 2 cm square is needed for each disc. The author finds it convenient to mark the bottom of the plate. Individual discs are removed from storage vials with sterile forceps and placed firmly in the appropriately marked area. Care should be taken to assure that the discs do not slide as they are put in place. The plates are returned to the incubator (37°C, high humidity). Antiserums to be used are determined according to the host from which the culture was isolated. If growth is not inhibited with antiserums against the host-specific strains, the remaining antiserums that are available are used in a second series of tests.

5. Most tests will be complete and ready for reading after two to five days of incubation, but if the culture grows slowly, it may take longer. A previously determined growth rate on the medium is helpful information. A clear zone with no colonies or a greatly reduced number of colonies around a disc similar to that observed with known homologous strains is a positive test. Zones of 0.5 mm or less should be ignored. Examination under a microscope ($20\times$ to $100\times$) will often facilitate interpretation of the results. At times, it will be helpful to stain colonies (a drop of Deines' stain diluted 1:50) to make them easier to view (29).

COMPLEMENT FIXATION TEST (CFT). One of the advantages of the CFT is that many diagnostic laboratories are already using the test for diagnostic purposes, and to use it for the identification of mycoplasma isolates requires only the production of antigens and the availability of known antiserums for serotyping. However, adapting some new isolates to growth in broth medium and the production of antigen that is suitably free of anticomplementary (AC) activity can be very time-consuming.

The procedures for the test are fairly simple if the strain grows readily in broth and if the antigen produced does not have AC activity. Depending on the number of serums to be tested and the density of growth in a given medium, 50–200 ml of broth will usually produce enough antigen for identification tests. The broth is centrifuged (27,000 G for fifteen minutes), and the pellet is washed three times by resuspension in and centrifugation out of phosphate-buffered saline (PBS) solution pH 7.0–7.5. After the final wash, the antigen is resuspended in PBS in 2–5 percent of the original broth volume. Two rows of doubling dilutions of antigen are made in veronal-buffered diluent (VBD) in tubes or in a microtiter tray. Dilutions of 1:2 through 1:128 are commonly used. An equal volume of VBD containing 2 units of rabbit antiserum, e.g. a 1:160 dilution of an antiserum that titers 1:320, is added to each tube or well in one row, and an equal amount of VBD is added to the other row (antigen AC control). Complement (2–5 CH_{50} units) is added to both rows, and after incubation overnight at 4°C, sensitized sheep red blood cells are added.

Fixation of complement in one or more tubes past the AC activity of the controls is an indication of homology, but in weak reactions, a block titration against several antiserum dilutions may have to be run to make a decision. The significance of weak reactions is easier to ascertain if an antiserum against medium components is also available and is run against a third row of antigen dilutions. If there is no indication of the fixation being contributed to by a reaction between medium components and antibodies against them, it is an indication of specificity and the weakness of the reaction is due to (a) antigen too dilute, (b) partial homology, or (c) mixture of strains. A new antigen should be prepared and/or the test antigen should be run against other antisera to help determine the reason for the weak reactions.

Metabolic Inhibition Test (MIT). As the name implies, this test is essentially a neutralization test in which known antiserum slows the growth of the unknown mycoplasma for a period of time or prevents it altogether. The slowed growth is detected by absence of color change. The given change depends on the substrate of mycoplasmal metabolism and indicator used. The MIT is less reliable for the identification of mycoplasmas than for serology. First, the accuracy of the test is highly dependent on the unknown strain being a pure culture. Then, too, precautions must be taken to make sure that the organism is growing well enough in broth to change the indicator within a reasonable time. The test is a simple one in concept, but in practice it often does not work well unless the diagnostic laboratory uses MIT routinely.

The first step of MIT is to determine which of the possible substrates is utilized by the unknown strain. After repeated cloning, the strain is introduced into base medium supplemented with either arginine (1%) and adjusted to a pH value of 7.3, or dextrose (0.5%) and adjusted to a pH of 7.8. Of course, the simpler the base medium, the better. Heart infusion broth plus fresh yeast extract and horse or other serum is often used as the base medium for this test.

The tetrazolium reduction inhibition (TRI) test can be used with those mycoplasmas that utilize neither arginine nor glucose but that reduce tetra-zolium (2,3,5-TTC). The addition of sodium thioglycollate (0.1%) to the base medium is desirable for obtaining the best results with the TRI test.

MISCELLANEOUS IDENTIFICATION TESTS. A growth precipitation test has been reported in which a well containing replicating mycoplasma organisms is surrounded by discs previously soaked with reference antisera. Precipitation lines indicate homology (32). In the radial growth precipitation test (33), the unknown mycoplasma is grown in an agar plate; known antiserums are placed in wells in the agar. Lines of precipitation indicate homology. Another method of identification that has been used considerably is poly-acrylamide gel electrophoresis of mycoplasmal whole cell or membrane proteins (34).

FLUORESCENT ANTIBODY TESTS. The previously described tests for the identification of mycoplasmas isolated from animals have been criticized for their failure to adequately detect mixtures of mycoplasmal species. Mixed cultures can be difficult if not impossible to purify. The direct or indirect fluorescent antibody (FA) technique for the staining of mycoplasmal colonies in agar plates is the best method for the recognition of mixed cultures and for identifying causal agents of mycoplasmosis in animals.

The production of antiserums to mycoplasmas and the preparation of immunofluorescent antibody have been described (30). The FA test was performed as follows on colonies in primary isolation plates or on cultures grown in medium. Cultures of mycoplasmas were diluted (usually 1:100 or 1:1000) in phosphate-buffered saline (PBS: 0.015M phosphate; 0.15M NaCl in distilled water), and 0.2 ml was spread on the surface of solidified medium. The inoculated plates were incubated in a humidified environment at 37°C. When distinct, well-isolated colonies appeared, they were flooded with ethyl alcohol (95%) and fixed for sixty minutes. The alcohol was replaced with PBS, and the plates were stored at 5°C. For the FA test, eight small agar blocks (4–5 mm) containing one to five colonies were cut and placed in each of the eight outlined areas on tissue culture slides* and treated with eight different conjugated antiserums. Each block was treated with a small drop of diluted (usually 1:20) FA. The slide was incubated at room temperature for twenty to thirty minutes in a moist chamber (a 100 mm plastic Petri dish containing saturated filter paper). Then, the plastic chamber was replaced on the slide to provide washing chambers for each of the agar blocks. They were washed in PBS containing merthiolate (1:10,000) until the background fluorescence was reduced to acceptable levels (two to forty-eight hours). The slide without chamber was placed under a dissecting microscope. If the mycoplasmal colonies were in the proper position (up), a cover slip (22 × 40

*Lab-Tex®, Miles Laboratories, Naperville, Illinois.

mm) was applied for observations of fluorescence by incident ultraviolet light at 125×. The degree of fluorescence was rated relative to that of the background: negative where background fluorescence was equal to that of the colony, or +1 to +4 when the colony was clearly brighter than the background (35).

DIAGNOSIS

Complement Fixation Tests have been used for many years to diagnose contagious bovine pleuropneumonia, but their use to diagnose mycoplasmosis in animals in North America is relatively recent. Workers at Iowa State University (36) have used CFT for the diagnosis of mycoplasmal pneumonia in swine for several years. Experimentally exposed pigs developed CFT titers three to six weeks postexposure; they persisted for several months. By means of CFT and slaughter of reactors, herds of swine were freed of the infection. They could be kept free by CFT on all replacements and breeding stock. These reports suggest that the techniques for the eradication of mycoplasmal pneumonia from swine may be at hand.

ANTIGEN PRODUCTION. The nonspecific reagents for CFT to detect mycoplasmal antibodies are available from commercial sources, but few antigens and control antiserums are available. Production of the antigens consists of (a) production of antigen in any medium to which the antigen strain was adapted; (b) three washes in buffered saline with resuspension of antigen in buffered saline at about 5 percent of the original broth volume; (c) tests for specific and anticomplementary activity; (d) to suitable suspensions of antigen, add glycerol (25%) and store at −20°C or below until used (37).

If the antigen has anticomplementary activity at dilutions beyond usefulness, one of several treatments may be used in an attempt to make it usable. Treatment with trifluorotrichloroethane (Genetron 113) is sometimes quite effective in reducing anticomplementary effect. One volume of antigen suspension is added to one volume of Genetron 113; the mixture is shaken vigorously for fifteen seconds and then centrifuged lightly. The aqueous top layer usually contains a large proportion of the original specific antigen, with less of the anticomplementary activity. Another means of fractionating antigen that decreases anticomplementary activity and is usable with some, but by no means all, *Mycoplasma* species is the chloroform-methanol lipid antigen extraction method described by Kenny and Grayston (38). Another procedure that is often used to decrease the anticomplementary effect of CFT antigens is heating at various temperatures up to boiling.

PERFORMANCE OF THE TEST. The microtechnique method for CFT was modified for use with porcine and bovine serums (37). The modification required for bovine serum was the addition of suitable, fresh, unheated calf serum (5%) to the diluent used for diluting guinea pig serum—the source of complement for the test. Until recently, complement was modified in the same way for use with swine serums. Slavick and Switzer rehydrated lyophilized guinea pig complement in undiluted, fresh, unheated swine serum and obviated the frequently encountered procomplementary effects of swine serums (36).

The indirect hemagglutination (IHA) test was adapted to the detection in cattle of antibodies to mycoplasmal species, particularly *M. bovis* and *M. bovigenitalium* (39). The antigen was stable for at least seven months at 5°C. The IHA test was sensitive, specific, and reproducible. It may have considerable usefulness for the detection of mycoplasmosis in animals, particularly if regulations should be promulgated concerning mycoplasmosis in animals intended for interstate or international transport or in bulls whose semen is widely disseminated for artifical insemination.

The enzyme-linked immunosorbent assay (ELISA) was adapted to demonstrate antibodies to *M. bovis* in the serum of artificially and naturally infected cattle (40). The test was rapid, reproducible, and sensitive; some cross-reactions with other mycoplasmas were reported. The fluorescent antibody procedure was used to identify *M. dispar* and ureaplasmas lining the bronchial epithelial of calves afflicted with pneumonia (41). The enzyme-linked immunoperoxidase technique was applied to frozen lung and bronchial smears from pigs with MPS (42). In the tissue sections, the reddish brown *M. hyopneumoniae* organisms lined the bronchial epithelium or occurred as pleomorphic spots in the smears. This diagnostic test does not require a fluorescent microscope.

UREAPLASMAS

Although ureaplasmas were associated with nongonococcal urethritis of humans thirty years ago, their role in calf pneumonia and vaginitis of the cow was established only recently by American (43), English (44), and Canadian workers (10). Their role as pathogens appears to be expanding rapidly, e.g. in chickens with chronic respiratory diseases and female monkeys with a high incidence of stillbirths and abortions. Ureaplasmas (so called because of their urease activity) are common in ruminant animals, in dogs and cats, and in bull semen.

CULTIVATION AND BIOCHEMICAL REACTION. Media suitable for mycoplasmas are inadequate for the propagation of ureaplasmas (43). Four requirements must be observed, or attempts to isolate and propagate ureaplasmas will be frustrated. First, the pH value of sterile medium should be 6.0 ± 0.5. Second, ureaplasmas are very sensitive to pH values above 7.4, and if exposed to alkaline environmental conditions for a short period of time, they will become nonviable. In the laboratory, it is convenient to transfer cultures at the end of a work day and subculture the ureaplasmas again the next morning. Serial tenfold dilutions of the inoculum should be made to avoid losing the cultures. Third, ureaplasmal growth is enhanced by the addition of urea to the medium even though it may not be an absolute growth requirement. The hydrolysis of urea with the production of ammonia is the characteristic biochemical reaction of ureaplasmas. None of the mycoplasmas is known to possess urease activity. Finally, ureaplasmas are sensitive to thallium acetate, and it should not be added to the medium.

Bull semen and milk may contain antibiotics or other substances with mycoplasmacidal and ureaplasmacidal activity. Serial tenfold dilutions of milk, semen, tissue suspensions of pneumonic lungs, or urine should be prepared for primary isolation in liquid and solid medium. A suspension of tissue prepared by using 1 g of tissue to 9 ml of diluent and grinding in a Ten Broeck tissue grinder is suitable for primary inoculation.

It should be emphasized that media employed for the isolation and propagation of *U. urealyticum*, i.e. Shepard's U-9 Broth, A-3 solid medium, and A-5 solid medium, usually are not satisfactory for ureaplasmas of ruminant origin. Modified Hayflick's medium and similar media should be utilized instead (see Appendix F). The A-7 plating medium of Shepard and Lunceford (Gibco), however, is satisfactory (45). To increase the number of successful isolations, three media should be inoculated during primary isolation attempts. Serums from individual horses vary in growth-promoting properties. They should be batch tested before being used and stored at $-20°C$ until needed. The serum does not need to be heat inactivated.

Optimal growth of ureaplasmas can be obtained on agar in an atmosphere containing 5% carbon dioxide in nitrogen or air. A candle jar or tissue culture incubator provides satisfactory growth conditions; growth is obtained under anaerobic conditions. Colonies can be detected twenty-four to forty-eight hours after inoculation using ×100 magnification. Colonies are very small (about 20–40 μm in diameter); they may develop the characteristic central-nippled appearance of mycoplasmas in three to four days if they are not crowded on the agar plate. The volume of agar medium in the plate will also affect the size of ureaplasmal colonies; 6–7 ml of molten agar medium should be added to each 50 mm Petri dish. Dienes' stain will be retained if applied to the colony. Older colonies (five to seven days of age) may show an irregular edge and may be multilobate with a "cauliflower head" appearance. Different isolates from the same species of animal may exhibit slightly different growth and colonial characteristics. Liquid cultures are usually clear. Usually, titers of 10^6 to 10^7 color changing units (CCU) are obtained routinely, but under optimum growth conditions, 10^9 CCU have been observed.

All ureaplasmal isolates of animal origin have phosphatase activity, do not reduce tetrazolium or methylene blue, and are digitonin sensitive.

Ureaplasmas do not ferment carbohydrates and apparently do not possess hexokinase activity.

The optimal temperature for growth of ureaplasmas is 36–37°C. Some isolates will grow at 30°C. Most ureaplasmal cultures become nonviable in three or four days at 37°C; most cultures remain viable for several weeks at 5°C. Ureaplasmal cultures were frozen at −20°C and remained viable for a year. At −76°C, they have remained viable with loss of less than 1 $\log_{10}$ titer for over five years. Ureaplasmas can be successfully freeze-dried (43).

UREASE STAIN. For differentiation of ureaplasmas, the urease test developed by Shepard and Howard (46) is the most practical and reliable procedure available. The addition of a mixture of equal parts of 10% urea and 0.8% manganese chloride directly applied to forty-hour-old colonies on solid medium results in an immediate color reaction on the surface of ureaplasmal colonies. The colonies at first become golden brown, then dark brown, and may eventually become black as a result of deposition of manganese on the surface of the colony. All serotypes of ureaplasma react similarly, and to date the author has not observed a single false positive reaction. The mixture of urea and manganese chloride may be stored at −20°C until used. A solid medium containing test reagent reacts similarly with the developing urea-plasmal colonies (45).

SEROLOGY. The serological identification of serotypes of *U. urealyticum* is not a practical procedure at the present time.

REFERENCES

1. Leach, R. H.: *J Gen Microbiol, 75*:135, 1973.
2. Kirchhoff, H., Kuwabara, T., and Barile, M. F.: *Infect Immun, 31*:445, 1981.
3. Robinson, I. M., Allison, M. J. and Hartman, P. A.: *Int J Sys Bacteriol, 25*:173, 1975.
4. Gourlay, R. N. and Wyld, S. C.: *Vet Rec, 97*:370, 1975.
5. Jasper, D. E., Ernø, H., Dellinger, J. D., and Christiansen, C.: *Int J Syst Bacteriol, 31*:339, 1981.
6. Stalheim, O. H. V., and Proctor, S. J.: *Am J Vet Res, 37*:879, 1976.
7. Ross, R. F.: In Leman, A. D., Glock, R. D., et al. (Eds.): *Diseases of Swine,* 5th ed. Ames, Iowa St U Pr, 1981, Chapter 51.
8. Gourlay, R. N.: *Isr J Med Sci, 17*:531, 1981.
9. Jasper, D. E.: *J Am Vet Med Assoc, 170*:1167, 1977.
10. Doig, P. H., Ruknke, H. L., and Palmer, N. C.: *Can J Comp Med, 44*:252, 1980.
11. Adler, H. E.: *Proc 82nd Meet US Anim Health Assoc,* 384, 1978.
12. Stalheim, O. H. V.: *J Am Vet Med Assoc,* (accepted for publication), 1982.
13. Koshimizu, K., and Ogata, M.: *Jap J Vet Sci, 36*:391, 1974.
14. Armstrong, D., Yu, B. H., Yagoda, A., and Kagnoff, M. F.: *J Infect Dis, 124*:607, 1971.
15. Rosendal, S.: *J Infect Dis, 138*:203, 1978.
16. Campbell, L. H., Snyder, S. B., Reed, C., and Fox, J. G.: *J Am Vet Med Assoc, 163*:991, 1973.
17. Switzer, W. P.: In Merchant, I. A., and Packer, R. A. (Eds.): *Veterinary Bacteriology and Virology,* 77th ed. Ames, Iowa St U Pr, 1967, Chapter 38.
18. Tan, R. J. S., Lim, E. W., and Ishak, B.: *Can J Comp Med, 41*:349, 1977.

19. Harasawa, R., Yasmamoto, K., and Ogata, M.: *Microbiol Immunol, 21*:179, 1977.
20. Tan, R. J. S., and Miles, J. A. R.: *Aust Vet J, 50*:142, 1974.
21. VonAmmar, A., Kirchhoff, H., Heitmann, J., Meier, C., Fischer, J., and Deegen, E.: *Berliner und Munchener Tierarztliche Wochenschrift, 93*:457, 1980.
22. Lemcke, R. M., Ernø, H., and Gupta, U.: *J Hyg Camb, 87*:93, 1981.
23. Armstrong, C. H.: *Proc 19th Meet Am Assoc Vet Lab Diagnosticians*, 75, 1976.
24. Armstrong, C. H.: *Am J Vet Res, 42*:1030, 1981.
25. Livingston, C. W., Jr., and Gauer, B. B.: *Proc 82nd Meet US Anim Health Assoc*, 389, 1978.
26. Cottew, G. S., and Yeats, F. R.: *Aust Vet J, 54*:293, 1978.
27. Langford, E. V.: *Proc 19th Meet Am Assoc Vet Lab Diagnosticians*, 106, 1976.
28. Anonymous: *Int J Syst Bacteriol, 22*:184, 1972.
29. Madoff, S.: *Ann NY Acad Sci, 79*:383, 1960.
30. Blackburn, B. O.: *Proc 19th Meet Am Assoc Vet Lab Diagnosticians*, 118, 1976.
31. Clyde, W. H., Jr.: *J Immunol, 92*:958, 1964.
32. Krogsgaard-Jensen, A.: *Appl Microbiol, 23*:553, 1972.
33. Howard, C. J.: *Vet Microbiol, 1*:23, 1976.
34. Stalheim, O. H. V., and Stone, S. S.: *J Clin Microbiol, 2*:169, 1975.
35. Stalheim, O. H., Foley, J. H., Stone, S. S., and Rhoades, K. R.: *Proc 18th Meet Am Assoc Vet Lab Diagnosticians*, 207, 1975.
36. Slavik, M. F. and Switzer, W. P.: *Ia State J Res, 47*:117, 1972.
37. Frey, M. L.: *Proc 19th Meet Am Assoc Vet Lab Diagnosticians*, 149, 1976.
38. Kenny, G. E., and Grayson, J. R.: *J Immunol, 95*:19, 1965.
39. Cho, H. J., Ruhnke, H. L., and Langford, E. V.: *Appl Microbiol, 28*:897, 1974.
40. Boothby, J. T., Jasper, D. E., Rollins, M. H., and Thomas, C. B.: *Am J Vet Res, 42*:1242, 1981.
41. Tinant, M. K., Bergeland, M. E., and Knudtson, W. U.: *J Am Vet Med Assoc, 175*:812, 1979.
42. Bruggmann, S., Engberg, B., and Ehrensperger, F.: *Vet Rec, 101*:137, 1977.
43. Livingston, C. W., Jr.: *Proc 19th Meet Am Assoc Vet Lab Diagnosticians*, 96, 1976.
44. Howard, C. J., Gourlay, R. N., Thomas, L. H., and Stott, E. J.: *Res Vet Sci, 21*:227, 1976.
45. Shepard, M. C. and Lunceford, C. D.: *J Clin Microbiol, 3*:613, 1976.
46. Shepard, M. C., and Howard, D. R.: *Ann NY Acad Sci, 174*:809, 1970.

SUPPLEMENTARY REFERENCES

Cassell, G. H. and Hill, A.: Murine and other small-animal mycoplasmas. In Tully, J. G., and Whitcomb, R. F. (Eds.): *The Mycoplasmas. II. Human and Animal Mycoplasmas*. New York, Academic Press, 1979, pp 235.

Jasper, D. E., Dellinger, J. D., Rollins, M. H., and Hakanson, H. D. Prevalence of mycoplasmal bovine mastitis in California. *Am J Vet Res, 40*:1043, 1979.

McMartin, D. A., MacOwan, K. J., and Swift, L. L.: A century of classical contagious caprine pleuropneumonia: From original description to aetiology. *Br Vet J, 136*:507, 1980.

Razin, S.: The Mycoplasmas. *Microbiol Rev, 42*:414, 1978.

Stalheim, O. H. (Ed.): *Laboratory Diagnosis of Mycoplasmosis in Food Animals*. Columbia, Missouri, Am Assoc of Vet Lab Diagnosticians, 1977.

Whitcomb, R. F.: The genus Spiroplasma. *Ann Rev Microbiol, 34*:677, 1980.

Whittlestone, P.: Enzootic pneumonia of pigs. In Brandly, C. A., and Cornelius, C. E. (Eds.): *Advances in Veterinary Science and Comparative Medicine*, Vol. 17. New York, Academic Press, 1973.

Chapter 28

MYCOLOGY: INTRODUCTION

M YCOLOGY has been a neglected field in many veterinary diagnostic laboratories. A number of important pathogenic fungi can be identified without difficulty. Those that cannot can be forwarded to a reference mycology laboratory. A duplicate culture of the submitted strain may be kept for study after an identification has been made. By this means, a collection of the more common pathogenic and nonpathogenic fungi can be acquired.

All diagnostic laboratories can carry out direct examination of clinical materials and inoculation of this material onto appropriate media if this is warranted. Histopathological examinations are of great value in all but the superficial mycosis.

The outline that follows is provided as an aid and is not meant to be a substitute for standard texts (see Supplementary References). The review *Fungal Diseases of Animals* (Ainsworth and Austwick) is especially useful, and the well-known *Manual of Clinical Mycology* (Conant et al.) provides, besides the basic elements of mycology, numerous excellent illustrations of common contaminating fungi as well as the well-known pathogens. The *Medical Mycology Manual* (Beneke and Rogers) contains a mine of information on practical procedures. Among a number of other useful books are *Practical Medical Mycology* (Koneman et al.) and *Veterinary Mycology* (Jungerman and Schartzman). The publication "Identification of Saprophytic Fungi Commonly Encountered in a Clinical Environment" is especially helpful.

It should be remembered that a number of fungi producing disease in animals are transmissible to humans. *Special care should be taken to prevent laboratory infections.*

Significance of Fungous Isolations

The great majority of fungi live in the soil or water, or as harmless commensals associated with humans and animals; e.g. *Histoplasma capsulatum* and *Blastomyces dermatitidis* occur in soil (geophilic), while *Candida albicans* lives in the alimentary canal. Many of the fungi such as *Aspergillus* spp., *Rhizopus* spp., and *Geotrichum* are widespread in nature and thus are frequently found in clinical materials. The significance of the isolation will depend on such considerations as (1) demonstration of fungi in tissue sections, (2) presence of clinical disease, (3) presence of pathologic lesions, and (4) repeated isolation of the same fungus. The widespread saprophytic fungi that occasionally cause disease, such as *Mucor* spp., *Aspergillus* spp., *Candida albicans*, etc., are frequently referred to as "opportunistic fungi." Production of disease by these fungi is thought to be related to such factors as "impaired resistance," prolonged steroid and/or antibiotic therapy, and various stresses, including terminal diseases and metabolic disturbances.

MYCOLOGICAL EXAMINATIONS

Procedures for a Mycological Examination

1. Direct examination of the clinical material. A small amount of material is added to a drop of 10% potassium hydroxide or lactophenol, and a coverslip is applied. The slide is gently heated to remove air bubbles and promote clarification. Some laboratories prefer 20% potassium hydroxide with glycerol (see Appendix D). Gram's (special procedure for clinical material), Giemsa, and Wright's stains are also frequently of value. Direct examinations are almost always negative in histoplasmosis and sporotrichosis. Tissue should be set aside for histopathologic study if this was not already done.

2. If there is evidence of a fungous infection, appropriate media are inoculated.

3. Examination of the growth both macroscopically and microscopically.

4. Those cultures that cannot be identified with certainty are forwarded to a mycology laboratory.

Material Required for Mycological Procedures

Almost all of the equipment required can be found in the diagnostic bacteriology laboratory. Straight dissecting needles are useful for breaking up colonies of fungi for examination. The inocula used to seed media are generally larger than those used in bacteriological work. Forceps and scalpels with small, sharp blades are especially useful. Bacterial contamination can be reduced by aseptic operations.

The reagents, stains, culture media, and other items required especially for mycological work are described in Appendix D.

Culture Media

The media recommended for isolation of the pathogenic fungi along with the incubation temperatures and duration of incubation are summarized in Table 28-I.

Blood agar, Sabouraud dextrose agar, and Sabouraud C and C agar are used routinely for primary fungous cultivation. The last mentioned medium is basically the same as the second except that it contains cycloheximide and chloramphenicol for the suppression of some saprophytic fungi and bacteria, respectively. Brain-heart infusion agar can usually be used instead of blood agar. Media are most useful in Petri dishes; the medium should be thicker than

TABLE 28-I

MEDIA FOR ISOLATION,* INCUBATION TEMPERATURE AND USUAL LENGTH OF INCUBATION FOR THE PATHOGENIC FUNGI

Disease	*Isolation Medium* *28°C*	*Incubation Temperature* *37°C*
Phycomycosis (Zygomycosis, Mucormycosis)	Sabouraud Dextrose Agar; chloramphenicol (not cyclo-heximide) can be used. (1–3 days)†	
Aspergillosis	Same as for Phycomycosis (1–3 days)	
Candidiasis	Sabouraud Dextrose Agar, Sabouraud C and C Agar‡ (1–3 days)	
Dermatophytosis (Ringworm)	Sabouraud C and C Agar (2–3 weeks)	
Blastomycosis	Sabouraud C and C Agar (2–3 weeks)	Brain-Heart Infusion Agar or Blood Agar (3–7 days)
Cryptococcosis	Sabouraud Dextrose Agar; chloramphenicol (not cyclo-heximide) can be used. (1–2 weeks)	
Histoplasmosis	Sabouraud C and C Agar: (2–4 weeks)	Brain-Heart Infusion or Blood Agar (2–4 weeks)
Coccidioidomycosis	Sabouraud C and C Agar (1–2 weeks)	
Sporotrichosis	Sabouraud C and C Agar (7–10 days)	Brain-Heart Infusion or Blood Agar (7–10 days)
Epizootic Lymphangitis	Blood or Serum Agar (2–8 weeks)	Blood or Serum Agar (2–8 weeks)
Geotrichosis	Sabouraud C and C Agar (1–2 weeks)	
Chromomycosis	Sabouraud C and C Agar (2–3 weeks)	
Maduromycosis	Sabouraud Dextrose Agar (2–3 weeks)	

*In some instances depending on source of the specimen and history it will be advisable to inoculate both kinds of Sabouraud media as well as blood agar.

†Usual incubation period.

‡Cycloheximide and chloramphenicol added (Mycosel, Mycobiotic, etc.).

usual, 25–35 ml per plate. It should be kept in mind that media containing cycloheximide should be incubated at room temperature (25–28°C) only.

In the notes that follow, either Sabouraud dextrose agar or Sabouraud C and C agar have been recommended. Generally speaking, it is advisable to inoculate each specimen onto both media. It is also frequently advisable to incubate media at 25–28°C and at 37°C. The media for isolation, the incubator temperature, and the usual time required for growth are summarized in Table 28-I. It is a good practice to incubate all plates for fungous isolation for at least four weeks.

If plate media are used, they should be sealed with a rubber band or tape, or placed in a sealed container to prevent dehydration. Media for fungi are often used in the form of slants rather than plates in order to minimize the dissemination of spores.

Examination of Cultures

After observing gross characteristics of the cultures, a portion of the colony is teased apart with needles, then transferred to a drop of lactophenol cotton blue. A coverslip is then added. An alternative useful procedure is to prepare a tape mount. Place a drop of lactophenol cotton blue on a microscope slide. The end of a transparent tape such as Scotch® brand cellophane tape is grasped with forceps, and a square piece smaller than a coverslip is cut off. The adhesive side of the piece of tape is applied carefully to an area of the colony where sporulation is expected. The tape is removed, and a corner of it is pressed to the slide near the lactophenol cotton blue. The forceps are detached, and the tape is pressed down gently to expel air bubbles from the mounting fluid. A drop of lactophenol cotton blue is then added to the tape, and a coverslip is applied. The edges of the coverslip may be sealed with nail polish. The tape carrying the fungal elements may alternatively be applied directly to a coverslip, thus obviating the disadvantage of looking at the fungus through the tape. If the culture has sporulated, the hyphae and characteristic reproductive elements can be seen. For verification of the morphologic observations, a slide culture may be prepared. The technique is described by Ajello and associates (3) (see also Appendix D).

Difficulty may be experienced in obtaining both growth phases of the dimorphic fungi. Methods are described (3) for the conversion from one phase to another.

Many of the saprophytic fungi can be identified by their characteristic fruiting bodies. Figure 28-I has been provided to aid in the identification of some of the more common contaminants.

Figure 28-1. Contaminants: *a, Alternaria* sp.; *b, Helminthosporium* sp.; *c, Hormodendrum* sp.; *d, Paecilomyces* sp.; *e, Penicillum* sp.; *f, Scopulariopsis* sp.; *g, Cephalosporium* sp.; *h, Gliocladium* sp.; *i, Trichoderma* sp.; *j, Fusarium* sp.; *k, Nigrospora* sp.; *l, Aureobasidium* sp. (Pullularia). (H. A. McAllister.)

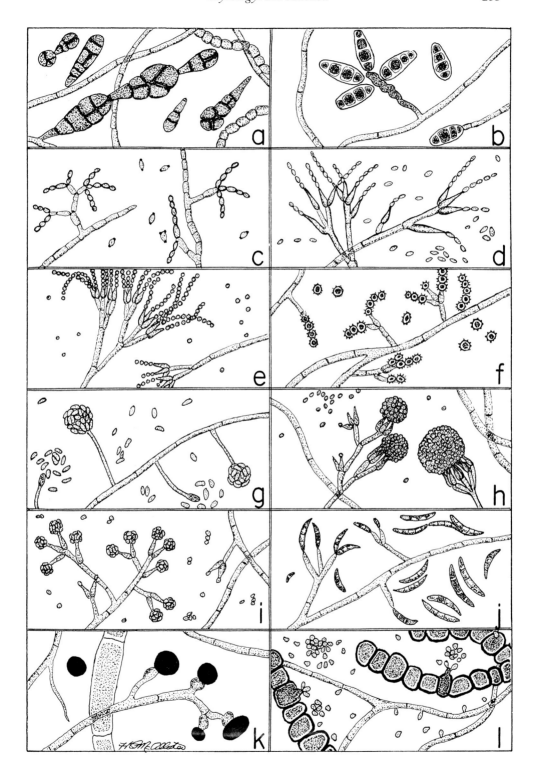

CLASSIFICATION OF FUNGI

Below are listed in barest outline the principal taxonomic categories of the fungi of medical and veterinary significance. Readers are referred to Alexopoulos and Mims (Supplementary References) for additional information.

Phylum Eumycophyta

Aseptate Mycelium	Septate Mycelium
Class: Zygomycetes	Classes: Ascomycetes (sac
(sexual and asexual cycles)	fungi, true yeasts, etc.)
Mucor	*Allescheria boydii*
Absidia	*Aspergillus*
Rhizopus	*Penicillium*, etc.
Mortierella	Basidiomycetes (cup
	fungi, truffles, rusts, smuts)
	Deuteromycetes (imperfect
	fungi; sexual stages not
	discovered)
	Includes many potential-
	ly pathogenic fungi.
	Dermatophytes
	Dirmophic fungi, etc.

FREQUENTLY USED MYCOLOGICAL TERMS

ARTHROSPORE. An asexual spore formed by the disarticulation of the MYCELIUM. Example: *Geotrichum candidum.*

ASCOSPORE. A sexual spore characteristic of the true YEAST and ascomycetes. They are produced in a saclike structure called an ASCUS. The ascospore results from the fusion of two nuclei. Examples: *Saccharomyces* spp., *Nannizzia* spp.

ASCUS: The specialized saclike structure characteristic of the true YEASTS in which ASCOSPORES (usually eight) are produced. Example: *Saccharomyces* spp.

BLASTOSPORE. A spore produced as a result of a budding process along the MYCELIUM or from a single spore. Examples: *Saccharomyces* spp., *Candida* spp.

CHLAMYDOSPORES. Thick-walled, resistant spores formed by the direct differentiation of HYPHAE. Examples: *Candida albicans, Histoplasma capsulatum.*

CLAVATE. Club-shaped.

COLUMELLA. The persisting, dome-shaped upper portion of the sporangiophore. Example: *Mucor* spp.

CONIDIUM. An asexual spore formed from HYPHAE by astriction, budding, or septal division. Example: *Penicillium* spp.

CONIDIOPHORE. A stalklike branch of the MYCELIUM on which CONIDIA develop either singly or in numbers. Example: *Penicillium* spp.

DEMATIACEOUS. Used to denote the dark brown or black fungi. Examples: *Philialophora* spp., *Hormodendrum* spp.

DIMORPHIC. Having two forms or phases, e.g. YEAST form and mycelial form. Example: *Blastomyces dermatitidis.*

ECTOTHRIX. Arthrospores appear outside the hair shaft. Example: *Microsporum* spp.

ENDOGENOUS. Originating or produced from within. Example: *Candida albicans* (usually).

ENDOTHRIX. Arthrospores appear within the hair shaft. Example: *Trichophyton* spp.

EXOGENOUS. Originating from without. Example: *Histoplasma capsulatum* infection.

GEOPHILIC. Denotes fungi whose natural habitat is the soil. Example: *Coccidioides immitis.*

GERM TUBE. Tubelike structure produced by germinating spores. They develop into HYPHAE. Example: *Candida albicans.*

GLABROUS. The smooth form. Example: the glabrous form of *Geotrichum candidum.*

HYPHAE. The filaments that compose the body of the thallus of a fungus.

MACROALEURIOSPORE. This is the larger of the two kinds of conidia that break from the attachment to hyphae by rupture through the cell wall. Example: *Dermatophytes.*

MACROCONIDIUM. A large, sometimes multicellular spore. Example: *Dermatophytes.*

MICROALEURIOSPORE. This is the smaller of the two kinds of conidia that break from the attachment to hyphae rupture through the cell wall. Example: *Dermatophytes.*

MICROCONIDIUM. A small, single-celled conidium borne laterally on *hyphae*. Examples: some species of *Microsporum* and *Trichophyton.*

MYCELIUM. A mat made up of the intertwining, threadlike HYPHAE.

NODES. The points on the STOLONS from which the RHIZOIDS arise. Example: *Rhizopus* spp.

PSEUDOHYPHA. Filament constituted by elongated budding cells that have failed to detach. Example: *Candida albicans.*

RACQUET HYPHAE. HYPHAE with terminal swelling of segments giving a shape resembling that of a tennis racquet. Example: *Dermatophytes.*

RHIZOID. Rootlike, branched HYPHAE extending into the medium. Example: *Absidia* spp.

SEPTATE. Having cross walls or septa in the HYPHAE.

SPORANGIUM. A closed, often spherical structure in which are produced asexual spores by cleavage. Example: *Rhizopus* spp.

STERIGMATA. Specialized structures, short or elongated, borne on a vesicle and producing conidia. Example: *Aspergillus* spp.

STOLON. A horizontal HYPHA or runner that sprouts where it touches the substrate. It forms RHIZOIDS in the substrate. Example: *Absidia* spp.

YEASTS. Unicellular fungi that reproduce by asexual budding or by sexually produced ascospores. Example: *Saccharomyces* spp.

ZYGOSPORE. A thick-walled, sexual spore of the true fungi that results from the fusion of two similar gametangia. Example: *Phycomycetes.*

SERODIAGNOSIS OF FUNGAL DISEASES

This area of laboratory diagnosis has received little attention in fungal diseases of animals. This is partly because these diseases are relatively infrequent and the required reagents have not usually been readily available to the veterinary microbiologist. The expense of the reagents has also been a deterrent. Some hospital and public health laboratories will make their serodiagnostic capability available to veterinarians, and some fungal diagnostic reagents are available commercially. Readers are referred to Kaufman (1) for details of procedures and some sources of reagents. The results of these human tests when used in animals may require different interpretations. Balows et al. (2) have reported on the value of serologic procedures in the diagnosis of canine histoplasmosis and blastomycosis.

SAFETY

All fungous cultures and specimens thought to harbor fungi should be considered potentially dangerous. Work with these materials should be carried out in a biological safety hood.

REFERENCES

1. Kaufman, L.: In Lennette, E. H. (Ed.-in-chief): *Manual of Clinical Microbiology,* 3rd ed. Washington, D.C., American Society for Microbiology, 1980, Chapter 61.
2. Balows, A., Ausherman, R. J., and Hopper, J. M.: *J Am Vet Med Assoc, 148*:678.

SUPPLEMENTARY REFERENCES

Alexopoulos, C. J., and Mims, C. W.: *Introductory Mycology,* 3rd ed. New York, John Wiley and Sons, Inc., 1979.

Beneke, E. S., and Rogers, A. L.: *Medical Mycology Manual,* 4th ed. Minneapolis, Burgess Publishing Company, 1980.

Haley, L. D., Trandel, J., and Coyle, M. B.: *Practical Methods for Culture and Identification of Fungi in the Clinical Microbiology Laboratory.* Cumitech 11. Washington, D.C., American Society for Microbiology, 1980.

Jungerman, P. F., and Schwartzman, R. M.: *Veterinary Mycology*. Philadelphia, Lea & Febiger, 1972.

Rogers, A. L. (Coordinator): Workshop—Identification of Saprophytic Fungi Commonly Encountered in a Clinical Environment. Washington, D.C., American Society for Microbiology.

Rippon, J. W.: *Medical Mycology*, 2nd ed. Philadelphia, W. B. Saunders Company, 1982.

Chapter 29

DERMATOPHYTES AND DERMATOPHYTOSES

HAROLD A. MCALLISTER

THE DERMATOPHYTOSES

Introduction

THE TERM RINGWORM is commonly applied to superficial fungal infections involving the keratinized layers of the skin and its appendages (hairs, nails, horns, and feathers) in animals and humans. Ringworm fungi are able to penetrate all layers of the skin, but they are generally restricted to the nonliving cornified portions, especially the stratum corneum (1,2,3). Invasion of the subcutaneous and deeper body tissues is not a feature of infection by these fungi (1), and for this reason, they tend to be self-limiting and rarely, if ever, lead to death (4). Nonetheless, ringworm can cause severe lesions and symptomatology with concomitant distress in the host, and since it can be unresponsive to treatment, significant economic losses may follow in its wake (4).

In medical and veterinary practice, ringworm has been known as tinea, dermatomycosis, or dermatophytosis. Considering that it is specifically due to a closely related group of mycelial keratinophilic fungi, the dermatophytes, there has been increasing acceptance of the term "dermatophytosis" in the scientific community (3,5,6,7,8,9). "Dermatomycosis" is deemed undesirable in this context because it would include any fungus infection of the skin, such as secondary spread from a systemic mycosis or infection by *Candida* spp. (3) and by agents of the superficial mycoses, such as *Trichosporon* spp. (1). "Dermatophytosis" eliminates ambiguity by emphasizing the etiologic relationship between the disease and its causative agents.

Tinea is an ancient Roman name pertaining to insects and their larvae, animals with which they associated the disease (3). Although it is retained in the human clinical literature, this term is technically inappropriate.

Origin and Distribution

There are at least thirty-eight known species of dermatophytes, many of which have only been recovered from soil (3). Traditionally, these organisms have been listed in the class Deuteromycetes (Fungi Imperfecti), but the perfect or ascus-bearing state has been described for sixteen of them (1,3,6,9).

These organisms are so closely related antigenically and physiologically that the asexual stages are placed in only three genera, namely *Microsporum, Trichophyton*, and *Epidermophyton*. In keeping with this fact, two genera from the class Ascomycetes (Euascomycetes), *Nannizzia* and *Arthroderma*, closely correspond to the traditional *Microsporum* and *Trichophyton* respectively. Moreover, the genus *Epidermophyton* is probably monotypic, i.e. constituted by the single *E. floccosum*, which is almost exclusively confined to humans (4).

Dermatophytes are categorized as geophilic, zoophilic, and anthropophilic depending on the habitat in which they are most likely to be found. Geophilic dermatophytes inhabit the soil and normally exist as free-living saprophytes. Zoophilic dermatophytes are primarily detected as parasites of animals other than humans, while humans serve as the main host for the anthropophilic types. All three groups nonetheless include species that can cause disease in both animals and humans. Transmission from animal to animal, from animals to humans, from one human to another, and from soil to either animals or humans are all possible. The mode of transmission depends on the type of dermatophyte involved and the individual circumstances surrounding the infection.

Dermatophytes are currently regarded as Ascomycetes in the family of soil keratinophiles, the Gymnoascaceae (1,3,9). Substrates such as soil, oatmeal, keratin, feathers, and others are sometimes necessary to bring about fruiting body formation, even if mated pairs are used in demonstrating the perfect phase (3). All dermatophytes may have originated from soil forms, but a significant number appear to have abandoned their saprophytic existence to become parasites (2,3). This adaptive process appears to entail losses in their sexual abilities, especially in the anthropophilic forms. Increasing affinity to the human host is believed to result in the gradual loss of both the sexual state and the ability to produce asexual spores (3). Certain species of dermatophytes have so far defied all efforts to induce a sexual state (3).

The perfect or sexual state of the dermatophytes is useful in the identification of certain species as well as in epidemiologic studies. Rebell and Taplin (10) suggest that adherence to rules of nomenclature requires the use of the sexual rather than the conidial state name. As long as the ascigerous state of many dermatophytes remains unknown, however, the use of dual designations is imperative.

Demonstrating the perfect state requires special mycologic techniques, and clinicians are more likely to be concerned with the conidial state seen in isolates from clinical specimens. Moreover, the fact that different species of Ascomycetes can produce the same conidial form may also introduce confusion. *Microsporum gypseum*, for example, is the conidial form of both *Nannizzia gypsea* and *N. incurvata, Trichophyton mentagrophytes* is the conidial form of both *Arthroderma benhamiae* and *A. vanbreuseghemii*, and *Trichophyton terrestre* is that of *Arthroderma insingulare, A. lenticularum,* and *A. quadrifidum* (1,3,9). The nomenclature applicable to the conidial state will therefore be largely adhered to throughout this chapter.

Diagnosticians with the need for inducing and studying the perfect state of the dermatophytes should consult mycology texts such as those by Rippon (3) or Emmons et al. (6) and manuals such as that by Beneke and Rogers (9) for specific laboratory techniques. These procedures are not routinely employed in the diagnosis of dermatophytosis by the smaller laboratories and are beyond the scope of this book.

Clinical Disease

There are no fundamental differences in the clinical manifestations of infections produced by different dermatophytes; ringworm or dermatophytosis should thus be regarded as a single clinical entity regardless of its causative agent (3,11,12,13). Dermatophytes have adapted to survival in the skin of particular hosts (2,4,6,9): *Microsporum nanum* in pigs, *M. canis* in the cat, *M. persicolor* in voles, *Trichophyton rubrum* in man, *T. verrucosum* in cattle, *T. erinacei* in hedgehogs, and *T. mentagrophytes* in rodents. Isolation of such host-adapted dermatophytes may therefore be of no clinical significance, unless there are lesions or there is transmission to susceptible individuals. In the cat, for example, *M. canis* infection is most often subclinical or inconspicuous, and attention is first drawn to the parasite following the development of ringworm in another host species after contact with the cat (3). Essentially all rodents carry *T. mentagrophytes* as normal flora with no significant signs of disease (3). When these animals act as vectors in transmitting their specifically adapted dermatophytes to other animal species, the recipients develop the eruptive symptomatology associated with ringworm (2). Human *T. verrucosum* infections are seen most often in farmers who are in contact with cattle (6).

As in all infectious processes, the lesions that develop are dependent not only on the invading agent but also on the host's reactivity to it (2). When the dermatophyte exceeds the limits of a balanced host-parasite relationship or the reactive threshold of the host is reached, clinically recognizable ringworm ensues (2). Since the organism is not usually capable of surviving the resulting inflammatory reaction, it tends to move away peripherally toward normal adjacent skin (2). This process tends to create the classical ringed lesions of alopecia with central healing and peripheral inflammation. Ringworm, however, manifests itself in many ways, ranging from the asymptomatic carrier state to the nodular or tumorous lesions called kerions; the classical ringed lesions may in fact be the exception rather than the rule (2).

The dermatophytes' ability to hydrolyze keratin may cause some damage to the epidermis and the hair follicles, but the mechanism whereby they actually produce disease is through hypersensitivity to their irritants and allergens (2,3,6,10,11,12). Multiple sterile vesicles known as dermatophytids or "id" lesions may appear anywhere in the epidermal cover as part of an allergic reaction to the hematogenous spread of fungal products (3,6,11,12). In humans, they are most often seen in the hands (12). "Id" reactions are not seen in most natural infections of animals, however (6). Secondary bacterial invaders such as *Staphylococcus aureus* may further contribute to the development of lesions in dermatophytosis, often generating pustules in the hair follicles (11).

In dogs and cats, lesions develop most frequently on the head and the extremities (11), but the infection can become generalized through the body, with weeping, crusted lesions reminiscent of a severe mange infestation (3). The disease tends to be more severe in dogs than in cats (3,4). In the horse, lesions are generally dry, raised, scaly, and most abundant in the saddle and girth area as well as in the hindquarters (3,4). These lesions may become small inflamed ulcers with a purulent exudate that tends to glue hairs together (3). In cattle, ringworm begins as scattered, discrete, circinate lesions with slight skin scaling and hair loss, followed by enlargement to circumscribed plaques up to 10 cm in diameter covered by a grayish white crust (3). These lesions are often severely inflamed and pruritic (3). Chronic, subclinical infections that cause little or no distress or economic losses are the rule in swine (3,4). Gallinaceous birds, including chickens and turkeys, are afflicted by white, moldy, patchy overgrowths of the comb and wattle; this clinical entity is known as favus (3,4). Thick white crusts often develop, and in severe cases of generalized infections, the base of the feathers is involved (3).

Dermatophytes grow upon or within follicular and epidermal cells, particularly in the stratum corneum (13,14,15). Scaling, hyperkeratosis, orthokeratosis, and acanthosis are apt to develop in infected areas, along with congestion and moderate lymphoplasmacytic cell infiltration of the underlying dermis (13,14,15). Intracorneal collections of neutrophils and edema of the superficial or papillary dermis are common findings (14). Degeneration of infected follicles leads to a foreign body reaction with the infiltration of the area by neutrophils and macrophages (14). The accumulation of necrotic debris at the site may obscure many of the hyphal elements. Secondary invasion by bacterial opportunists can locally enhance a pattern of multifocal abscessation.

Although the septate hyphae of the fungus and arthrospores within hair fragments are often seen in routinely stained hematoxylin-eosin histopathology sections, their presence is best demonstrated utilizing special stains. The Periodic Acid-Schiff reaction, Gridley's modification thereof, and Gomori's methenamine silver stains are most useful in this respect (13). The demonstration of dermatophyte-type fungal elements in skin biopsy specimens featuring lesions is a valuable adjunct to culture procedures. It can help to confirm clinical dermatophytosis, although it cannot distinguish the agents of the disease from each other.

Identification of the Etiologic Agents

Studies of colony characteristics, microscopic morphology, and nutritional requirements make possible the differentiation of dermatophyte species. Large, multicellular spores known as macroconidia or macroaleuriospores are numerous in the genera *Epidermophyton* and *Microsporum* (Fig. 29-1). The size and shape of these structures and the thickness and character of their walls so differ from one species to the other that they constitute an important basis for identification. Unfortunately, macroaleuriospores are few or absent from *Trichophyton* mycelia; although helpful when seen (Fig. 29-2), they cannot be consistently used in the identification of species in the genus. Nutritional requirements and colony characteristics are used instead.

Another criterion useful in the identification of dermatophytes is offered by the arrangement and type of growth of fungal elements along infected hairs. A mosaic of arthrospores outside the hair shaft is said to be ectothrix (Fig. 29-3), while arthrospores in roughly parallel chains inside the hair shaft are said to be endothrix. *Microsporum* hair infections are generally characterized by a small-spored ectothrix pattern of growth (6). The fungus invades the hair shaft by growing downward in it, and on the surface of the hair it forms a sheath of spores that are 2–3 µ in diameter each (6). Different species of *Trichophyton* produce either endothrix or ectothrix arthrospores. The ectothrix types grow into the hair follicle, surround the shaft and penetrate it. Growth of hyphae continues both within and on the surface of the shaft, producing rows of arthrospores by septation at both sites. There are both small-spored and large-spored types in the genus; the latter feature spores that are 3–5 µ in diameter (6). In endothrix infections, the fungus grows from the epidermis into the hair follicle, penetrates the hair, and extends down into it. After this initial period there is no substantial growth on the external surface of the hair shaft (6).

Speciation of *Trichophyton* spp. usually requires the expertise of a reference mycology laboratory because a series of test media with differing basal composition and variously enriched with vitamins must be kept on hand for the purpose; in addition, acid-cleaned glassware should be used for these nutritional tests (9). Laboratories with a need to develop such expertise are advised to consult specialized manuals in laboratory medical mycology such as those by Al-Doory (16), Beneke and Rogers (9), Haley and Callaway (17), and Rebell and Taplin (10).

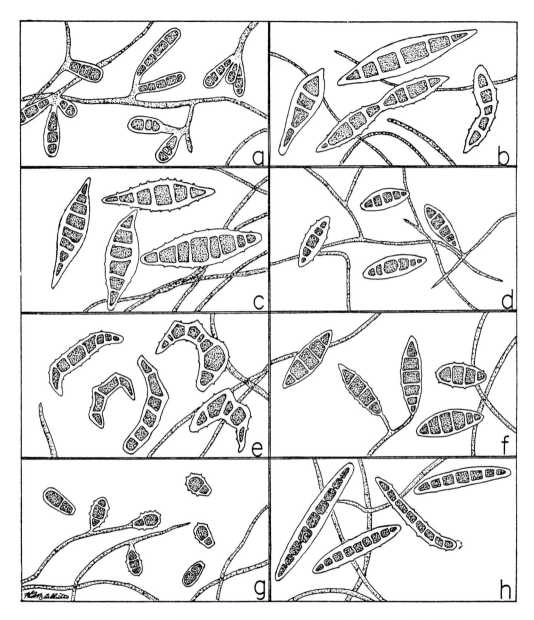

Figure 29-1. Macroconidia: *a, E. floccosum; b, M. audouinii; c, M. canis; d, M. cookei, e, M. distortum; f, M. gypseum; g, M. nanum; h, M. vanbreuseghemii* (H. A. McAllister).

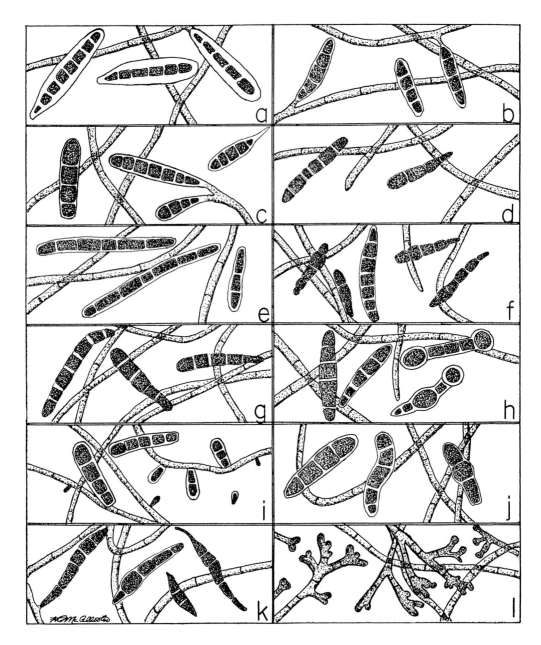

Figure 29-2. Macroconidia: a. *T. ajelloi*; b, *T. equinum*; c. *T. gallinae*; d, *T. gourvilii*, e, *T. megninii*, f, *T. mentagrophytes*, g, *T. rubrum;* h, *T. simii*, i, *T. terrestre*, j, *T. tonsurans;* k, *T. verrucosum;* Favic chandeliers: 1, *T. schoenleinii* (H. A. McAllister).

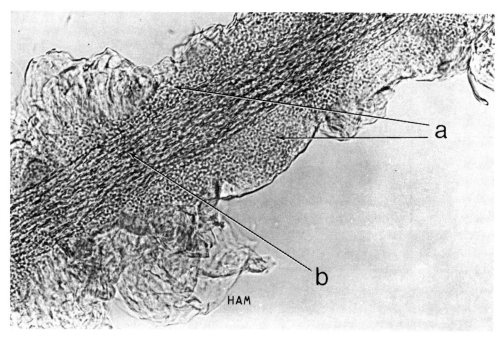

Figure 29-3. Sheath of ectothrix arthrospores around a cat hair: *a*, arthrospores; *b*, hair shaft. ×445 (H. A. McAllister).

Useful identifying characteristics for common dermatophytes of medical importance are provided for the reader's convenience in the figures and in Table 29-I. Many, but not all, dermatophytes should be identifiable down to the species level with the information provided in this chapter. Confirmation by trained mycologists whenever feasible is recommended, however.

LABORATORY PROCEDURES

Preliminary Examination

If possible, examine the patient's lesions in a darkened room using ultraviolet light rays of 3,660 angstrom units (Wood's lamp). Since hairs infected by certain *Microsporum* species produce a marked yellowish green fluorescence under such light, the precise localization of infected areas may be ascertained. Infections caused by zoophilic dermatophytes such as *M. canis* and *M. distortum* can be detected by this means, but almost all *Trichophyton* species and many *Microsporum* species are not fluorescent in hair.

Specimen Collection

1. Skin: wash the infected area with 70% alcohol to remove surface contaminants and collect scrapings from the borders of the lesion.

Table 29-I

COMMON DERMATOPHYTES OF MEDICAL SIGNIFICANCE

Species	Stimulatory or Required Factors	Habitat Preference	Hair Invasion	Hair Fluorescence
E. floccosum	None	Anthropophilic	Hair not invaded	None
M. audouinii	Yeast extract is stimulatory	Anthropophilic	Ectothrix	Bright greenish yellow
M. ferrugineum	None	Anthropophilic	Ectothrix	Bright greenish yellow
M. canis	None	Zoophilic	Ectothrix	Bright greenish yellow
M. distortum	None	Zoophilic	Ectothrix	Bright greenish yellow
M. gallinae	Yeast extract and thiamine stimulatory	Zoophilic	Ectothrix	None
M. gypseum	None	Geophilic	Ectothrix	None
M. cookei	None	Geophilic	Hair not invaded	None
M. nanum	None	Zoophilic	Ectothrix (sparse)	None
T. concentricum	Thiamine is stimulatory	Anthropophilic	Hair not invaded	None
T. megnini	L-histidine on a NH_4NO_3 basal medium are required for growth	Anthropophilic	Ectothrix	None
T. rubrum	None	Anthropophilic	Ectothrix, hair rarely invaded	None
T. schoenleinii	Generally none, thiamine by some strains	Anthropophilic	Endothrix	None or dull
T. tonsurans	Thiamine is greatly stimulatory	Anthropophilic	Large spored endothrix	None
T. violaceum	Thiamine is required	Anthropophilic	Endothrix	None
T. equinum	Nicotinic acid is required (except by T. equinum var. autotrophicum)	Zoophilic	Large spored ectothrix	None
T. mentagrophytes	None	Zoophilic (except var. interdigitale)	Small spored ectothrix	None or dull
T. verrucosum	Thiamine or both thiamine and inositol	Zoophilic	Large spored ectothrix	None

The most common zoophilic species are *M. canis* from cat and dog; *M. gallinae* from chickens and turkeys; *M. nanum* from swine; *T. equinum* from the horse; *T. mentagrophytes* from rodents; and *T. verrucosum* from cattle. Note special requirements for the isolation of some of the species of medical significance. Refer to text for additional information on all species.

2. Nails: scrapings or clippings of nails should be collected from spots near the bed of the nail.
3. Hair: remove hairs from the edges of the lesion by plucking them out with tweezers. The basal portions of hairs usually contain the best material.
4. Place skin scrapings, nails, and hair in covered containers. Cardboard specimen boxes, envelopes, or disposable Petri dishes may also be used. It is important to make sure that the hairs and other debris are not spilling out of the container.
5. Skin biopsies: fix the specimen in 10% neutral buffered formalin equal to at least ten times the tissue volume and allow twenty-four hours for proper fixation. The amount of formalin can be significantly reduced after fixing for convenience in shipping. Forward the specimen to the laboratory in a small amount of formalin within a spill-proof and crush-proof container. Formalin kills the fungi, so separate specimens are required for culture. Neutral buffered formalin is prepared as follows:

Sodium chloride	8.5	g
Sodium phosphate, monobasic, anhydrous	4.0	g
Sodium phosphate, dibasic, anhydrous	6.5	g
Distilled water	900	ml
37–40% formaldehyde solution	100	ml

Dissolve the reagents in water with the aid of heat. Cool, then add the formaldehyde solution. Unbuffered formalin is acidic and promotes artifactual precipitates in histopathology slides. It is not recommended.

Direct Observation of Clinical Material

Thin nail scrapings, skin scabs, and hairs from suspect cases should be examined after the partial digestion of proteinaceous debris with potassium hydroxide. To do this, some of the material is placed on a microscope slide and one or two drops of 10% KOH solution are added. A coverslip is placed over the drops, and then the specimen is allowed to stand for a few minutes with occasional gentle warming (no boiling). Nail scrapings may require several hours for clearing. Microscopic examination should reveal hyphal fragments and arthrospores in infected material, especially around or within hair shafts.

A rapid stain reported by Swartz and Lamkins (18) greatly enhances visualization of the fungus. It is prepared as follows:

1. Wetting agent: dissolve 0.15 g of sodium benzoate and 0.85 g of dioctyl sodium sulfosuccinate in 100 ml of distilled water.

2. Add together and mix: one part 10% KOH solution, one part wetting agent solution, and two parts of permanent blue-black ink.

Staining by this and by comparable alternative mixtures of ink and KOH is less intense than that of lactophenol cotton blue, but the fungi take up the dye selectively and the specimen is cleared with minimal shrinkage (6).

A significant problem with direct observation of clinical material is the presence of large numbers of artifacts in the preparation. Pigment granules (melanosomes) and degenerate keratinocytes within hair shafts are often mistaken for rounded arthrospores, resulting in false positive reports of dermatophyte infection. A network of debris and cholesterol crystals deposited around epidermal cells and known as "mosaic" can effectively mimic septate hyphae (6). Even experienced workers can have difficulty with artifacts. Moreover, various fungi, such as *Alternaria*, can be present in healthy animal fur as incidental contaminants. For these reasons, culture attempts and the demonstration of fungi and lesions in biopsy specimens are needed to confirm clinical dermatophytosis.

Culture

Whether arthrospores have been detected or not, the materials collected should be inoculated on Sabouraud agar with cycloheximide (Actidione®) and chloramphenicol. Mycobiotic agar (Difco) or Mycosel® agar (BBL) may be used instead. These inhibitory media are very useful in the isolation of dermatophytes from clinical materials heavily contaminated with bacteria and saprophytic fungi, although they sometimes cannot prevent the growth of *Alternaria* species and of other resistant bacteria and fungi.

It should be kept in mind that some dermatophytes (2,3,7,9,10,12,16,17,19) have special nutritional requirements (i.e. *T. equinum*, nicotinic acid; *T. megninii*, L-histidine; *T. tonsurans*, thiamine; *T. verrucosum*, thiamine or both inositol and thiamine; *T. violaceum*, thiamine) and that these organisms may fail to grow on ordinary culture media. For this reason, failure to recover a dermatophyte in culture cannot automatically preclude a diagnosis of dermatophytosis. Appropriate commercially available *Trichophyton* agars, as listed in the species descriptions, should be on hand in attempts to culture dermatophytes from horses and from cattle.

Some authors (2,6,9) recommend agar slants in cotton-plugged test tubes for fungus cultures. Petri dishes are advised against because spores may escape into the air when the lid is lifted (2). Although this criticism is valid, primary cultures, especially those from primary sources, almost always yield a mixed flora that cannot be effectively resolved into its components when it develops in small tubes and bottles. In a tube, the contaminant fungi can completely overgrow dermatophytes, preventing their detection. In addition, the narrow neck makes it difficult to reach in with forceps to prepare tape mounts.

Isolated colonies of fungi are best obtained by scattering hairs, nails, and scabs throughout the surface of an agar plate and pressing them down gently with forceps or a swab. The size of the inoculum should not be excessive, and about 10 mm should be allowed between individual crusts or hairs. The plates are sealed with masking tape to delay drying and to prevent the escape of spores; they are then incubated at 25°C (except for *T. verrucosum*), lid up, for not less than two weeks. The plates must be handled without inverting them, since this maneuver will dislodge spores from the aerial mycelium; the spores fall on the lid and back on the plate, producing additional growth on the agar. Due care and possibly the use of a hood are advisable when opening and closing plates. A good precaution is to place a paper towel soaked in 2% Amphyl® disinfectant (or a similar fungicide) over the work area of the table and to fold and discard it upon completion of the work (2). If growth is fast and the aerial mycelium reaches the lid, then the plate must not be opened at all except under a biologic safety hood.

Dermatophyte test medium (Fungassay®, Pitman-Moore; DTM®, Charles Pfizer & Co.; Dermocult®, Medical Technology Corp.) may be used as a primary culture medium. It can also be used as a differential agar. In the latter case, a suspicious isolate from the original plate is transferred with teasing needles to dermatophyte test medium to confirm identity as a dermatophyte. Such fungi are easily recognizable on this agar because on it they show a white or near-white aerial mycelium and they change the medium's color from a dark yellow to red. Unfortunately, this characteristic color change can be produced by some saprophytic fungi, yeasts, and bacteria. These contaminants, however, do not produce a white cottony mycelium.

Final Identification

Plates with colonies of fungi should be examined after the mycelia attain a diameter of 10 mm or more but before individual colonies of different types begin to merge with each other. Colonial characteristics such as texture, pigment, and rate of growth must be recorded. The mycelium can be examined effectively using a modification of Endo's (20) tape mount procedure. Perfectly transparent tape such as Scotch® brand cellophane tape or Scotch® brand double-sided cellophane tape is used to mount the hyphal elements with minimal disturbance of the spatial relationships between the reproductive structures.

The procedure is as follows:
1. Place a drop of lactophenol cotton blue fluid on a microscope slide and another one on a coverslip; neither drop should be too large.

2. Cut a piece of tape smaller than the coverslip and grasp one corner with a forceps, keeping the tape free of dirt and fingerprints.

3. Place the adhesive side of the tape against the fungus and apply gentle pressure with a dissecting needle. Caution: excess pressure will tear off a thick mycelial mass unsuitable for viewing, and spores may be scattered in the air.

4. Gently transfer the tape to the coverslip. Place a corner against dry glass and the fungus-bearing portion directly over the mounting fluid.

5. Using the dissecting needle, pry the tape loose from the forceps and press the dry corner gently against the glass.

6. Invert the coverslip and place it over the mounting fluid on the microscope slide. Gently press out any air bubbles trapped under the coverslip. Fingernail polish may be used to seal the preparation and prevent quick drying. Note that the only material between the fungus and the microscope objective is the coverslip glass.

Longer-lasting preparations with a more perfect rendition of the spatial relationships are possible with the slide cultures discussed elsewhere in this book, as well as by Beneke and Rogers (9).

A statement of identity for a fungal isolate recovered in a laboratory, when issued, constitutes an opinion. Definitive identification, especially for publication purposes, requires concurrence with that opinion by other qualified workers. To secure such concurrence, the record offered for independent objective scrutiny should include a thorough description of the gross mycelial appearance and drawings or photographs of all reproductive structures observed. The result of ancillary nutritional tests, if any, should be provided as well. Failure to record such data and failure to include it in published accounts very seriously compromise the validity of any alleged final identification.

THE DERMATOPHYTES

Epidermophyton (1 species)

The only widely accepted species in this genus, *E. floccosum*, infects only skin and nails, and it is virtually confined to humans (4). Reports of infections in domestic animals are essentially nonexistent (2,9) but there are some of isolations from mice and from a dog (4,7,9). Since hair is not attacked, there is no fluorescence under Wood's lamp. The perfect state has not been reported. Colonies are usually greenish yellow, olive, or khaki, powdery, with radial furrows and a yellow to tan reverse. Macroaleuriospores are numerous in young cultures, but sterile hyphae may overgrow strains maintained in serial cultures. The macroaleuriospores are two to six celled, fingerlike in shape, and arranged in clusters of two, three, or more. Microaleuriospores are absent.

Some workers (1) accept a second and possibly geophilic species in the genus, *E. stockdaleae*.

Microsporum (15 species)

M. amazonicum

This dermatophyte produces distinctive macroaleuriospores. It is rare, having been described only from the apparently normal hairs of certain native Brazilian rats (10). The colonies are fluffy to powdery with an olive gray buff color (10). Most of the macroaleuriospores have four cells, but there may be as many as eight; the cell walls and septa are thick, there is an oil-droplet-like inclusion in each cell, and the surface of the spore is echinulate (10). The sexual state is *Nannizzia borelli*.

M. audouinii

This anthropophilic fungus rarely attacks animals, although isolates have been reported from the dog, the cat, the guinea pig, the rabbit, the gibbon, and a monkey. The correctness of such reports has been questioned (6). The organism produces a yellow-greenish fluorescence in hair, where small ectothrix spores can be demonstrated. The perfect state has not been reported. The colonies are usually light tan to brown with a buff salmon to orange brown reverse. Macroaleuriospores are rare, irregular or spindle shaped, with two to nine cells; microaleuriospores are sessile or on short stalks, clavate, and single celled. Unlike the closely related *M. canis*, this fungus grows poorly on rice (9). The addition of yeast extract to the medium stimulates the formation of macroaleuriospores by some strains (9).

M. boullardii

M. boullardii is geophilic, and it has been isolated from soil but not from animals. Its macroaleuriospores resemble those of *M. fulvum*, but unlike the latter, it fails to produce ascospores with *Nannizzia fulva* (10).

M. canis

M. canis is generally regarded as a zoophilic dermatophyte, but it often attacks humans. Most human infections are acquired from animals. It is the etiologic agent of roughly 98 percent of the cases of feline ringworm and about 70 percent of the cases of canine ringworm in North America (2); in cats, the infection is most often subclinical, whereas in dogs, the disease is more obvious (3,4). The head is the most common site of infection, with areas of alopecia around the nose, eyes, and ears, but the infection can become generalized (3). In addition to humans, cats, and dogs, numerous animals can serve as hosts: bats, canaries, cattle, chimpanzees, chinchilla, donkeys, foxes, gibbons, goats, gorillas, guinea pigs, horses, jaguars, lions, lynx, monkeys, orangutans, pigs, rabbits, sheep, tigers, and others (4).

M. canis produces a yellow-greenish fluorescence in hair, and it is associated with small ectothrix spores. Its colonies are white to buff in color with a characteristic yellow to orange brown reverse. The macroaleuriospores are numerous, spindle shaped, with thick walls and six to fifteen cells; the microaleuriospores are rare, small, clavate to elongate, and single celled. The ascomycetous state of *M. canis* is *Nannizzia otae*.

M. cookei

This is a geophilic dermatophyte rarely recovered from animals. It has been isolated in the absence of clinical lesions from a variety of hosts, including baboons, dogs, cats, humans, marsupials, monkeys, rabbits, reptiles, and various wild animals, including rodents (4,7,9,10). It is not known to invade hair and it produces neither fluorescence nor arthrospores. *Nannizzia cajetani* represents its perfect state. Colonies are powdery yellowish or dark tan with a deep purplish red reverse. The macroaleuriospores are numerous, ellipsoidal, thick walled, with four to six cells; the microaleuriospores are abundant and obovoid.

M. distortum

M. distortum is a zoophilic dermatophyte that rarely infects humans. It has been isolated from ringworm in dogs, cats, horses, swine, guinea pigs, monkeys, rabbits, and humans (4,7). It produces a yellow-greenish fluorescence in hair, where it is associated with small ectothrix spores. The perfect state has not been reported. Colonies are white to tan, fluffy, with a yellow to tan reverse. The macroaleuriospores are numerous, thick walled, with six to fifteen cells, bent, and distorted; the microaleuriospores are clavate and sessile.

M. ferrugineum

This is an anthropophilic dermatophyte not yet isolated from animals (9). It produces a yellow-greenish fluorescence in hair, where it forms small ectothrix spores. The perfect state has not been reported. Colonies are waxy, slow growing, deep yellow to orange, with many deep furrows and may be completely lacking in both macro- and microaleuriospores.

M. fulvum

M. fulvum is a geophilic species occasionally involved in dermatophytosis; it has been recovered from humans, pigs, jaguars, lions, and certain types of goats and monkeys (4). Hair invasion is by ectothrix spores without fluorescence (3), and the perfect state is *Nannizzia fulva*. The mycelium resembles that of the *M. gypseum* complex, and in fact, this organism may often have been recorded as *M. gypseum* in the past (4). The colony has a dense downy to floccose chamoislike surface, and it is tawny buff in color with a white periphery. There are large numbers of macroaleuriospores similar to those of *M. gypseum* but longer, more clavate or bullet shaped, and frequently lateral rather than conspicuously clustered (10). Microaleuriospores are also numerous.

M. gallinae (Synonym: Trichophyton gallinae)

This is a zoophilic dermatophyte that rarely attacks humans. It is primarily a cause of ringworm (favus or white comb) in gallinaceous birds such as chickens and turkeys, but it has also been recovered from cats, dogs, laboratory mice, monkeys, pigeons, and quail (4,9). Favus is characterized by a white, moldy, patchy overgrowth on the comb and wattle (3). White crusts may develop, and the infection may become generalized, involving the base of feathers; in chickens, the inflammatory reaction can lead to focal necrosis (3).

In mammals, hair invasion is ectothrix without fluorescence. The perfect state has not been reported. Colonies are heaped, radially folded, with a deep red color that diffuses throughout the medium. Macroaleuriospores may be abundant. They are generally spatulate or slipper shaped, i.e. elongate with a blunt tip, smooth walled, and with two to ten cells. Microaleuriospores are pyriform to clavate, single or in clusters, and few in number. The addition of thiamine or yeast extract to the medium may increase sporulation (6,9).

M. gypseum

M. gypseum is geophilic, but it often attacks humans and animals. It has been isolated from baboons, buffaloes, cats, cattle, chickens, chimpanzees, chinchillas, dogs, fowl, guinea pigs, horses, leopards, monkeys, mice, parrots, rabbits, rats, squirrels, tapirs, tigers, and other species (4,9). Large ectothrix spores are typical but few in number, and fluorescence is absent or dull. The perfect state is represented by both *Nannizzia incurvata* and *N. gypsea*. The colonies grow fast, producing a flat, powdery, buff to cinnamon brown surface with a pale yellow to tan reverse. Upon microscopic examination, the powder is revealed as virtually solid masses of macroaleuriospores. These are large, rough, ellipsoid, thin walled, ranging from three- to nine-celled types, but preponderantly four to six celled. The microaleuriospores are clavate and sessile.

M. nanum

This dermatophyte is a zoophilic species that rarely infects humans, and it has been recovered from the soil only in pigyards. The organism periodically causes ringworm in swine, infecting up to a third of the herd; it is most commonly seen in Yorkshires, but all breeds can be affected (3). There have been no isolations of this fungus in animals other than pigs and humans (4,6,9). The lesions caused by *M. nanum* in swine are usually mild (3). They may cover large areas of the body, but the initial reaction subsides to leave only inconspicuous scaling and discoloration without alopecia or systemic disturbances (3). The disease becomes chronic and subclinical, and once established, there is little tendency for a spontaneous cure (3). Ectothrix spores with little or no fluorescence are produced in hair. The perfect state is *Nannizzia obtusa*. Colonies are white to buff or yellow, cottony, with a red to brown reverse. Characteristic ovoid, clavate, or pear-shaped macroaleuriospores with thin walls and one to three cells are produced in large numbers; microaleuriospores are rarely produced on Sabouraud's agar.

M. persicolor

M. persicolor is a zoophilic dermatophyte very rarely isolated from humans. It is a frequent mild pathogen of small wild rodents, particularly bank voles, where it produces tail lesions (4,6,10). It has also been isolated from field voles, bats, dogs, guinea pigs, shrews, and mice (4). It does not parasitize hair (10), and its perfect state is *Nannizzia persicolor*. Colonies are flat to gently folded, fluffy, yellowish buff to pale pink, with a reverse that ranges from peach or rose to a deep shade of ochre. Clavate, fusiform, or globose microaleuriospores in clusters similar to those of *T. mentagrophytes* are abundant, while macroaleuriospores are rare, elongate, fusiform to clavate, thin walled, and usually six celled. The organism may be confused with *T. mentagrophytes*.

M. praecox

This is a geophilic dermatophyte once isolated from a pustular vesicular lesion in a human wrist (10). No perfect state is known. Colonies are powdery buff with a yellow orange undersurface, similar to *M. gypseum*. The macroaleuriospores are also similar to those of *M. gypseum*, but they are longer, narrow and lanceolate in profile, and have six to nine cells.

M. racemosum

M. racemosum is a geophilic species isolated from the soil and from rats (4,10). The perfect state is *Nannizzia racemosa*. Colonies are flat, powdery, cream white, fast-growing, and with a grape red undersurface. The macroaleuriospores are thin walled, with five to ten cells, tapered, and frequently bear a terminal filament. Distinctive club-shaped, mostly stalked microaleuriospores are found in large wandlike clusters (10).

M. vanbreuseghemii

This dermatophyte is geophilic, and it rarely attacks humans or animals. It has been recovered from a dog, a squirrel, and humans (3,4,6,10). In infected hair, it produces ectothrix spores and no fluorescence or very poor fluorescence. The perfect state is *Nannizzia grubyia*. Colonies are flat, powdery, creamy yellow to pink, with a colorless or yellow reverse. Numerous long cylindrofusiform, rough, thick-walled macroaleuriospores with seven to ten cells are produced in culture, along with pyriform or obovate microaleuriospores.

Trichophyton (22 species)

T. ajelloi

T. ajelloi is a geophilic saprophyte of very low pathogenicity that rarely infects humans or animals. Isolations have been made from baboons, dogs, cattle, horses, marsupials, monkeys, mice, guinea pigs, squirrels, and humans (3,4). It does not attack hair, and the perfect state is *Arthroderma uncinatum*. Colonies are downy, cream to orange tan, with a colorless, reddish, or bluish black reverse. The macroaleuriospores are smooth, cylindric, with tapering ends and five to twelve cells. The microaleuriospores are abundant, sessile, and pyriform to ovate.

T. concentricum

This is an anthropophilic species not known to infect animals (9). It does not attack hair, and the perfect state has not been reported. Colonies are white to cream colored, deeply folded, with a cream to brown reverse. No macroaleuriospores or microaleuriospores are produced. The growth of at least half of the isolates is stimulated by the addition of thiamine to the medium (6,9).

T. equinum

T. equinum is a zoophilic species that rarely attacks humans; it causes ringworm in horses and donkeys, and, occasionally, in dogs (4,9). Colts and yearlings are most susceptible, at first developing swellings that can be felt through the hair. These can progress to small inflamed ulcers with an exudate, a condition often known as girth itch (3). Alopecia develops as the lesions enlarge and turn chronic; crusts may fall off from healed lesions, leaving bald areas with a "motheaten" appearance (3). Hair invasion is by large ectothrix spores; there is no fluorescence, and the perfect state has not been reported. Colonies are white to cream colored with a bright yellow to dark pink or brown reverse. Macroaleuriospores are extremely rare, but variable numbers of thin, elongate to pyriform stalked microaleuriospores are usually seen. All strains require nicotinic acid (niacin) for growth except those from Australia and New Zealand (3). The organism will grow on commercially available *Trichophyton* agar #5 (nicotinic acid-casein agar) but not on ordinary Sabouraud's (16).

T. erinacei

T. erinacei is a zoophilic dermatophyte primarily associated with ringworm in hedgehogs, but occasionally it has been isolated from the mouse and rat, the dog, and humans (4,16). Hair invasion is normally ectothrix (10). The perfect state has not been reported. Colonies are flat, powdery, white to ivory white, with a clear yellow diffusing pigment underneath (16). The macroaleuriospores are club shaped, with fewer cells than those of *T. mentagrophytes*. The microaleuriospores are also club shaped and produced abundantly on terminal hyphal branches. Forms intermediate in appearance between microaleuriospores and macroaleuriospores are numerous (10).

T. georgii

This geophilic species is most likely a soil saprophyte, but it has been isolated from the opossum (4). It does not attack hair, and the perfect state is *Arthroderma ciferri*. Its colonies are pale brown with a spotted brown reverse. Microaleuriospores are abundant and variable in size and shape, but no macroaleuriospores are produced.

T. gloriae

T. gloriae is also geophilic, and it has not been isolated from humans nor animals (9,10). It does not attack hair, and the perfect state is *Arthroderma gloriae*. Colonies are flat, downy to powdery, and they vary from white, cream, or yellowish to cinnamon in color. They are chrome yellow underneath. Abundant narrow macroaleuriospores vary in size and shape from one-celled rod- or spindle-shaped types to long cylindrical ones with up to ten cells (10). They occur mainly on terminal hyphal branches where they are arranged laterally in groups or clusters (10). Small pyriform microaleuriospores are present laterally on the hyphae; they are usually distinguishable from the one-celled macroaleuriospores (10).

All strains are partially dependent on thiamine for growth and enrichment enhances sporulation; chlamydospores and hyphal swellings are the usual structures seen on Sabouraud's agar (9).

T. gourvilii

This anthropophilic dermatophyte has not been isolated from infections in animals. Hair invasion is endothrix without fluorescence, and the perfect state has not been reported. Some workers regard it as a soil saprophyte (9).

Colonies are folded, heaped, waxy, with a lavender to deep red pigmentation (16). Macro- and microaleuriospores are usually present in very small numbers.

T. longifusum

T. longifusum is probably geophilic. It has been isolated from soil and from one human (10). The perfect state has not been reported. Colonies are fluffy to powdery, white to pale yellowish, with a yellowish brown undersurface. There are no microaleuriospores, but long, cylindrical, distinctive macroaleuriospores are seen. They develop in clusters and characteristically arise from each other, merging into the terminal hyphae (10).

T. megninii

This anthropophilic dermatophyte is rarely isolated from animals, but it has been recovered from dogs, cats, cattle, mice, and chickens (4,7). Hair invasion is ectothrix without fluorescence, and the perfect state is not known. Colonies are white to pink with a nondiffusible rose to red pigment on the reverse side (9). L-Histidine is required for growth. The organism grows on commercially available *Trichophyton* agar #7 (histidine-ammonium nitrate agar). Macroaleuriospores are rare, clavate, two to ten celled, with thin smooth walls, while pyriform to clavate microaleuriospores are very numerous.

T. mentagrophytes

T. mentagrophytes is primarily a zoophilic dermatophyte that often attacks humans and may also survive saprophytically in the soil (9). One variety, *T. mentagrophytes* var. *interdigitale*, is anthropophilic (1). Infections by *T. mentagrophytes* have been reported in a large number of wild and domestic animals, including baboons, buffaloes, cats, cattle, chinchillas, chickens, chimpanzees, dogs, foxes, guinea pigs, goats, horses, hedgehogs, kangaroos, kiwis, mice, monkeys, marsupials, nutrias, opossums, polecats, porcupines, parrots, rabbits, rats, rodents, swine, sheep, squirrels, tapirs, tigers, and others (4,7,9,10). *T. mentagrophytes* is probably the commonest dermatophyte (3).

Hair invasion by *T. mentagrophytes* is usually small-spored ectothrix and without fluorescence, but some strains may be endothrix with a dull fluorescence (3). The latter (var. *quinckeanum*) have been implicated in mouse favus, which produces numerous white crusty lesions throughout the body. Most rodents, however, carry *T. mentagrophytes* as normal flora without signs of disease (3). Two perfect states are known: *Arthroderma benhamiae* and *A. vanbreuseghemii*. Colonies are powdery or granular, light buff to rose tan with a buff to deep wine or brown reverse. Macroaleuriospores are three to five celled, thin walled, clavate, and not too abundant. The most consistent microscopic feature is the production of large numbers of microaleuriospores in grapelike clusters, especially in the zoophilic strains.

T. phaseoliforme

This species has been isolated from soil and from normal fur in rodents (10). The perfect state has not been reported. Colonies are white to bright cinnamon, eventually developing white or ivory colored, spherical, ascocarp-like structures on the surface (10). Cylindrical macroaleuriospores are borne in terminal clusters, and they include one-celled types resembling micro-aleuriospores. Large numbers of microaleuriospores develop laterally on both vegetative hyphae and in the ascocarplike structures; they are typically curved and cashew nut shaped (10).

T. proliferans

T. proliferans has been isolated from humans but may be present in the soil (10). No perfect state is known. Colonies are white, fluffy, with surface folds, and the reverse may develop a greenish yellow diffusing pigment that may intensify in color (10). The macroaleuriospores range from small unicellular rods to well-developed cylindrical structures of the *Trichophyton* type, but clearly differentiated microaleuriospores are absent (10). The most distinctive feature is the presence of "propagules"; these are lateral, multicelled, cylindrical, or spindle-shaped structures giving rise to multiple frondlike hyphae (10).

T. rubrum

T. rubrum is an anthropophilic species rarely isolated from animals (3). So far, it has been recovered from a baboon, a dog, a cat, some cattle, the chimpanzee, a rabbit, a guinea pig, sheep, and a mouse (4,7,9,10). This species has never been isolated from soil (3). It appears to be dependent on humans for dissemination, and no perfect phase is known (3). Hair is rarely invaded, but when it is, ectothrix spores in chains are the rule (3,6,7). There is no fluorescence. Most colonies are white and cottony with a reddish to rose purple reverse. Macroaleuriospores are absent or rare, but "teardrop" micro-aleuriospores lateral to the hyphae or in "pine tree" clusters may be seen.

T. schoenleinii

This anthropophilic dermatophyte is rarely isolated from animals, but it has been recovered from dogs, cats, hedgehogs, mice, cattle, horses, rabbits, and guinea pigs (4,7,9,10). Hair invasion involves rare endothrix spores without fluorescence or with a dull grayish to yellow fluorescence (6).

Longitudinal tunnels produced within the hair shaft are filled with air bubbles after disintegration of the hyphae (1,6). No perfect state has been reported. Colonies are slow growing and waxy with many irregular folds and a yellowish to light brown color. Macroaleuriospores are absent and micro-aleuriospores rare, but numerous "favic chandeliers" are present; these structures are broadened hyphal tips with short lateral and terminal branches. Favic chandeliers do occur in other *Trichophyton* spp., but they are generally rare (1,6,9). The growth of some isolates is stimulated by thiamine (1).

T. simii

T. simii may be a soil saprophyte, but it is presently regarded as a zoophilic species that rarely attacks humans. It causes ringworm in poultry and in monkeys; it has also been isolated from chimpanzees, dogs, guinea pigs, horses, rabbits, gerbils, mice, rats, shrews, and squirrels (4,7,9,10). Hair invasion is both ecto- and endothrix, frequently with a vivid green fluorescence. The perfect state is *Arthroderma simii* (6,9). Colonies are white to pale buff in color with a colorless to vinaceous reverse. Macroaleuriospores are numerous, cylindrical in shape, with four to ten cells. Microaleuriospores are rare at first, but they increase in number as the culture ages.

T. soudanense

This anthropophilic species is not known to attack animals (9). Hair invasion is endothrix without fluorescence, and the perfect state is not known. Colonies are slow growing, lemon yellow to apricot colored, with a yellow to orange reverse. The species produces no macroaleuriospores and variable numbers of ovoid, clavate, or pyriform microaleuriospores (9,16).

T. terrestre

T. terrestre is a geophilic species of very low pathogenicity. It has been isolated from the hair of animals including badgers, cats, dogs, hedgehogs, horses, mink, moles, monkeys, mice, opossums, polecats, rats, and others as well as humans, but it does not invade hair (4,9). Lesions have not been demonstrated in these cases (9,10). The perfect states are *Arthroderma insingulare*, *A. lenticularum*, and *A. quadrifidum* (9). Colonies are powdery, velvety, pale lemon to buff, with a yellow or reddish reverse. Clavate to pyriform microaleuriospores are abundant, as well as thin-walled, cigar-shaped, two- to twelve-celled smooth macroaleuriospores. Transitional forms from micro- to macroaleuriospores are also abundant (9).

T. tonsurans

This anthropophilic species so far has been isolated from only two animals, a horse and a dog (4,7,9). Hair invasion is large-spored endothrix without fluorescence (6). The perfect state has not been reported. Colonies are folded, velvety to powdery, with considerable color variation. Macroaleuriospores are rare, thin walled and club shaped, but microaleuriospores are numerous, attached to the hyphae by short sterigmata. Growth is greatly stimulated by the addition of thiamine to the medium (6,9).

T. vanbreuseghemii

T. vanbreuseghemii is a geophilic species of doubtful pathogenicity so far isolated only from the dog and the cat (4). Animal infections are rare (9). It has also been implicated in dermatophytosis in the hand of a forestry worker (10). Hair invasion is ectothrix (10). The perfect state is *Arthroderma gertleri*. Colonies are buff colored, with the texture of fine glove leather and central folds (10). The macroaleuriospores are plump, thin walled, cylindrical or club shaped, with rounded ends, and they appear in small clusters; they are reminiscent of *T. ajelloi* (10). The individual cells of the macroaleuriospores usually separate from each other as they mature. Microaleuriospores are club shaped and stubby, and they develop mainly along the hyphae.

T. verrucosum

This zoophilic dermatophyte often attacks humans; it has been isolated from buffalo, canaries, cats, cattle, dogs, donkeys, dromedaries, fowl, goats, horses, mules, pigs, sheep, and zebu (4). Presently, *T. verrucosum* is regarded as the primary cause of ringworm in cattle (6,9).

Bovine ringworm is more frequently seen in younger animals, with up to 40 percent of the calves and yearlings infected, and it is more frequent in winter than in summer (3). These facts seem to be accounted for by crowding within buildings, where there is increased contact between animals and with spore-laden debris (3). Cattle ringworm begins as scattered circular lesions with slight scaling and alopecia. The disease may become stationary and chronic, or the lesions may enlarge into plaques covered by thickened crusts that are firmly attached to the animal (3). Their removal leaves a weeping, bleeding, erythematous base (3). Spontaneous healing is the rule, leaving dry, scaly patches with alopecia and scar formation (3). There is no satisfactory topical treatment for ringworm in cattle; good hygiene and sanitation are important for the control of this disease (3).

T. verrucosum produces very large ectothrix spores in chains. Infected hair in humans is not fluorescent, but some fluorescence has been noted in cattle (10). The perfect state has not been reported. Colonies are slow growing, heaped, deeply folded, and white to yellow. The organism may fail to grow in routine media because thiamine is required for growth and many strains also require inositol (9,16). All strains grow on commercially available #3 *Trichophyton* agar and some also grow on #4 agar (9,16). Unlike other dermatophytes, the organism grows best at 37°C (6,9). Macroaleuriospores are rarely seen; they are variable in size and shape with three to five cells and a "rattail" appearance (9). Microaleuriospores are tearshaped, abundant only in media enriched with thiamine.

T. violaceum

T. violaceum is an anthropophilic fungus rarely isolated from animals. It has been reported from buffaloes, horses, cattle, cats, dogs, mice, sheep, a pigeon, and a mule (4,7,9,10). Hair invasion is endothrix without fluorescence, and the perfect state has not been reported. Colonies are heaped, folded, waxy, slow growing, and violet in hue, but growth is minimal in the absence of thiamine. The organism grows well on commercial #4 *Trichophyton* agar (casein-thiamine agar) (9,16). Some pyriform microaleuriospores may develop on thiamine-enriched media. Generally, no macroaleuriospores are seen (9,16).

T. yaoundei

This is an anthropophilic dermatophyte also found as a soil saprophyte; no infections in animals have been reported (9). Hair invasion is endothrix without fluorescence, and the perfect state is not known. Colonies are slow growing, white to cream in color initially, changing to a chocolate brown hue as they mature. Macroaleuriospores are not produced; microaleuriospores are rare (16).

REFERENCES

1. Ajello, L., and Padhye, A.: In Lennette, E. H., Balows, R., Hausler, W. J., and Truant, J. P. (Eds.): *Manual of Clinical Microbiology*, 3rd ed. Washington, D.C., American Society for Microbiology, 1980, p. 543.
2. Jungerman, P. F., and Schwartzman, R. M.: *Veterinary Medical Mycology*. Philadelphia, Lea & Febiger, 1972.
3. Rippon, J. W.: *Medical Mycology*. Philadelphia, Saunders, 1974.
4. Ainsworth, G. C., and Austwick, P. K. C.: *Fungal Diseases of Animals*, 2nd ed. Farnham Royal, Slough, England, Commonwealth Agricultural Bureaux, 1973.
5. Finegold, S. M., Martin, W. J., and Scott, E. G.: *Bailey and Scott's Diagnostic Microbiology*, 5th ed. St. Louis, Mosby, 1978.
6. Emmons, C. W., Binford, C. H., Utz, J. P., and Kwon-Chung, K. J.: *Medical Mycology*, 3rd. ed. Philadelphia, Lea & Febiger, 1977.
7. Dvořák, J., and Otčenášek, M.: *Mycological Diagnosis of Animal Dermatophytoses*. The Hague, Dr. W. Junk N. V., Publishers, 1969.
8. Merchant, I. A., and Packer, R. A.: *Veterinary Bacteriology and Virology*, 7th ed. Ames, Iowa St U Pr, 1967.
9. Beneke, E. S. and Rogers, A. L.: *Medical Mycology Manual*, 4th. ed. Minneapolis, Burgess, 1980.
10. Rebell, G. and Taplin, D.: *Dermatophytes — Their Recognition and Identification*, rev. ed. Coral Gables, Florida, U of Miami Pr, 1970.
11. Carter, G. R.: *Essentials of Veterinary Bacteriology and Mycology*, rev. ed. East Lansing, Mich St U Pr, 1982.

12. Conant, N. F., Smith, D. T., Baker, R. D., and Callaway, J. L.: *Manual of Clinical Mycology*, 3rd ed. Philadelphia, Saunders, 1971.
13. Smith, H. A., Jones, T. C., and Hunt, R. D.: *Veterinary Pathology*, 4th. ed. Philadelphia, Lea & Febiger, 1972.
14. Ackerman, A. B.: *Histologic Diagnosis of Inflammatory Skin Diseases*. Philadelphia, Lea & Febiger, 1978.
15. Muller, G. H. and Kirk, R. W.: *Small Animal Dermatology*, 2nd. ed. Philadelphia, Saunders, 1976.
16. Al-Doory, Y.: *Laboratory Medical Mycology.* Philadelphia, Lea & Febiger, 1980.
17. Haley, L. D. and Callaway, C. S.: *Laboratory Methods in Medical Mycology*, 4th. ed. Atlanta, U.S. Department of Health, Education and Welfare, Center for Disease Control, 1978.
18. Swartz, J. H., and Lamkins, B. E.: *Arch Dermatol, 89:*89, 1964.
19. Vanbreuseghem, R.: *Guide Pratique de Mycologie Médicale et Vétérinaire*. Paris, Masson et Cie., 1966.
20. Endo, R. M.: *Mycologia, 58:*655, 1966.

SUPPLEMENTARY REFERENCES

Ainsworth, G. C., and Austwick, P. K. C.: *Fungal Diseases of Animals*, 2nd ed. Farnham Royal, Slough, England, Commonwealth Agricultural Bureaux, 1973.

Beneke, E. S., and Rogers, A. L.: *Medical Mycology Manual*, 4th ed. Minneapolis, Burgess, 1980.

Dvorák, J., and Otcenásek, M.: *Mycological Diagnosis of Animal Dermatophytoses*. The Hague, Dr. W. Junk N. V., Publishers, 1969.

Jungerman, P. F., and Schwartzman, R. M.: *Veterinary Medical Mycology*. Philadelphia, Lea & Febiger, 1972.

Rebell, G. and Taplin, D.: *Dermatophytes — Their Recognition and Identification*, rev. ed. Coral Gables, Florida, U of Miami Pr, 1970.

Chapter 30

YEASTS CAUSING INFECTION

T HE YEASTS CONSTITUTE A LARGE GROUP, a number of species of which are found as commensals in the gastrointestinal tract of humans and animals. Only a small number of species have been implicated as causes of disease in animals. A wider variety have been involved in infections in humans, perhaps because of the intensive and prolonged use of antibiotics.

The normal vegetative forms of yeast and yeastlike fungi are round or oval cells with a diameter usually in the range of 2.5–6 µm. Their mode of reproduction is by budding (blastospores), and some possess capsules. Pseudohyphae and true hyphae are produced by various species. *Candida* spp. produce chlamydospores of different shapes, and *Saccharomyces* produce ascospores.

Geotrichum candidum will be discussed in this chapter, although it resembles a mold more than a yeast. Its early colony morphology resembles that of the yeasts. The principal pathogenic yeasts of animals are *Candida albicans* and *Cryptococcus neoformans*.

Yeasts are recovered occasionally from clinical materials. Most often they are contaminants or are derived from the normal flora on mucous membranes, e.g. *Candida* spp. The kind of disease and/or lesion will usually lead the clinician or pathologist to suspect a yeast infection. Also, the recovery of a potentially pathogenic yeast in nearly pure culture or in large numbers may indicate disease due to a yeast. The repeated demonstration and recovery of a yeast from a lesion provides strong evidence of its significance. In instances of what appear to be contaminating yeasts, efforts at identification are not usually made.

Yeasts will frequently be seen on primary media such as blood agar and Sabouraud agar. The colonies are usually small initially and can easily be confused with colonies of micrococci or staphylococci; however, they have a yeasty odor and on further incubation they may display pigmentation, as evidenced by a cream, salmon, or tan appearance. The colonies of *Pityrosporum* and *Trichosporon* may become membranous after prolonged incubation. Identification of the colonies as those of yeasts is made by demonstration of the typical yeast forms in wet mounts or in Gram-stained smears. Table 30-I is provided to aid in the identification of the principal genera of yeasts encountered in clinical specimens.

Both carbohydrate fermentation and assimilation tests are used in the

TABLE 30-I
DIFFERENTIATION OF GENERA OF YEASTS FOUND IN CLINICAL SPECIMENS

	Microscopic morphology	Pseudomycelium	Capsule	Chlamydospores	Urease	Other
Candida	Spherical	+	−	*Candida albicans* *C. stellatoidea* *C. tropicalis**	most negative	*Candida albicans* produces germ tubes
Cryptococcus	Spherical, oval, or elongate	Rudimentary or none	+	−	+	Colonies often very mucoid
Pityrosporum	Bottle-shaped, monopolar budding	Pseudomycelium rare	−	−	+	Growth stimulated by fatty acids
Rhodotorula	Spherical, oval, or elongate; multilateral budding	−	+	−	+	Orange to red pigment
Saccharomyces	Spherical, oval, cylindrical, or elongate; multilateral budding; ascospores, usually 4	−	−	−	−	
Torulopsis	Spherical, oval, or elongate; multilateral budding	−	−	−	−	Predominantly nonfermentative (except glucose) non-assimilatory
Trichosporon	Various shapes (barrel-shaped arthrospores)	+ and true hyphae	−	−	some species +	Resistant to cycloheximide

*Chlamydospores produced rarely

identification of species of yeasts. The former tests can be carried out in conventional carbohydrate broths, and the latter can be performed with carbohydrate discs applied to yeasts in pour plates (1). Several commercial systems are available to facilitate the identification of yeasts and yeastlike organisms recovered from clinical specimens. Three that have been widely used are the API 20 C® Strip,* the Minitek™ yeast carbon assimilation system,† and the Uni-Yeast-Tec® system.‡ All employ carbohydrate assimilation and/or fermentation tests.

For a detailed discussion of the yeasts associated with disease in man and animals, including identification, readers are referred to the Supplementary References.

CANDIDIASIS (MONILIASIS, THRUSH, CANDIDOSIS)

CAUSE. *Candida albicans.*

PATHOGENICITY. Infections most frequently involve the mucous membranes and are often sufficiently severe to produce diphtheritic membranes. Systemic infections are infrequent.

Chickens, Turkeys, and Other Birds: Infection of the mouth, esophagus, and crop.

Dogs and Cats: Mycotic stomatitis; white to gray patches on the oral mucosa. Enteritis of kittens.

Calves and Foals: Infections of the oral and intestinal mucosa. Systemic candidiasis has been described in calves on prolonged antibiotic therapy (2).

Cattle: An infrequent cause of bovine mastitis. Genital infections are rare.

Horses: Genital infections occur occasionally in the male and female.

Swine: Infections of the lower esophagus and esophagogastric region of the stomach (3). Cutaneous candidiasis has been described in swine (4).

Other Species: Infections have been reported in rodents, primates, dolphins (5), and other animals. In humans, the mucous membranes of the mouth, tongue, and vulva are more commonly involved than the skin and nails. The oral form with the characteristic white patches is seen frequently in infants. The lungs and bones may be infected, and endocarditis may be seen in the systemic form.

DIRECT EXAMINATION. Yeastlike cells and hyphae can be demonstrated in wet mounts (10% NaOH) of scrapings from lesions. The thin-walled oval and budding yeast cells are 2–6 μm in diameter.

---

*Analytab Products, Inc., Plainview, New York.

†Flow Laboratories, Inc., McLean, Virginia.

‡BBL Microbiology Systems, Cockeysville, Maryland.

The examination of tissue sections is of great value in the diagnosis of candidiasis in that the actual invasion of the tissue by the organism can be demonstrated.

Gram-stained smears disclose Gram-positive oval and budding yeast cells (Fig. 30-1).

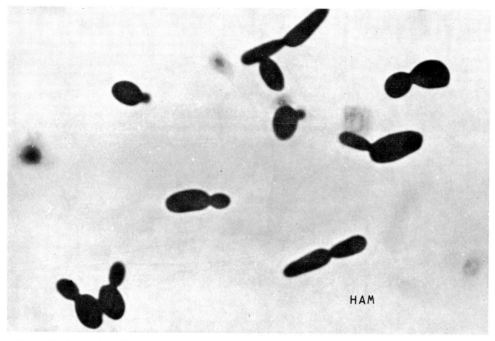

Figure 30-1. *Candida albicans* from Sabouraud dextrose agar. Similar forms are seen in clinical specimens. Gram's stain, ×2250 (H. A. McAllister).

ISOLATION PROCEDURES. If candidiasis is suspected, the specimen is inoculated onto Sabouraud agar and Sabouraud C and C agar, then incubated at 25°C. It is also advisable to inoculate blood agar and incubate at 37°C.

In one to three days, colonies are cream colored, pasty, and smooth with a yeastlike odor. The colonies do not look unlike those of staphylococci and micrococci.

IDENTIFICATION. *Candida* spp. produce pseudohyphae and do not possess ascospores. Typical oval and budding yeast cells are seen in wet mounts and in Gram-stained smears. These are of little value in identification. Some differential features of genera of yeastlike organisms are given in Table 30-I.

Demonstration of the large, round, thick-walled chlamydospores (Fig. 30-2) characteristic of *C. albicans* is essential for identification. These are produced on chlamydospore agar (see Appendix D) and other media, including cornmeal agar with Tween 80. Plates of this medium are inoculated by cutting through the agar along the line of inoculation. The cut in the agar should be at an acute angle to the bottom of the plate to facilitate subsequent microscopic examination. After incubation at 25°C for two to four days, the plates are examined under the low power objective of the microscope. Production of chlamydospores is favored by lowered oxygen tension, and for this reason, they are more apt to be seen below the surface of the medium. The chlamydospores appear blue in the chlamydospore agar as a result of absorption of the trypan blue, whereas the filaments are colorless.

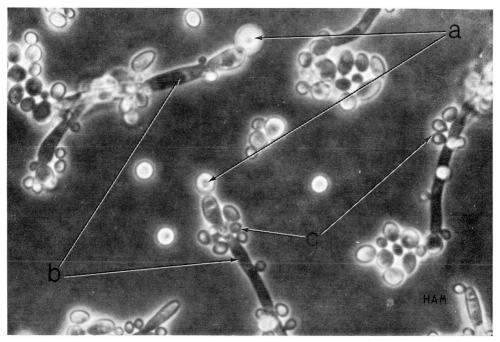

Figure 30-2. *Candida albicans* on chlamydospore agar: *a*, terminal chlamydospores; *b*, pseudo-hyphae; *c*, blastospores. Dark-phase illumination, ×975 (H. A. McAllister).

Another useful medium for the identification of *C. albicans* is Levine's eosin-methylene blue agar. On this medium, characteristic germ tubes are produced after incubation for twenty-four hours at 37°C in a candle jar. The germ tubes of *C. albicans* may be demonstrated after incubating yeast cells in a small amount of serum for two to four hours at 37°C. They appear as small sprouts developing from yeast cells.

It should be kept in mind that *C. tropicalis* and *C. stellatoidea* can produce small numbers of chlamydospores on rare occasions.

A small percentage of *C. albicans* cannot be identified by the above procedures. These strains and other species of the genus can only be identified with certainty by means of cultural and biochemical criteria (6) (see Table 30-II).

ANIMAL INOCULATION. Strains of *C. albicans* are pathogenic for mice and rabbits. One milliliter of a saline suspension (Brown or McFarland tubes 5 to 6; twenty-four- to forty-eight-hour culture) administered intravenously to a rabbit or intraperitoneally to a mouse will usually kill within four to ten days. Numerous small abscesses are found in the swollen kidneys. The typical yeastlike cells can be seen in smears from the lesions (Fig. 30-3). Other *Candida* species do not as a rule kill experimental animals.

OTHER *CANDIDA* SPECIES

The other *Candida* species are identified by the criteria listed in Table 30-II.

C. tropicalis and *pseudotropicalis*. These species are uncommon causes of mastitis in cows.

C. parapsilosis. Cases of bovine mastitis (7) and abortion in a cow (8) have been attributed to this organism.

C. guilliermondii and *C. krusei*. Gedek (7) recovered these additional species from the milk of cows with mastitis.

C. rugosa. This organism was incriminated as a cause of pyometra in a mare (9) and can also cause bovine mastitis.

CRYPTOCOCCOSIS (TORULOSIS)

CAUSE. *Cryptococcus neoformans.*

This yeastlike organism, which is the only pathogenic species of the genus, occurs widely in nature.

PATHOGENICITY. Cryptococcosis is a subacute or chronic infection that frequently involves the central nervous and respiratory systems.

Dogs and Cats: Cryptococcosis is seen most frequently in dogs and less commonly in cats. Infections usually begin in the paranasal sinuses, with later extension to the brain and meninges or lungs. Subcutaneous granuloma are sometimes encountered around the head and feet.

Horses: Infections, which begin usually as nasal granuloma, may extend to the lungs and viscera.

Cattle: A cause of sporadic cases of mastitis.

Other Animals: Infections have also been reported in the fox, dolphin (10), monkey, civet, ferret, guinea pig, and cheetah.

DIRECT EXAMINATION. A strongly presumptive diagnosis of cryptococcosis can frequently be made by the demonstration of the capsulated organism

TABLE 30-II

DIFFERENTIATION OF *CANDIDA* SPECIES

	Growth 37°C	Pellicle: broth	Chlamydospores	Germ tubes	Fermentation: Glucose	Maltose	Sucrose	Lactose	Galactose
C. albicans	+	−	+	+	+	+	−	−	+
C. stellatoidea	+	−	rare	+	+	+	−	−	−
C. tropicalis	+	+	rare	−	+	+	+	−	+
C. pseudotropicalis	+	−	−	−	+	−	+	+	+
C. parapsilosis	+	−	−	−	+	−	−	−	+
C. krusei	+	+	−	−	+	−	−	−	−
C. guilliermondii	+	−	−	−	+	−	+	−	+
C. rugosa	−	−	−	−	−	−	−	−	−

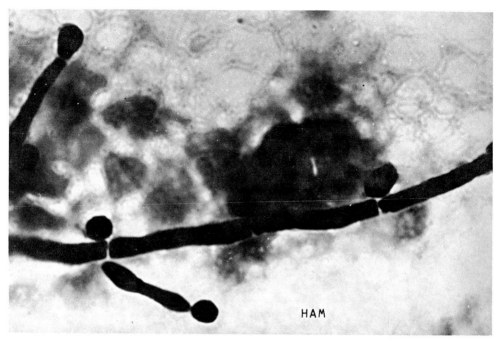

Figure 30-3. *Candida albicans* showing spherical forms and pseudohyphae. Gram-stained kidney impression smear from an experimentally infected mouse, ×2250 (H. A. McAllister).

in the nasal discharge in the paranasal form and in various clinical specimens such as pus, milk, exudate, and cerebrospinal fluid. The yeast cells are 5–20 μm in diameter. The addition of India ink to wet preparations facilitates demonstration of the characteristic thick capsules (Fig. 30-4).

ISOLATION PROCEDURES. Blood agar and Sabouraud dextrose agar are inoculated, the former being incubated at 37°C and the latter at 25°C. Cycloheximide inhibits the growth of this organism.

CULTURAL CHARACTERISTICS. Growth at 25°C (seven to fourteen days); wrinkled, whitish, granular colonies become slimy, mucoid, and cream to brownish in color on further incubation. No mycelium is present, and the colony flows to the bottom of the slant. At 37°C (seven to fourteen days); colonies are essentially the same as those described above. After shorter incubation periods, colonies resemble those of staphylococci and micrococci.

IDENTIFICATION. Budding cells, usually with large capsules, can be seen in wet mounts containing India ink. Some saprophytic strains of *Cryptococcus* spp. will not grow at 37°C.

A presumptive identification of *C. neoformans* is usually based upon the demonstration and recovery of yeast cells with a large capsule from an animal showing clinical signs of cryptococcosis. Because *Rhodotorula* and various species of cryptococci also produce capsules, definitive identifica-

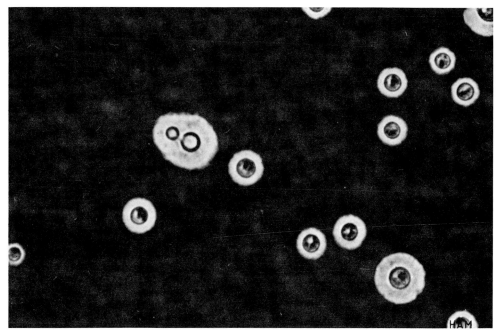

Figure 30-4. Encapsulated yeast cells of *Cryptococcus neoformans* from an experimentally infected mouse brain. India ink preparation, ×975 (H. A. McAllister).

tion requires additional tests. In addition to the features listed in Table 30-I, differentiation of *C. neoformans* from other species of cryptococci can be accomplished on the basis of (a) ability to grow at 37°C; (b) pathogenicity for the mouse (see below); (c) production of brown colonies on birdseed agar (see Appendix D); (d) typical carbohydrate assimilation pattern (Table 30-III).

The commercial kits referred to earlier in this chapter can be used to identify species of *Cryptococcus*.

Members of the genus *Cryptococcus* produce urease on Christensen's urea agar while *Candida* spp. do not. Strains of the true yeast *Saccharomyces* can be distinguished from the cryptococci by the presence in the former of ascospores. The ascospores stain well with methylene blue. Species of *Geotrichum* and *Trichosporon* both produce true mycelia.

MOUSE INOCULATION. The pathogenic strains of *Cryptococcus neoformans* kill mice. Dose: 1 ml intraperitoneally or 0.02 ml intracerebrally of a heavy saline suspension prepared from a Sabouraud agar slant.

Pathogenic strains inoculated intraperitoneally may produce lesions in the brain in about three weeks, while those injected intracerebrally may develop brain lesions in about a week. Those mice that do not die within two weeks are necropsied. Gelatinous masses are found in the abdominal cavity

TABLE 30-III

DIFFERENTIATION OF *CRYPTOCOCCUS* SPECIES

	Glucose Fermentation	Urease	Growth 37°C	Sugar Assimilation:					Nitrate Assimilation
				Gluc.	Malt.	Suc.	Lact.	Galact.	
C. neoformans	−	+	+	+	+	+	−	+	−
C. albidus	−	+	−	+	+	+	+	+	+
C. albidus var. *diffluens*	−	+	−	+	+	+	−	+	+
C. luteolus	−	+	−	+	+	+	−	+	−
C. laurentii	−	+	− or +	+	+	+	+	+	−
C. terreus	−	+	−	+	+	−	−	+	− or +

and associated with the brain and lungs. The typical budding, encapsulated organism can be demonstrated and cultured from the material.

GEOTRICHOSIS

CAUSE. *Geotrichum candidum.*

As mentioned previously, this organism is a mold rather than a yeast, but in its early colonial growth, it appears yeastlike. This fungus is found widely in nature, and its isolation is not necessarily significant. Two cultural forms occur. They are referred to as the glabrous, or yeastlike, and the fluffy form. The latter is sometimes given the name *Oospora.* The glabrous form of *G. candidum* is the form usually associated with disease.

PATHOGENICITY. Infections due to this fungus in animals are rare. They have been reported from cattle, dogs, fowl, and humans. The bronchi, lungs, udder, and the mucous membranes of the alimentary tract are most frequently affected. The disease is usually mild and is characterized by the formation of granulomas that may suppurate.

DIRECT EXAMINATION. Purulent material or scrapings from lesions are examined in wet mounts. The organism appears as rectangular (4–8 μm) or large spherical (4–10 μm) arthrospores. They are thick-walled, nonbudding, and in stained smears are strongly Gram-positive.

ISOLATION PROCEDURES. If there is microscopic evidence of *G. candidum*, material is inoculated onto blood agar, which is incubated at 37°C, and onto Sabouraud C and C agar and regular Sabouraud agar, which are incubated at 25°C.

CULTURAL CHARACTERISTICS. 25°C (one to two weeks): Colonies grow fairly rapidly, are membranous with radial furrows, and soft with a dry granular surface. 37°C: The fungus does not grow well at this temperature. The colonies, which appear early, are small, and the mycelium penetrates the subsurface of the medium.

IDENTIFICATION. 25°C and 37°C: The mycelium is made up of septate hyphae that fragment, producing chains of rectangular to round arthrospores (see Fig. 30-5).

Identification is based upon cultural characteristics and the demonstration of the characteristic arthrospores (see Fig. 30-5). This fungus can be distinguished from *Coccidioides immitis* and *Blastomyces dermatitidis* by the fact that these species produce cottony, filamentous colonies at room temperature. *G. candidum* produces a soft yeastlike colony at this temperature.

MALASSEZIA (PITYROSPORUM)

Malassezia is now considered the correct genus name for those lipophilic yeasts that reproduce by unipolar budding. The species *Malassezia furfur* is associated with tinea versicolar, dandruff, seborrhea, and blepharitis in humans.

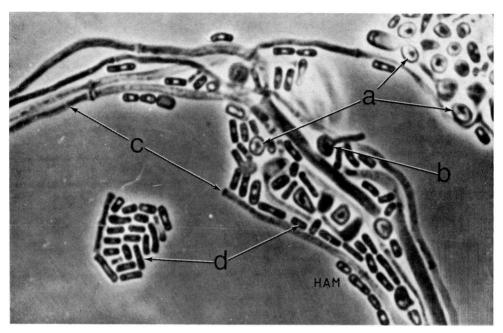

Figure 30-5. Lactophenol cotton blue preparation of *Geotrichum candidum* from a Sabouraud dextrose agar slide culture: *a*, yeast cells; *b*, yeast cell with germ tube; *c*, hyphae; *d*, rectangular athrospores. Dark-phase illumination, × 1000 (H. A. McAllister).

The yeasts that have been called *Pityrosporum canis* and *P. felis* and similar organisms from other animals have been given the name *M. pachydermatis*. The latter species is thought to occur often as a commensal on the oily areas of the skin and ears of dogs, cats and probably other animals. It is frequently recovered from the ears of dogs with chronic otitis externa and is thought to have at least a secondary role in the etiology of some cases (11).

M. pachydermatis consists of bottle-shaped, small budding cells 1–2 by 2–4 μm, reproducing by a process known as bud fission in which the bud detaches from the mother cell by a septum (Fig. 30-6).

DIRECT EXAMINATION. Wet mounts (10% sodium hydroxide) are made from material taken from the affected ear in the inflammatory area. After the mount has cleared, it is examined for clusters of thick-walled, round, and budding forms (3–8 μm in diameter) surrounded by short, straight, and angular mycelial fragments. The short hyphae are generally observed in clinical materials only. Although a presumptive identification can be made on a direct examination, it is advisable to attempt isolation.

ISOLATION PROCEDURES. Isolations have been made from cases of otitis externa employing Sabouraud agar at 25°C with an incubation period of two weeks (12). Growth was increased by the addition of sterile olive or coconut

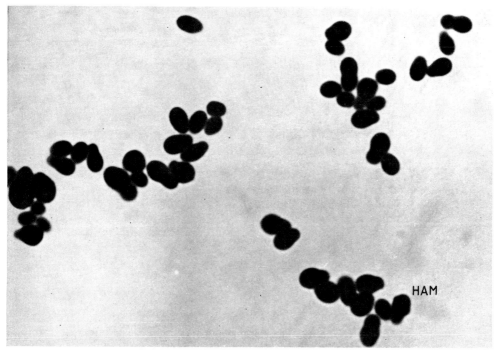

Figure 30-6. *Malassezia pachydermatis* (*Pityrosporum canis*). Gram-stained smear from an olive oil enriched Sabouraud dextrose agar culture, ×2440 (H. A. McAllister).

oil to the Sabouraud plates prior to inoculation. Small creamy, punctiform colonies give rise with additional incubation to confluent, membranous colonies.

For definitive identification, readers are referred to Lodder and Lennette et al. (Supplementary References).

ADDITIONAL YEASTS

Torulopsosis (now called candidiasis) is an infrequent disease of humans and has only been reported on several occasions in animals. The species involved in most infections is *Torulopsis glabrata* (now called *Candida glabrata*), which occurs as a commensal in animals and is also found in the soil. It has been reported as causing bovine mastitis and abortion, and systemic infections in dogs and monkeys (10).

Species of the genera *Rhodotorula* and *Trichosporon* have on rare occasions been associated with disease in animals. Readers are referred to the Supplementary References for the identification of these and other widespread but usually nonpathogenic yeasts. Ainsworth and Austwick (13) have reviewed a number of reports of diseases caused by less well known yeasts.

REFERENCES

1. Mickelson, P. H., McCarthy, L. R., and Propst, M. A.: *J Clin Microbiol, 5*:297, 1977.
2. Mills, J. H. L., and Hirth, R. S.: *J Am Vet Med Assoc, 150*:862, 1967.
3. Kadel, W. L., Kelley, D. C., and Coles, E. H.: *Am J Vet Res, 30*:401, 1969.
4. Reynolds, I. M., Miner, P. W., and Smith, R. E.: *J Am Vet Med Assoc, 152*:182, 1968.
5. Cordes, D. O.: *New Zealand Vet J, 30*:46, 1982.
6. Silva-Hunter, M., and Cooper, B. H.: In Lennette, E. H., (Ed.-in-chief): *Manual of Clinical Microbiology.* 3rd ed. Washington, D.C. American Society for Microbiology, 1980, Chapter 55.
7. Gedek, B.: *Tierärztl Wochenschr, 82*:241, 1969.
8. Bisping, W., Refai, M., and Trautwein, G.: *Tierärztl Wochenschr, 77*:260, 1964.
9. Abou-Gabal, M., Hogle, R. M., and West, J. K.: *J Am Vet Med Assoc, 170*:177, 1977.
10. Rippon, J. W.: *Medical Mycology.* Philadelphia, Saunders, 1974, Chapter 9.
11. Smith, J. M. B.: *Aust Vet J, 44*:414, 1968.
12. McAllister, H. A., and Carter, G. R.: Unpublished observations, 1971.
13. Ainsworth, G. C., and Austwick, P. K. C.: *Fungal Diseases of Animals*, 2nd ed. Farnham Royal, Slough, England. Commonwealth Agricultural Bureaux. 1973, Chapter 18.

SUPPLEMENTARY REFERENCES

Ainsworth, G. C., and Austwick, P. K. C.: *Fungal Diseases of Animals*, 2nd ed. Farnham Royal, Slough, England, Commonwealth Agricultural Bureaux, 1973.

Beneke, E. S., and Rogers, A. L.: *Medical Mycology* (Laboratory Manual), 4th ed. Minneapolis, Burgess, 1980, Chapter 9.

Conant, N. F., Smith, T. L., Baker, R. D., Callaway, J. L., and Martin, D. S.: *Manual of Clinical Mycology*, 3rd ed. Philadelphia, Saunders, 1971.

Lennette, E. H. (Ed.-in-chief): *Manual of Clinical Microbiology*, 3rd ed. Washington, D.C. American Society for Microbiology, 1980, Chapter 55.

Lodder, J. (Ed.): *The Yeasts.* Amsterdam, North-Holland Publishing Co., 1970.

Rippon, J. W.: *Medical Mycology.* 2nd ed. Philadelphia, Saunders, 1982.

Chapter 31

FUNGI CAUSING SUBCUTANEOUS INFECTIONS

INCLUDED IN THIS CHAPTER are those fungi that produce infections that principally involve the skin and subcutis. The two species *Sporothrix schenckii* and *Histoplasma farciminosum* are dimorphic fungi. A variety of "monomorphic" fungi cause the diseases called chromomycosis and maduromycosis. The remaining disease, rhinosporidiosis, is caused by a fungus that has not yet been satisfactorily cultivated on artificial media.

SPOROTRICHOSIS

CAUSE. *Sporothrix schenckii.*

PATHOGENICITY. Infections in humans and some animals are characterized by the formation of subcutaneous nodules or granulomas. The organisms frequently enter through wounds in the skin and are spread via the lymphatics. The nodules eventually ulcerate and discharge pus. The disease is probably seen most commonly in practice in the horse, in which it is an ascending infection involving the leg. Involvement of bones and visceral organs with fatal termination has been reported from humans, swine, dog, horse, donkey, mule, cattle, fowl, camel, and rodents.

DIRECT EXAMINATION. In pus and tissue, the organism appears as a single-celled cigar-shaped body, usually within neutrophils (Fig. 31-1). These structures are very difficult to demonstrate in smears and wet mounts of pus and tissue scrapings; however, they can usually be demonstrated in clinical material by fluorescent antibody staining (1).

ISOLATION PROCEDURES. Pus is taken aseptically from, preferably, an unopened lesion and inoculated onto brain-heart infusion agar or blood agar and Sabouraud agar, with and without inhibitors. Incubate the former two media at 37°C and the Sabouraud media at 25°C.

CULTURAL CHARACTERISTICS. 25°C (seven to ten days): Colonies appear early, but the characteristic structures are not evident until the aerial mycelium is produced. Colonies are white and soft at first, then become tan to brown to black. The texture is leathery, wrinkled, and coarsely tufted.

37°C (seven to ten days): Colonies develop rapidly and are yeastlike, smooth, soft, and cream to tan in color.

IDENTIFICATION. 25°C: The mycelium consists of fine, branching septate hyphae that bear pyriform or ovoid microconidia 2–5 μm. The latter are

337

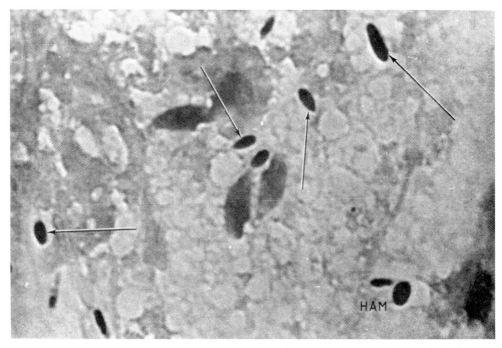

Figure 31-1. *Sporothrix schenckii.* Gram-stained impression smear from experimentally infected mouse testes. The elongated yeast cells are commonly called "cigar bodies." ×2440 (H. A. McAllister).

borne in clusters from the ends of conidiophores or as sessile forms directly on the sides of hyphae (Fig. 31-2). Thick-walled, large chlamydospores may be seen in old cultures.

37°C: There is no mycelium. Colonies are composed of the same elements that occur in pus and tissue, the cigar-shaped cells (1–4 µm by about 1 µm) and spherical or oval budding cells (2–3 µm) (Fig. 31-3). Some large pyriform cells (3–5 µm) may be seen.

EPIZOOTIC LYMPHANGITIS (AFRICAN FARCY, JAPANESE GLANDERS)

CAUSE. *Histoplasma farciminosum* (Synonyms: *Cryptococcus farciminosus, Zymonema farciminosum*).

The name *Histoplasma farciminosum* will be used, although according to Ajello (2) the organism does not belong in the genus *Histoplasma*.

PATHOGENICITY. This organism causes epizootic lymphangitis, a disease of horses, mules, and donkeys in Mediterranean countries, Asia, Africa, and parts of the U.S.S.R. Although infections usually involve the skin and lymphatics of the legs and neck, they may involve mucous membranes and internal organs. The lesions are granulomatous and frequently ulcerative. A

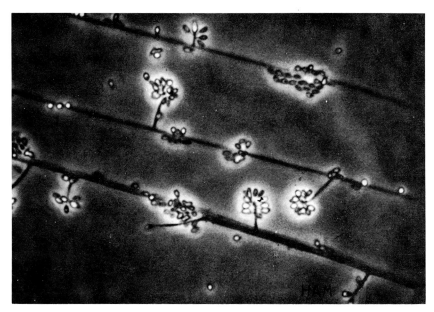

Figure 31-2. *Sporothrix schenckii:* mycelial phase with small conidia in "flowerette" arrangements. Lactophenol cotton blue preparation from slide culture at 24°C. Dark-phase illumination, ×840 (H. A. McAllister).

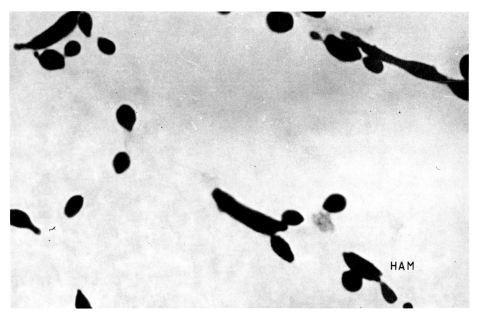

Figure 31-3. *Sporothrix schenckii.* Yeast cells from a blood agar plate culture (37°C). Gram's stain, ×2250 (H. A. McAllister).

primary pulmonary form of the disease in horses without lymph node involvement has been described (3).

DIRECT EXAMINATION. A diagnosis is usually based upon demonstration of typical yeastlike, globose or pear-shaped, double-contoured cells (2.5–4 µm) in wet mounts of the pus and discharges from lesions. Like other fungi, the organisms are Gram-positive. A fluorescent antibody procedure has been used for identification (4).

ISOLATION PROCEDURES. Material for inoculation should be taken aseptically from unruptured nodules. Cultivation is difficult and slow, and various media have been used. Nègre and Boquet (5) used a medium composed of macerated horse manure, agar, and lymph gland extract. Small colonies were obtained after incubation at 25–30°C for three to four weeks. Other workers have used Sabouraud agar with success. Bullen (6) obtained primary cultures in Hartley's digest agar (with 10% horse serum) incubated at 37°C.

IDENTIFICATION. 37°C (two to eight weeks): Colonies appear as minute gray flakes and are composed of yeastlike cells and some hyphae.

25°C (two to eight weeks): Colonies appear as minute gray flakes that later become dry, scaly, and wrinkled like earthworm casts. They are composed of septate mycelia 1–4 µm thick with characteristic, irregular thickening of hyphae and thick-walled chlamydospores varying 5–10 µm in diameter.

CHROMOMYCOSIS (CHROMOBLASTOMYCOSIS)

Chromomycosis is a clinical entity caused by several species of dematiaceous fungi. Although many cases have been described in humans, there have only been several reported from animals, viz. from horses (7) and dogs (8). The infections, which result from wounds or abrasions, produce nodular and frequently ulcerating lesions of the feet or legs with regional granulomatous lymphadenitis. The following species of fungi have been incriminated:

Hormodendrum dermatitidis
Hormodendrum pedrosoi (syn. *Fonsecaea pedrosoi*)
Hormodendrum compactum (syn. *Fonsecaea compacta*)
Phialophora verrucosa
Curvularia spp.
Cladosporium carrionii

DIRECT EXAMINATION. Material from granulomatous and ulcerous lesions is examined in 10% sodium hydroxide under a coverslip. Organisms are single celled or clustered, spherical (6–12 µm), thick walled, with a black or dark brown pigment. They multiply by cross wall formation or splitting rather than by budding. These structures are also demonstrable in histopathological sections.

ISOLATION PROCEDURES. Material is seeded on Sabouraud C and C agar.

These are incubated at room temperature for up to three weeks, as growth is usually slow.

IDENTIFICATION. Until considerable experience has been acquired in the identification of these dark brown or black (dematiaceous) fungi, they should be submitted to a reference mycology laboratory for confirmation of identification. Detailed descriptions of these fungi are given in standard texts (see Supplementary References).

MYCETOMAS (MADUROMYCOSIS)

A mycetoma is a localized swollen lesion consisting of granulomas and abscesses that suppurate and drain through sinus tracts. Although there are a number of agents that can cause mycetomas, in this discussion, reference is made to mycetomas caused by the true fungi. Although many species of fungi have been associated with mycetomas in man, the disease in animals has been reported infrequently. The cases reported have mostly been from horses, cattle, dogs, and cats (9, 10). The lesions occur most frequently on the extremities but may be found in the nasal mucosa, on the peritoneum, and involving the skin in various locations.

Incision of the lesions discloses discrete brown or black fungous microcolonies embedded in a large mass of granulation tissue. The following species of fungi have been recovered from mycetomas in animals:

Dog:	*Allescheria boydii*
	Curvularia geniculata
Horse:	*Brachycladium (Curvularia) spiciferum*
	Allescheria boydii
	Helminthosporium sp.
Cattle:	*Helminthosporium* sp.
Cat:	*Brachycladium speciferum*

The first mentioned species is a "hyaline" nonpigmented fungus, while the other three are dematiaceous (black or brown pigment) fungi. The first grows rapidly, while the others require several weeks.

DIRECT EXAMINATIONS. Scrapings or biopsy tissues are examined grossly for the characteristic microcolonies, which are small (0.5–3 mm), irregularly shaped, and variously colored. These colonies or "grains" are examined in 10% sodium hydroxide, then pressed out by means of a coverslip and observed microscopically. The "grains" of maduromycosis reveal mycelia usually 2–4 µm in width, in contrast to the narrower filaments found in the actinomycotic granule. Also of importance is the presence in the "grains" of chlamydospores.

ISOLATION PROCEDURES. Material containing microcolonies or grains may be washed in sterile physiological saline. Contamination can be overcome by leaving overnight in contact with a combination of penicillin (1000 units/ml) and streptomycin (100 µm/ml) or chloramphenicol (0.05 mg/ml). The fungi of this disease are sensitive to cycloheximide. Sabouraud dextrose agar is seeded and incubated at 25°C.

IDENTIFICATION. Fungi are identified according to their morphological and physiological characteristics. Detailed descriptions are given in texts (see Supplementary References). Cultures may be submitted to a reference laboratory for final identification.

RHINOSPORIDIOSIS

CAUSE. *Rhinosporidium seeberi.*

PATHOGENICITY. A disease of cattle, horse, mule, dog (11), and humans characterized by polyps on the nasal and ocular mucous membranes. These growths are considered to result from the tissue response to infection with *Rhinosporidium seeberi*. It is of interest that most infections involve the nose of male animals. That the disease may occur in avian species is suggested by the report of a case in a wood duck (12).

DIRECT EXAMINATION. Wet mounts are prepared from tissue taken from the polyps or from nasal discharge. The large sporangia (up to 200–300 µm) are readily distinguishable from the thin-walled *Haplosporangium parvum* and the smaller spherical cells of *Coccidioides immitis*.

The sporangia develop from small globose cells, 6–8 µm. They continue to grow, and when they reach approximately 100 µm, they become sporangia and their contents are transformed into thousands of spores. The sporangia may reach a size of 200–300 µm before the spores are released.

The typical sporangia can be readily seen in stained sections of the polyps.

ISOLATION PROCEDURES. This organism has not yet been successfully propagated on artificial media.

REFERENCES

1. Kaufman, L.: In Lennette et al. (Eds). *Manual of Clinical Microbiology*, 4th ed. Washington, D.C., American Society for Microbiology, 1980, Chapter 61.
2. Ajello, L.: *Mycosen, 11*:507, 1968.
3. Fawi, M. T.: *Sabouraudia, 9*:123, 1971.
4. Fawi, M. T.: *Br Vet J, 125*:231, 1969
5. Nègre, L. and Boquet, A.: *Ann Pasteur, 33*:269, 1919.
6. Bullen, J. J.: *J Pathol Bacteriol, 61*:117, 1949.
7. Simpson, J. G.: *Vet Med Small Anim Clin, 61*:207, 1966.
8. Hoskins, H. P. et al. (Eds.): *Canine Medicine*, rev. ed. Santa Barbara, California, American Veterinary Publications, 1959, p. 644.
9. Bridges, C. H.: *Am J Pathol, 33*:411, 1957.
10. Bridges, C. H., and Beasley, J. N.: *J Am Vet Med Assoc, 137*:192, 1960.
11. Stuart, B. P., and O'Malley, N.: *J Am Vet Med Assoc, 167*:942, 1975.

12. Davidson, W. R., and Nettles, V. F.: *J Am Vet Med Assoc, 171*:989, 1977.

SUPPLEMENTARY REFERENCES

Ainsworth, G. C., and Austwick, P. K. C.: *Fungal Diseases of Animals*, 2nd ed. Farnham Royal, Slough, England, Commonwealth Agricultural Bureaux, 1973.

Beneke, E. S., and Rogers, A. L.: *Medical Mycology Manual*, 4th ed. Minneapolis, Burgess, 1980, Chapters 7 and 11.

Conant, N. F., Smith, T. L., Baker, R. D., Callaway, J. L., and Martin, D. S.: *Manual of Clinical Mycology*, 3rd ed. Philadelphia, Saunders, 1971.

Jungerman, P. F., and Schwartzman, R. M.: *Veterinary Medical Mycology*, Philadelphia, Lea & Febiger, 1972, Chapter 4.

Lennette, E. H., et al. (Eds.): *Manual of Clinical Microbiology*, 4th ed. Washington, D.C., American Society for Microbiology, 1980, Section VII.

Rippon, J. W.: *Medical Mycology*, 2nd ed. Philadelphia, W. B. Saunders Company, 1982.

Chapter 32

FUNGI CAUSING SYSTEMIC OR DEEP INFECTIONS

INCLUDED in this section are a variety of fungi that cause infections of one or several organs. The most widely disseminated of these infections is histoplasmosis, a disease of the reticuloendothelial system.

PHYCOMYCOSIS (MUCORMYCOSIS)

Phycomycosis includes diseases caused by the "perfect fungi," *Mucor, Absidia, Rhizopus,* and *Mortierella.* Strains of these genera occur widely in nature. Several species of different genera referred to below have been recovered from phycomycosis of horses.

Mortierella hygrophila and *M. polycephala* have been recovered from phycomycosis in fowl and cattle, respectively. *M. wolfii* has been shown to cause abortion (mycotic placentitis) in cattle. This can occasionally be followed, in twenty-four to forty-eight hours, by acute pneumonia and death (1). *A. ramosa,* occasionally cited as a cause of bovine abortion, was found to be a common contaminant of placentas submitted for microbiological examination (2). Readers are referred to Ainsworth and Austwick (3) for additional information on phycomycosis.

PATHOGENICITY. Lesions are granulomatous or ulcerative; the former type is more often generalized. These fungi frequently infect lymph nodes of the respiratory and alimentary tracts. Lymph nodes enlarge and become caseocalcareous. Ulceration of stomach and intestine has been attributed to mucormycosis.

Pigs: Lesions are found in mediastinal and submandibular lymph nodes; embolic "tumors" are seen in the liver and lungs. Fungi of this group are found in gastric ulcers.

Cattle: Lesions are found in the bronchial, mesenteric, and mediastinal lymph nodes and in nasal and abomasal ulcers; abortions are attributed to these fungi.

Horses: There are several reports of mucormycosis in this species.

Infections have also been reported in dogs, mink, guinea pigs, and the mouse. The disease occurs occasionally in fowl and exotic birds.

DIRECT EXAMINATION. Coarse, nonseptate, branching hyphae with rounded sporangia at the terminal portion are seen. The coarseness or thickness of the pieces of hyphae is especially significant.

ISOLATION PROCEDURES. Blood agar and Sabouraud dextrose agar are inoculated and incubated at 37°C and 25°C respectively.

The mucoraceous fungi are characterized by nonseptate hyphae and the formation of spores by cleavage within a sporangium. Most species are saprophytic, but at least seven species have been shown to be potentially pathogenic. They all grow rapidly both at room and incubator temperature on blood and Sabouraud agar. Their growth is inhibited by cycloheximide. Generally speaking, those species capable of causing disease have a greater capacity for growth at 37°C than the strictly saprophytic strains.

MUCOR

Several species of *Mucor* have been incriminated in fungal infections (3), but *M. pusillus* is the only definitely pathogenic species. Members of the genus produce a thick, colorless mycelium, without rhizoids, from which simple or branched sporangiophores arise, bearing terminal globose sporangia. The small globose spores of *M. pusillus* are 2–4 µm in diameter.

This rapidly growing fungus will fill the Petri dish in four to five days with abundant growth of floccose aerial mycelium. It is white at first, then turns a dark gray.

IDENTIFICATION. Structures characteristic of the genus are shown in Figure 32-1. It may be necessary to submit the culture to a reference laboratory for species identification.

ABSIDIA

The two pathogenic species are *A. ramosa*, with oval spores 2–5 µm in diameter, and *A. corymbifera*, with globose spores 2–4 µm in diameter.

They are rapidly growing fungi whose growth resembles *Rhizopus*. Long aerial stolons are formed that arch over the advancing mycelium. Rhizoids form where the stolons contact the medium. The sporangiophores are produced at intervals along the stolon and bear pyriform terminal sporangia.

IDENTIFICATION. Structural elements resemble those of *Rhizopus* spp. Some of the morphological elements that characterize the genus are shown in Figure 32-2. Submit to a reference laboratory for species identification.

Ainsworth and Austwick (3) differentiate the genera *Mucor, Absidia,* and *Rhizopus* as follows: "In *Mucor* the globose sporangia occur singly or in groups along prostrate hyphae whereas in *Absidia* they are borne in groups by arching, stoloniferous hyphae which arise from behind the advancing margin. *Rhizopus* resembles *Absidia* but the sporangiophores arise only at points of contact of the stolons with the medium."

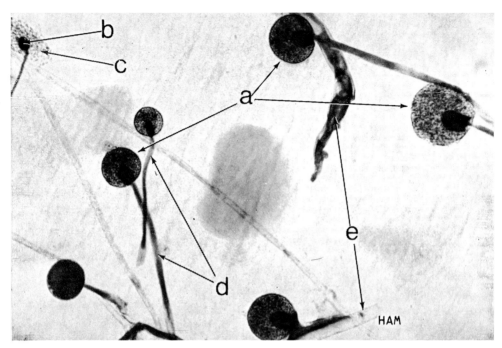

Figure 32-1. *Mucor* sp.: *a*, Mature, spore-filled sporangia; *b*, columella; *c*, spores from ruptured sporangium; *d*, sporangiophores; *e*, nonseptate hyphae. Tape mount, lactophenol cotton blue, ×220 (H. A. McAllister).

RHIZOPUS

Two species have been reported as pathogenic for animals: *R. microsporus* and *R. cohnii.*

Stolons are characteristic of this genus, but the sporangiophores arise only at points of contact with the medium. Rhizoids can usually be seen in mycelial mounts.

This fast growing group will fill the Petri dish in five days with a dense, cottony aerial mycelium. At first their growth is white, then it turns gray.

MORTIERELLA

Field strains of *M. wolfii* grow on blood agar and Sabouraud agar at 27°C and 37°C. However, the fungus does not compete well if there is heavy bacterial growth (blood agar) or on mycological agars with fungal contaminants. The colonies on Sabouraud and blood agars are white, velvetlike, dense, and characteristically lobulated (Fig. 32-3) but sterile. The hyphae are hyaline and 2.5–5.0 μm in width. Sporangia are produced on hay and potato-carrot agar. For definitive identification based on the morphology of sporangia and spores, readers are referred to di Menna et al. (4).

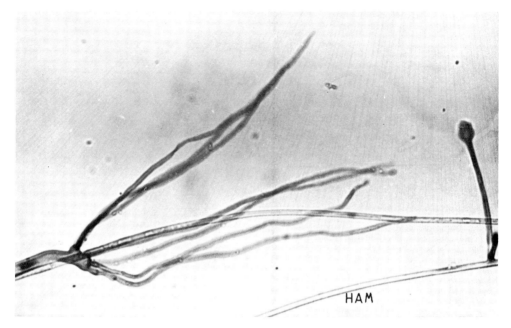

Figure 32-2. *Absidia* sp. showing rhizoids. Slide culture, lactophenol cotton blue, ×250 (H. A. McAllister).

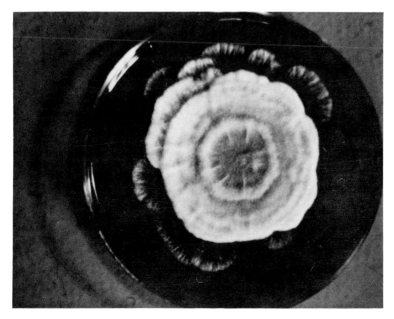

Figure 32-3. *Mortierella wolfii* colony on blood agar, after seventy-two hours incubation at 37°C.

FLORIDA HORSE LEECH

This name has been given to a phycomycosis in the horse caused by several fungi within the same class but of a different order from those usually associated with phycomycosis.

The fungi causing this disease are in the same class but of a different order than those producing the zygomycoses. These infrequent infections of horses are characterized by the formation of granulomatous processes of varying size. The foci of proliferative inflammation may be found anywhere on the body but usually about the limbs, abdomen, neck, and alae nasi.

At least three fungal species have been implicated: *Hyphomyces destruens, Entomophthora coronata,* and *Basidiobolus ranarum.* The disease and the mycology have been dealt with comprehensively by Bridges and associates (5, 6).

ASPERGILLOSIS

The most prevalent pathogenic species is *Aspergillus fumigatus.* Other potentially pathogenic species are *A. flavus, A. nidulans,* possibly *A. niger,* and *A. phialisepticus.* Some workers are of the opinion that the differences among these species are small and that they should all be included in the species *A. fumigatus.* Many strains of aspergillus occur as contaminants.

PATHOGENICITY. Fowl: Several manifestations—(a) a diffuse infection of the air sacs, (b) a diffuse pneumonic form, and (c) a nodular form involving the lungs. The disease may be limited to one organ or become generalized. "Brooder pneumonia" is an important form of the disease in chicks and poults.

Cattle: Infections of the uterus, fetal membranes, and fetal skin resulting in abortion.

Horses: Infections involving the skin and mucous membranes have been reported. *Aspergillus* spp. appear to be the sole cause of guttural pouch mycosis, a serious disease of stabled horses.

Dogs: Infections of the nasal chambers.

Other Animals: Infections occur occasionally in many animal species. The respiratory tract of the penguin in captivity is frequently infected.

Humans: Primary and secondary infections in a wide variety of tissues and locations: lungs, skin, nasal sinuses, external ear, bronchi, bones, and meninges.

DIRECT EXAMINATION. Small pieces of tissue or deep scrapings are examined in 10% sodium hydroxide. Short pieces of thick, septate, sometimes almost spherical hyphae are characteristic. The typical conidial heads are only seen in the lung and air sacs where there is access to oxygen.

ISOLATION PROCEDURES. Material is inoculated onto blood agar and Sabouraud dextrose agar and incubated at 37°C and 25°C respectively. Streptomycin and penicillin can be added to these media to suppress bacterial contaminants.

CULTURAL CHARACTERISTICS. 25°C and 37°C (one to three days). Growth is rapid, and colonies are white at first but later turn green to dark green, flat, and velvety.

IDENTIFICATION. The identifying microscopic structure seen at both temperatures is the conidiophore with large terminal vesicles bearing many sterigmata from which chains of spores are produced (Figs. 32-4 and 32-5).

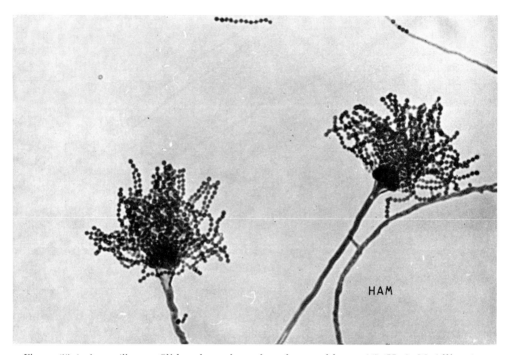

Figure 32-4. *Aspergillus* sp.: Slide culture, lactophenol cotton blue, × 445 (H. A. McAllister).

The structural elements that characterize the genus are indicated in Figure 32-5. Identification of the many species of aspergillus requires a specialist. Because of the ubiquity of the aspergilli, typical structural elements should be demonstrable in tissues or isolated in heavy culture more than once before pathogenic significance is inferred.

The agar gel double diffusion test is claimed to be a reliable procedure for the diagnosis of nasal aspergillosis in dogs (7).

BLASTOMYCOSIS
(CAUSE: *BLASTOMYCES DERMATITIDIS*)

PATHOGENICITY. This disease, which is characterized by the formation of granulomatous nodules, occurs principally in humans and dogs. A case has been described in the horse and the cat. In dogs, the granulomatous lesions

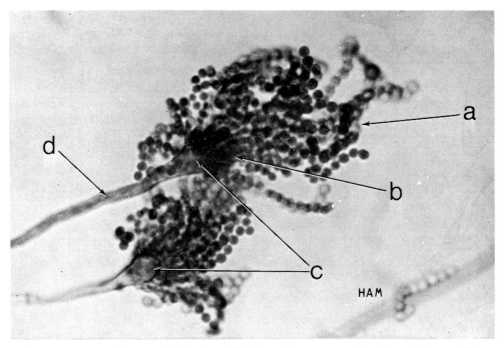

Figure 32-5. *Aspergillus* sp. *a*, chains of conidia; *b*, sterigmata; *c*, vesicle; *d*, conidophore. Slide culture, lactophenol cotton blue, ×975 (H. A. McAllister).

are usually found in the lungs and/or the skin and subcutis. The latter frequently ulcerate.

DIRECT EXAMINATION. The large, spherical, thick-walled cells (5–20 µm in diameter) are readily seen in wet mounts (Fig. 32-6). A single bud is frequently seen connected to the larger mother cell by a wide base. Some cells give a double contoured effect.

ISOLATION PROCEDURES. Blood or brain-heart infusion agar and Sabouraud C and C are inoculated. The former two are incubated at 37°C and the latter at 25°C.

CULTURAL CHARACTERISTICS. 25°C (twenty-one days): Moist, grayish, yeastlike colonies that develop white, cottony mycelia. As the growth ages, it becomes tan, then dark brown to black.

37°C (three to five days): Creamy, waxy, wrinkled colony, cream to tan in color.

IDENTIFICATION. Microscopic, 25°C: septate hyphae bearing small, oval (2–3 µm) or pyriform (3–5 µm) conidia laterally close to the point of septation (Fig. 32-7). Older cultures form chlamydospores (7–15 µm) with thickened walls.

37°C: Thick-walled budding yeast cells (7–15 µm in diameter) similar to

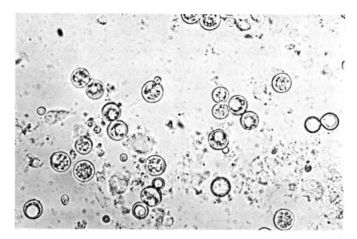

Figure 32-6. Wet mount of exudate from lung of a dog with *Blastomyces dermatitidis* infection. Single and budding spheres with double contoured, refractile walls. ×410 (F. K. Ramsey and G. R. Carter).

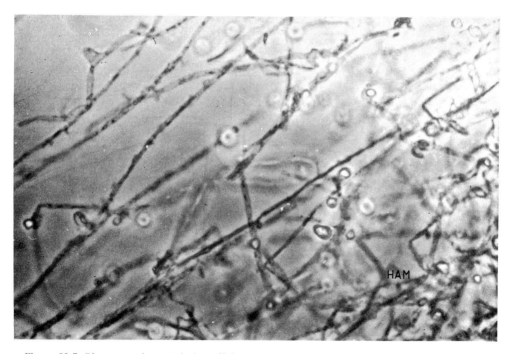

Figure 32-7. *Blastomyces dermatitidis.* Mycelial mass showing hyphae and conidia from a Sabouraud dextrose agar slide culture (25°C). Dark-phase illumination, lactophenol cotton blue. ×975 (H. A. McAllister).

those seen in tissue sections and exudate (Fig. 32-8). Identification is based on cultural and morphological characteristics. The finding of typical lesions and organisms in tissue sections or clinical specimens supports the identification.

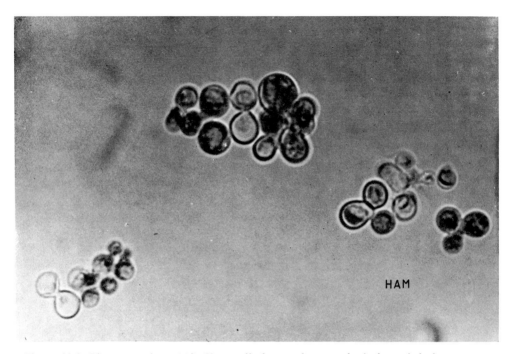

Figure 32-8. *Blastomyces dermatitidis.* Yeast cells from culture on brain-heart infusion agar at 37°C. Wet mount, lactophenol cotton blue, ×975 (H. A. McAllister).

HISTOPLASMOSIS
(CAUSE: *HISTOPLASMA CAPSULATUM*)

This fungus is found in nature and has a worldwide distribution. It is more heavily concentrated in the soil of certain geographic areas. The areas of heaviest concentration in the United States are the northeast, central, and south central states.

PATHOGENICITY. Histoplasmosis is a generalized disease involving the reticuloendothelial system. Infections have been reported from dogs, cats, sheep, swine, horses, and humans. Some of the lesions seen in dogs and cats are ulcerations of the intestinal canal; enlargement of the liver, spleen, and lymph nodes; necrosis and tubercle-like lesions in the lungs, liver, kidneys, and spleen.

DIRECT EXAMINATION. Because *H. capsulatum* is small and rarely found extracellularly, it is extremely difficult to demonstrate in clinical materials.

Smears are made from scrapings of ulcers and from cut surfaces of lymph nodes and other tissues. They are stained by Giemsa's or Wright's method and examined under the oil immersion objective. The organisms occur intracellularly as small, round or oval, yeastlike single or budding cells. A clear halo is seen around the darker-staining central material. The characteristic small yeast cells can be seen in stained sections of affected tissue (Fig. 32-9). They are most commonly seen in monocytic cells.

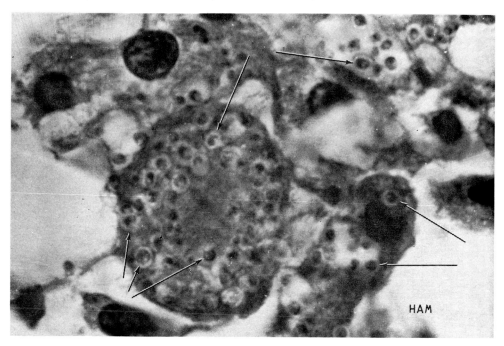

Figure 32-9. *Histoplasma capsulatum.* Small intracellular yeastlike cells in a monocyte. Tissue section, hematoxylin-eosin, ×2250 (H. A. MacAllister).

ISOLATION PROCEDURES. Material is inoculated onto Sabouraud C and C agar and blood or brain-heart infusion agar. The former is incubated at 25°C and the latter at 37°C. To reduce contaminants, material may be incubated for three hours with penicillin and streptomycin prior to inoculation. Growth is not always obtained at 37°C.

CULTURAL CHARACTERISTICS. 25°C (two to four weeks): Colonies are cottony, white to cream at first, later becoming tan to brown.

37°C (two to four weeks): Colonies are small, white, and yeastlike.

IDENTIFICATION. 25°C: A mycelium is produced with septate hyphae on which are borne two kinds of spores: (a) small, smooth, round to pyriform microconidia, either on short lateral branches or attached directly by the base; (b) large macroconidia or chlamydospores (7–18 µm in diameter), round, thick walled, and covered with knoblike or tuberculate projections (Fig. 32-10).

37°C: Yeastlike forms are seen (Fig. 32-11).

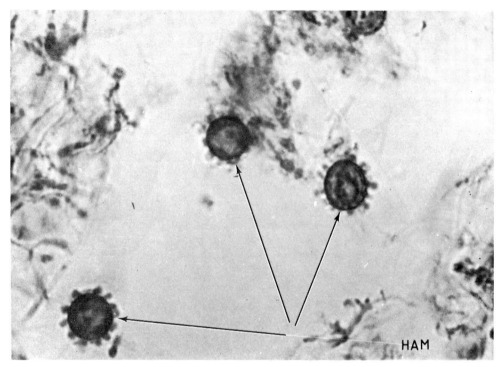

Figure 32-10. *Histoplasma capsulatum*. Mycelial phase showing tuberculate chlamydospores. Sabouraud dextrose agar slide culture (24°C). Lactophenol cotton blue, × 1140 (H. A. McAllister).

Identification is based on the cultural and morphological characteristics of the organism. Demonstration of organisms in macrophages and mouse pathogenicity are confirmatory.

MOUSE INOCULATION. The organism can be recovered from grossly contaminated specimens by mouse inoculation. Material is ground in physiological saline containing chloramphenicol (0.05 mg/ml) by means of a tissue grinder, or in a mortar and pestle with sterile alundum. Each of four mice are inoculated intraperitoneally with graded doses, e.g., 0.2 ml, 0.5 ml, and 1.0 ml. After four weeks, the mice are sacrificed and material from the livers is inoculated onto media as described above. Portions of liver are taken for histopathological examination.

The mycelial phase may be converted to the yeast phase by mouse inoculation. Mycelial elements are suspended in gastric mucin (see Appendix D) prior to inoculation.

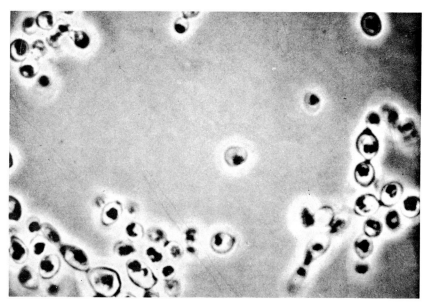

Figure 32-11. *Histoplasma capsulatum.* Yeast phase showing yeastlike forms from a culture at 37°C. Lactophenol cotton blue wet mount, dark-phase illumination, ×1900.

COCCIDIOIDOMYCOSIS
(CAUSE: *COCCIDIOIDES IMMITIS*)

This fungus occurs widely in the soil of certain arid areas of the southwestern United States and South America.

PATHOGENICITY. The disease is characterized by the formation of nodules or granulomas. It has been reported from human, horse, cattle, sheep, dog, cat, and some captive wild animals. The gross lesions in cattle resemble those of tuberculosis and are usually seen in the bronchial and mediastinal lymph nodes, and less frequently in the lungs. Lesions have been found in lung, brain, liver, spleen, and kidney of dogs.

DIRECT EXAMINATION. In unstained wet mounts, the organism is seen as a nonbudding, thick-walled sporangium having diameters varying 5–50 μm (Fig. 32-12). These large sporangia contain numerous endospores, 2–5 μm in diameter. The large sporangia burst, releasing the endospores, and leave empty "ghost" spherules.

ISOLATION PROCEDURES. *Because of the highly infectious nature of this organism, great care must be exercised in handling cultures and infectious materials.* Growing cultures should be covered with sterile saline before introducing the inoculating needle to prevent dispersion of arthrospores.

Clinical material should be inoculated onto blood agar, Sabouraud C and

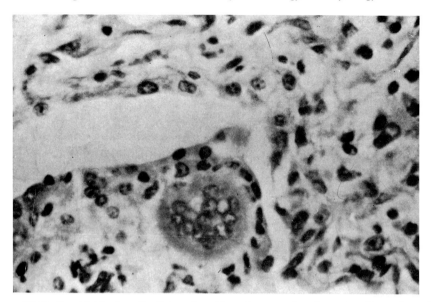

Figure 32-12. *Coccidioides immitis.* Spherule with endospores in a hemotoxylin-eosin stained section, ×840.

C agar, and regular Sabouraud agar. The former medium is incubated at 37°C and the latter at 25°C.

CULTURAL CHARACTERISTICS. 25°C: (one to two weeks): Colonies are flat, moist, and membranous, developing later a coarse, cottony aerial mycelium, the color of which varies from white to brown.

37°C: Colonies are similar in appearance to those seen at room temperature.

Only the mycelial form develops on artificial media. The tissue or yeastlike form described above is not seen, but it can be obtained by animal inoculation (4) (see below).

IDENTIFICATION. This is based upon cultural and morphological characteristics. 25°C: The mycelium consists of branching septate mycelia that form chains of thick-walled, barrel-shaped arthrospores (2–3 µm long). When stained by lactophenol cotton blue, the chains of arthrospores are stained in a characteristic alternate fashion ("collarettes") (Fig. 32-13).

The following are important differential considerations:

1. *Geotrichum* spp. (glabrous or yeastlike form) remain yeastlike in appearance on Sabouraud agar (25°C).
2. *Oospora* spp. (fluffy form of *Geotrichum*) do not produce alternate stained arthrospores and are not pathogenic for animals.
3. *Coccidioides immitis* produces typical spherules filled with endospores.

ANIMAL INOCULATION. Saline or broth suspensions of arthrospores from cultures or clinical materials are injected into mice intraperitoneally. Seven

Figure 32-13. *Coccidioides immitis.* Barrel-shaped arthrospores from a Sabouraud dextrose agar slide culture. Lactophenol cotton blue tease mount, ×2250 (H. A. McAllister).

to ten days after inoculation, the mice are sacrificed and the peritoneum, lungs, and spleen are examined for nodules or other lesions. Material from the lesions and nodules is examined in lactophenol cotton blue under a coverslip. If the preparation is sufficiently thin, spherules of varying size, including mature ones with endospores, are usually evident.

ADIASPIROMYCOSIS (HAPLOMYCOSIS)

This is principally a respiratory infection of many species of rodents and other wild mammals, including insectivores, herbivores, and carnivores. There is one report of the infection in a dog (8) and several reports of the disease in humans. The causal agents are the soil-borne fungi *Emmonsia parva* or *E. crescens.*

Light gray to yellowish focal lesions are found in the lungs of apparently healthy animals. The organism occurs as thick-walled spherical cells or adiaspores in the granulomatous lesions. They have been confused in histopathological sections with *Coccidioides immitis.*

As this disease is not apt to be seen at all frequently in the veterinary diagnostic laboratory, those desiring detailed information on cultivation and identification should consult Ainsworth and Austwick (3) and Emmons et al. (Supplementary References).

REFERENCES

1. Carter, M. E., Cordes, D. O., and di Menna, M. E.: *American Association of Veterinary Laboratory Diagnosticians,* 24th Annual Proc., 1981, p. 127.
2. Carter, M. E., Cordes, D. O., di Menna, M. E., and Hunter, R.: *Res Vet Sci, 14*:201, 1973.
3. Ainsworth, G. C., and Austwick, P. K. C.: *Fungal Diseases of Animals,* 2nd ed. Farnham Royal, Slough, England, Commonwealth Agricultural Bureaux, 1973.
4. Di Menna, M. E., Carter, M. E. and Cordes, D. O. *Res Vet Sci, 13*:439, 1972.
5. Bridges, C. H., and Emmons, C. W.: *J Am Vet Med Assoc, 138*:579, 1961.
6. Bridges, C. H., Romane, W. M., and Emmons, C. W.: *J Am Vet Med Assoc, 140*:673, 1962.
7. Lane, J. G., and Warnock, D. W.: *J Small Anim Pract, 18*:169, 1977.
8. Al-Doory, Y., Vice, T. E., and Mainster, M. E.: *J Am Vet Med Assoc, 159*:87, 1971.

SUPPLEMENTARY REFERENCES

See Supplementary References in Chapter 28.

Chapter 33

LABORATORY DIAGNOSIS OF INFECTIOUS ABORTION

C. A. KIRKBRIDE

A NUMBER of microbial agents have been incriminated as causes of abortion in the principal farm animals. Fetuses and their membranes are frequently submitted to the diagnostic laboratory for microbial examination. The procedures followed in such an examination are essentially similar for the various animals. Because abortion in the bovine is of great economic significance, the procedure outlined will deal with this species. Some of the important agents that are implicated in abortions in the domestic animals are listed below. The list does not include all of the bacterial species, fungi, protozoa, and viruses that are occasionally isolated from the aborted fetuses of all four species. For a comprehensive discussion of the causes of infectious abortions, readers are referred to the Supplementary References.

Bovine:	*Brucella abortus*
	Campylobacter fetus
	Leptospira
	Corynebacterium pyogenes
	Listeria monocytogenes
	Fungi: *Aspergillus, Absidia, Mucor, Mortierella*, etc.
	Infectious bovine rhinotracheitis virus
	Bovine virus diarrhea virus
	Chlamydiae
	Trichomonas fetus
	Toxoplasma gondii
Equine:	Rhinopneumonitis virus
	Equine arteritis virus
	Salmonella abortus-equi
	Streptococcus spp.
	Klebsiella
	Leptospira
	Fungi: *Mucor, Aspergillus*, etc.
Ovine:	*Campylobacter fetus*
	Brucella spp.

359

Chlamydiae
Listeria monocytogenes
Salmonella abortus-ovis, other *Salmonella* spp.
Corynebacterium pyogenes
Toxoplasma gondii
Porcine: *Brucella suis*
Leptospira
Erysipelothrix rhusiopathiae
Hog cholera virus
Pseudorabies virus
Parvovirus

In addition to the pathogens listed, numerous agents with low virulence that are not generally considered pathogens of adult animals have been associated with abortions. These organisms include *Escherichia coli, Bacillus cereus, Serratia marcescens, Pseudomonas* sp., alpha-hemolytic streptococci, and *Aeromonas hydrophila.* Some of these organisms are ubiquitous in the environment of livestock and are frequently isolated from abortion specimens. In these instances, a question arises as to whether these organisms should be considered a possible cause of the abortion. If the organism is present in large numbers and nearly pure culture, if lesions compatible with a bacterial infection are present, and if no other significant etiologic agent is found, the organism may be considered to have contributed to the cause of the abortion. It is always possible that the primary infectious agent had been eliminated by the time of abortion or was not discovered by the diagnostic methods used. Bacteria of low virulence may also be present concurrent with a viral infection.

In order to recover the agent from the fetus or its membranes, it is imperative that the laboratory receive fresh or properly preserved specimens. The procedures outlined below refer specifically to the bovine fetus, but they are equally appropriate for other fetuses. The only difference is in the agents being sought.

It is important that tissues be taken for histologic examination. Some viruses produce characteristic lesions, and the confirmation of a fungus as a cause of abortion is dependent on the demonstration of fungal elements in lesions.

EXAMINATION OF THE PLACENTA

1. Remove dirt and debris with running water and spread in a specimen tray.
2. Examine cotyledons and membranes for evidence of necrosis and exudate.

3. Prepare moist smears from necrotic areas of the cotyledons, digest five to ten minutes with 10% KOH, and examine microscopically for fungal elements.
4. Prepare smears from lesions and normal cotyledons and stain by Gram's method, Gimenez or Koster's technique, and acid-fast procedures if indicated.
5. If organisms are observed, they will provide guidance in the isolation procedures to be carried out. It is often difficult to isolate the causal organisms on culture media because of contaminants.
6. One or more cotyledons, especially any with lesions, should be fixed in 10% formalin for histologic examination.

EXAMINATION OF THE FETUS

1. The fetus is washed if necessary and examined externally for congenital defects, mycotic skin lesions, etc.
2. The fetus is then opened and examined internally. Fluid from the body cavities or heart blood may be taken for serologic examination. Tissues are taken for histologic and microbiologic examinations. If the fetus is well preserved, brain, heart, lung, liver, spleen, thymus, kidney, and any other tissue containing possible lesions may be included. If postmortem autolysis is extensive, at least lung and liver should be examined histologically. Tissues for virus isolation may be pooled or cultured individually.
3. The stomach wall is punctured with a hypodermic needle and 2–3 ml of the content is withdrawn into a sterile syringe. The stomach content is generally the best fetal specimen for bacteriologic examination, followed by lung, then liver.
4. Cultures are prepared as follows:

Aerobic
Stomach content, fetal
tissues, and placenta   Blood agar   37°C
Stomach content, suspected
mycotic skin lesions,
cotyledons   Sabouraud's   30°C
Microaerophilic (10% carbon dioxide)
Stomach content, fetal
tissues, and placenta   Blood agar   37°C

To isolate *Brucella* sp. from contaminated specimens, 6,000 Units Polymyxin® B, 25,000 Units Bacitracin®, 100 mg Actidione®, and 20 ml filter-sterilized, brucella-negative bovine serum may be added per liter of tryptose or trypticase soy agar, or 1.25 ml of a 1:1000 solution of ethyl violet or crystal violet may be added to a liter of serum enriched agar. (Biotype 2 *Brucella abortus* is sensitive to all dyes and will not grow in this medium.) Inoculated plates of these media should be incubated in an atmosphere containing 10% carbon dioxide.

A special selective medium is of value for the isolation of *Campylobacter fetus* (see Appendix B).

While successful fluorescent antibody techniques have been described for such abortifacient pathogens as *Campylobacter fetus, Brucella* sp., and *Listeria monocytogenes*, standardized fluorescent antibody conjugates are not available, and diagnosis of these infections is usually done through isolation and identification of the organism. Cold enrichment techniques are not necessary to isolate *L. monocytogenes* from fetal stomach content or tissues. Darkfield microscopic examination of a drop of fetal calf or lamb stomach content often reveals organisms with the characteristic shape, size and motility of *C. fetus*. The tentative diagnosis made this way should be confirmed by isolation and identification of the organism. The subspecies of *Campylobacter fetus*, especially those isolated from bovine fetuses, should be identified (Chapter 6).

LEPTOSPIROSIS

It is generally not practical to attempt isolation of leptospires from aborted bovine fetuses because of the time, effort, and low success rate involved. Leptospires can be isolated from aborted porcine fetuses more easily, but two to four weeks are usually required (Chapter 5). Leptospires can be seen frequently by darkfield examination of body fluids of porcine fetuses and much less commonly in body fluids of bovine fetuses. Great care must be used in identifying leptospires by direct examination, because many morphologically similar objects are present in body fluid.

Leptospires are more readily isolated from the urine of aborting cows than from aborted calves (Chapter 5), and their presence in the urine of a cow that has aborted in the past week or two provides significant diagnostic information.

Leptospires have been demonstrated in fetal tissues by microscopic examination of silver impregnated tissues (Levaditi stain). Kidney is the preferred tissue. Fluorescent antibody techniques using an incident light microscope have the advantage of more specificity. Kidney smears or cryostat sections may be used. In porcine fetuses, the leptospires tend to retain their morphologic characteristics and may be easily recognizable. In bovine fetuses, the morphologic characteristics are often lost and identification depends upon specific fluorescence. Standardized fluorescent conjugates are not readily available.

Care must be taken in diagnosing leptospiral abortion by serologic methods. Microscopic agglutination titers 1:100 to 1:400 may persist years following an infection, and up to one-half the cows infected with serovar *hardjo* may have microscopic agglutination titers less than 1:100 (1). Vaccination often results in microscopic agglutination titers as high as 1:6400 within two weeks. These titers usually decline to less than 1:400 within eight weeks, but the titers of cattle vaccinated after recovering from an infection often rise above 1:6400 and persist for several months above 1:400. Cows that abort from serovar *pomona* infection often have titers greater than 1:12,800 at the time of abortion. Cows that abort because of leptospiral infection have reached or passed their maximum titer at the time of abortion, and one should not expect an increase in leptospiral titer in serum taken after abortion. To determine if leptospiral infection is active in a herd, paired serum samples taken two to three weeks apart from a minimum of ten animals or 10 percent of the herd should be examined serologically. A four-fold increase in titer in an unvaccinated animal indicates a recent infection.

VIRUS INFECTIONS

For details of the laboratory diagnosis of virus infection, workers are referred to standard texts.

Direct fluorescent antibody examination of cryostat sections of fetal kidney is a highly successful technique for diagnosing abortion caused by the IBR virus (bovine herpesvirus, Group I) (2). Calves aborted because of IBR infection consistently have microscopic lesions of focal hepatic necrosis. In about one-third of the positive cases, IBR virus can be isolated from fetal tissues or placenta using standard cell culture techniques. The identity of the isolated virus is established using specific fluorescent antibody conjugates or immunoelectron microscopy with specific antiserum.

The BVD virus may be identified by direct fluorescent antibody examination of cryostat sections of fetal kidney or lung. Virus isolation from fetal tissues or placenta using standard cell culture techniques is quite successful. Most BVD viruses associated with abortion are noncytopathic, and their presence in second or third passage cell cultures is detected by fluorescent antibody examination using specific BVD conjugate. BVD virus does not consistently produce specific lesions in fetal tissues, and in some cases the virus may be present in the fetus without resulting in abortion.

Porcine parvovirus infection of sows pregnant up to day 56 causes death and resorption of part or all of the embryos, or death and mummification of part or all of the fetuses in a litter. If delivery occurs, generally it is at or after normal term. The most successful method of diagnosis is direct fluorescent antibody examination of cryostat sections of lung from mummified fetuses

(3). Parvovirus in fetal tissues may also be identified by specific hemagglutination activity. Parvovirus can be isolated from tissues of many mummified fetuses and some normal fetuses. Fetuses infected after seventy days gestation may produce antibody to parvovirus that can be detected by serum neutralization or specific hemagglutination inhibition tests run on serum or body fluids.

Pseudorabies (Aujeszky's, porcine herpesvirus) virus infection of pregnant sows often causes fetal death and partial mummification. The virus can sometimes be identified by direct fluorescent antibody examination of fetal lung, liver, or spleen. The virus is readily isolated from fetal tissues and causes herpesvirus-type cytopathic effect on cell cultures. There may be focal necrosis of fetal liver and spleen, and intranuclear inclusion bodies may be present in the margins of these lesions. Serum neutralization titers of 1:2 or greater in unvaccinated aborting sows provides evidence of infection and presumptive evidence of the cause of the abortion.

Gross and microscopic lesions are of value in the diagnosis of equine rhinopneumonitis, and the virus can be readily isolated from fetal lung and liver. The equine arteritis virus can be cultivated in cell cultures of equine kidney.

Numerous other viruses have been associated with livestock abortions, but other than virus isolation techniques, specific diagnostic procedures have not been described.

Standard fluorescent antibody conjugates for IBR, BVD, porcine parvovirus, and pseudorabies virus are available from the National Animal Disease Center, Ames, Iowa.

CHLAMYDIAL ABORTION

Chlamydiae may be isolated from infected fetal tissues or placenta by inoculation of chick embryos or cell cultures (4). The isolation success rate from bovine fetuses is rather low. Chlamydia-infected ovine and bovine placenta characteristically have necrotic gray-brown cotyledons and leathery intercotyledonary chorion. Microscopically, there is necrosis of the trophoblasts and inclusions. Examination of smears of these affected cotyledons stained by the Gimenez method (Appendix E) reveals chlamydial elementary bodies that appear as tiny red single or aggregated intracellular dots against a blue-green background. Chlamydial elementary bodies can be detected by fluorescent antibody techniques, but standard antichlamydial conjugate is not presently available. Cows aborting due to chlamydial infection characteristically have a significant antibody titer rise two to three weeks after termination of pregnancy when examined by the complement fixation test.

MYCOTIC ABORTION

Fungus infections of fetal calves consistently produce inflammation and necrosis of the placenta, and in about one-third of all cases also produce skin lesions. Occasionally, fungi infect fetal lungs producing bronchopneumonia. Diagnosis is made by histologically demonstrating hyphae associated with inflammation and necrosis. It is often necessary to apply special stains (Gomori's methenamine-silver nitrate, Periodic Acid-Schiff, or Gridley's) to tissues to make hyphae visible. Fungi often contaminate fetal stomach content and placentas, and isolation of them from these specimens does not necessarily implicate them as the cause of abortion. If a histologic diagnosis of mycotic infection is made, culture may enable identification of the agent(s).

Mycotic abortion occurs rarely in livestock species other than cattle.

TRICHOMONAS FETUS

Trichomonas infection usually causes bovine abortion before four months of gestation. Aborted fetuses are often badly macerated. Trichomonads can be seen by low power microscopic examination of a drop of fetal stomach content, and the organisms can be cultured in special media (5).

TOXOPLASMA GONDII

Toxoplasma infection is a common cause of ovine and caprine abortion in some areas. Other livestock are affected less commonly. The organisms may be seen by histologic examination in infected cotyledons, and more rarely, in fetal tissues. Isolation of toxoplasma requires mouse inoculation (6).

REFERENCES

1. Ellis, W. A., et al.: Bovine leptospirosis: Serological findings in aborting cows. *Vet Rec,* *110*:178, 1982.
2. Reed, D. E., et al.: Infectious bovine rhinotracheitis virus-induced abortion: Rapid diagnosis by fluorescent antibody technique. *Am J Vet Res, 32*:1423, 1971.
3. Mengeling, W. L., et al.: Fetal mummification associated with porcine parvovirus infection. *JAVMA, 166*:993, 1975.
4. Storz, J., and Whiteman, C. E.: Bovine chlamydial abortions. *Bovine Practitioner, No. 16,* pp. 71–75, November 1981.
5. Kinsey, P. B., et al.: Bovine trichomoniasis: Diagnosis and treatment. *JAVMA, 177*:616, 1980.
6. Dubey, J. P., and Schmitz, J. A. Abortion associated with toxoplasmosis in sheep. *JAVMA, 178*:675, 1981.

SUPPLEMENTARY REFERENCES

Dennis S. M.: *JAVMA, 155*:1913, 1969.

Dunne, H. W.: In Dunne, H. W., and Leman, A. D. (Eds.): *Disease of Swine*, 4th ed. Ames, Iowa St U Pr, 1975, Chapter 48.

Faulkner, L. C. (Ed.): *Abortion Disease of Livestock*. Springfield, Thomas, 1968.

Kirkbride, C. A. (Ed.): *Laboratory Diagnosis of Bovine Abortion*. Roanoke, Virginia, Proceedings 17th Annual Meeting American Association of Veterinary Laboratory Diagnosticians, 1974.

Chapter 34

BOVINE MASTITIS

F. H. S. NEWBOULD

THE BACTERIA and fungi listed below, and others, have been implicated as causes of bovine mastitis.

Streptococcus agalactiae	*Pasteurella haemolytica*
Str. dysgalactiae	*P. multocida*
Str. uberis, Str. bovis	*Bacillus* spp.
Fecal streptococci	*B. anthracis*
Group G streptococci	*Mycoplasma* spp.
Staphylococcus aureus	*Nocardia* spp.
Staph. epidermidis	*Trichosporon* spp.
Micrococci	*Candida* spp.
Escherichia coli	*Cryptococcus neoformans*
Klebsiella spp.	*Saccharomyces*
Enterobacter spp.	*Torulopsis*
Corynebacterium pyogenes	*Mycobacterium* spp.
C. bovis	*Fusobacterium necrophorum*
C. ulcerans	*Leptospira* spp.
Pseudomonas spp.	*Brucella* spp.
Serratia marcescens	*Clostridium perfringens*

Despite the large number of genera and species of microorganisms that have been implicated as causal organisms, it is still true that approximately 90 percent of infections are due to *Staph. aureus, Str. agalactiae, Str. dysgalactiae,* and *Str. uberis,* when *Str. uberis* is classified by the usual methods, i.e. on a Camp plate. It should be kept in mind, however, that streptococci giving no Camp reaction and that hydrolyse esculin may be *Str. uberis,* esculin-positive *Str. dysgalactiae,* or fecal streptococci.

ROUTINE PROCEDURES

Because of the large variety of potential pathogens, it is essential for accurate diagnosis that all samples submitted for laboratory examination be taken carefully and aseptically into sterile vials or tubes. Immediately after the samples are taken, they should be cooled to 4–5°C and maintained at that temperature until cultured in the laboratory. Frequently, milks kept longer than twenty-four hours and/or at ambient temperature are useless for diagnostic purposes. In such circumstances, because of differences in growth rates between genera or species, the actual causative organisms may be overgrown by others less significant. Only if it is quite impossible to obtain another one should such samples be cultured. If rapid delivery to the laboratory cannot be made, freeze the samples in a food freezer and keep frozen until delivery to the laboratory. This will have no effect on bacterial diagnosis but usually results in breakup of leukocytes present in the milk.

In the laboratory, first examine and record gross appearance of each milk sample. Then, proceed with culture preparation as detailed below.

Cultural Procedures

Primary

Blood Agar Medium
Either bovine (preferably calf) or ovine blood may be used for preparation of the blood agar. Staphylococcal antihemolysins, present in the blood of certain animals, may inhibit zones of alpha- and beta-hemolysis, which aid in the identification of *Staphylococcus aureus*. Therefore, each new lot of blood should be tested. When washed erythrocytes are used, testing the blood for antihemolysins is unnecessary. Rabbit, horse, or human blood is not satisfactory for use in the blood agar medium.

Routine Plating of Milk Samples
SAMPLE MIXING. In samples that have been refrigerated, some of the bacteria will have been swept up with the fat into the cream layer. Refrigerated samples should be warmed to 25°C. Shake the samples thoroughly twenty-five times up and down a distance of one foot in not more than seven seconds, or use an equivalent mixing procedure.

INOCULUM SIZE. Plating methods in general use for the microbiological diagnosis of bovine mastitis are not strictly quantitative but are based on the recognition of specific microorganisms that are reported as present or absent.

Transfer the inoculum to the medium surface with a wire loop such as the standard 0.01 ml loop. On occasion, milk samples from infected quarters may contain less than 200 bacteria per milliliter of milk (two colonies or less per 0.01 ml). To lessen the likelihood of the worker mistakenly considering such samples to be negative, a volume of 0.025 ml may be cultured. (A loop of 8 mm ID made with 20 gauge Chromel-A wire will deliver approximately 0.025 ml.)

SURFACE AREA. The area of surface over which the inoculum is spread will depend on the inoculum volume. A common practice is to divide a Petri dish into two parts and spread milk from two quarters of a cow on a single plate.

SPREADING INOCULUM. Streak the loopful of milk over the designated area in such a manner that at least some single, well-isolated colonies will result.

INCUBATION OF CULTURES. Normally, incubation temperature should be 35–37°C. Plates should be examined after eighteen to twenty-four hours incubation for colonies of streptococci, staphylococci, or other organisms, then returned to the incubator for at least another twenty-four hours incubation before being reexamined. If a stained smear indicates the presence of an anaerobe, incubate some plates under appropriate conditions (see Chapter 15).

NOTE: Some laboratories also inoculate Edward's medium. This is an additional expense, although it may facilitate the rapid presumptive identification of *Str. agalactiae* and *Str. dysgalactiae*.

Although usually unnecessary, it may be helpful to inoculate a plate of medium selective for staphylococci (see Appendix B). Sodium azide incorporated in blood agar at a concentration of 1:2000 provides a selective medium favoring the growth of streptococci and other Gram-positive organisms.

It may be useful to incubate, for six to fifteen hours at 37°C, those milk samples from which no isolations are made on direct plating. After incubation, a small loopful (.001 ml) should be streaked on a blood agar plate. Incubate twenty-four hours at 37°C. This should only be done when the sample is known to be free of contaminants.

If milks are very abnormal and no bacteria are isolated, it may be worthwhile to reculture for the presence of mycoplasma (see Chapter 27).

Secondary

Identification of organisms isolated
Examine the colonies that have developed on the incubated blood plates. Record colony form, hemolysis, etc. Stain using Gram's method and, depending on the types of organisms, proceed as follows.

GRAM-POSITIVE COCCI. Perform a catalase test (see Appendix C) on a slide, being careful not to pick up any blood cells with the organisms.

   Catalase positive — staphylococci or micrococci
   Catalase negative — streptococci
Staphylococcus. Do a slide coagulase (clumping factor) test (Appendix C).
   Coagulase positive — *Staph. aureus*
   Coagulase negative — *Staph. epidermidis*, *Staph. saprophyticus*, or *Micrococcus*
                  spp.
Streptococcus. Set up a Camp plate and test all streptococci found (see Appendix B). Camp plates must be read after eighteen to twenty-four hours incubation.

TABLE 34-I

DIFFERENTIATION OF SOME STREPTOCOCCI FREQUENTLY ISOLATED FROM
BOVINE MASTITIS

Organism	Lancefield Group	Hemolysis: Blood agar	Camp Test	Esculin	Sodium Hippurate	0.1% Methylene Blue Milk
Str. agalactiae	B	α (narrow) β or none	+	—	+	—
Str. dysgalactiae	C	α or none	—	—†	—	—
Str. uberis	Var. Some E	α or none	—*	+	+	—
Str. bovis	D	α or none	—	+	V	—
Enterocci	D	α or none	—	+	V	+
Str. spp.	G	α (wide)	—	+	—	—

*Approximately 12 percent *Str. uberis* strains give a positive Camp test.
†Some (up to 30%) may be positive.

GRAM-POSITIVE RODS. Non-sporeforming, short, pleomorphic. These are probably corynebacteria. If milk is very abnormal, probably *C. pyogenes*. If milk appears normal, probably *C. bovis*, or *C. ulcerans* is sometimes isolated from moderately acute cases of mastitis. These species can be differentiated by biochemical tests (see Chapter 17). *C. ulcerans* inhibits the formation of staphylococcal β toxin (reverse Camp reaction); microscopically, this organism appears as a coccus in all but very young cultures.

SPORE-FORMING RODS. A number of species of *Bacillus* can be isolated from infected quarters. *B. cereus* is quite common. Identify as described in Chapter 18.

GRAM-NEGATIVE RODS. These should be streaked on MacConkey's agar plates and identified by appropriate biochemical tests.

OTHER ORGANISMS. When other organisms such as mycoplasmas, fungi, anaerobes, and prototheca, are found on the original plate, isolate and identify by methods outlined in the appropriate chapters. The identification of mastitis pathogens is described in a National Mastitis Council publication (2).

Diagnosis

It is usual to define an infected quarter as one shedding milk from which bacteria are recovered in pure culture and an increased number of leukocytes (as compared to a normal quarter in the same cow) are found on one or more occasions. In cases of acute clinical mastitis, bacterial culture is usually

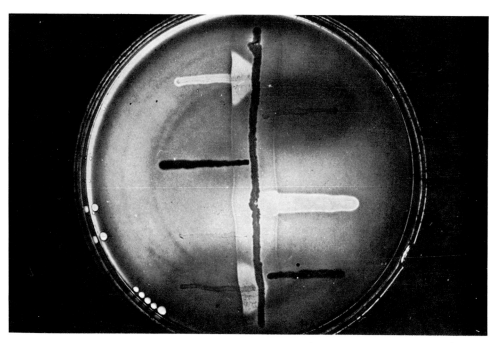

Figure 34-1. Camp test. Completion of the partial hemolysis along the *Staphylococcus aureus* streak by two cultures of *Streptococcus agalactiae*. The "arrowhead" effect is shown above and below, adjacent to the staphylococcal streak. The other streaks represent other streptococcal species (H. A. McAllister).

adequate. For diagnosis of subclinical mastitis, which accounts for up to 70 percent of all infected quarters, an estimate of the somatic cell count is usual. Cell counts may be estimated indirectly by the California Mastitis Test (CMT), by the Direct Microscopic Somatic Cell Count (1), or by an electronic counter. The direct count method, which is time-consuming and requires more expensive equipment, sometimes can be helpful in a preliminary identification of some causal organisms, e.g. yeasts, fungi, bacilli, etc. A 10 to 20 field count is usually sufficient in examining individual quarter samples. Detailed directions for both counting methods may be found in *Standard Methods for the Examination of Dairy Products* (1). Milks with cell counts in excess of 500,000 per milliliter are definitely abnormal and usually from infected glands. A proportion of glands shedding milk with cell counts between 150,000 and 500,000 per milliliter can be infected.

Particularly with the more common mastitis pathogens, the isolation in more or less pure culture of the same organism from two of three properly taken samples one week apart may be accepted as evidence of subclinical infection (3).

It should be kept in mind that micrococci, staphylococci (both coagulase positive and coagulase negative), streptococci, mycoplasmas, and diphtheroids, including *Corynebacterium bovis*, may be found in milk from apparently normal glands.

SUSCEPTIBILITY TESTS. These are carried out in the routine manner (see Chapter 35). Because of the consistent susceptibility of *Str. agalactiae* to penicillin, sensitivity tests are not usually carried out on this species. Some laboratories use a set of discs specially selected to represent the antimicrobial agents in current use for the treatment of mastitis.

REFERENCES

1. *Standard Methods for the Examination of Dairy Products*, 13th ed. New York, American Public Health Association, Inc., 1971.
2. Brown, R. W., et al.: *Microbiological Procedures for Use in the Diagnosis of Bovine Mastitis*, 2nd ed. Washington, D.C. (30 F Street, N.W.), National Mastitis Council, 1981.
3. Neave, F. K.: Diagnosis of mastitis by bacteriological methods alone. *Proc Seminar on Mastitis Control*. I.D.F. document 85. Brussels, Belgium, Int., Dairy Fed., 1975, pp. 19–36.
4. Baird-Parker, A. C.: *Ann NY Acad Sci, 236*:7–13, 1974.

SUPPLEMENTARY REFERENCES

Brown, R. W., Morse, G. E., Newbould, F. H. S., and Slanetz, L. W.: *Microbiological Procedures for the Diagnosis of Bovine Mastitis*. Washington, D.C., National Mastitis Council, U of N H Pr, 1969.
Schalm, O. W., Carrol, E. J., and Jain, N. C.: *Bovine Mastitis*. Philadelphia, Lea & Febiger, 1971.

Chapter 35

ANTIMICROBIAL AGENTS AND SUSCEPTIBILITY TESTS

Harold A. McAllister

INTRODUCTION

THE objective of this chapter is to offer veterinary practitioners and operators of small microbiology laboratories the essential background on antimicrobial susceptibility testing. Armed with this information, it should be feasible for them to prepare clinically relevant and therefore cost-effective antibiograms. The antibiogram is a statement delineating the pattern of susceptibility and resistance of a selected bacterial isolate to various antimicrobial agents. The material provided, despite appearances, is far from being exhaustive.

It should go without saying that directors of state-run veterinary diagnostic laboratories and their technologists, as well as those of roughly comparable facilities within colleges of veterinary medicine, should be aware of the explosion of information on the subject in the last half dozen years. Since they are expected to set the highest standards in the profession, more advanced sources are grouped for their convenience under supplementary references at the end of the chapter. That literature offers detailed instructions on numerous pertinent but specialized procedures such as the determination of antibiotic levels in body fluids.

A general impression of the author in the course of teaching veterinary students and of participating in an active diagnostic microbiology laboratory is that veterinarians are often hampered by a considerable lack of familiarity with the pitfalls and shortcomings of the antibiogram. An equally strong impression is that they are most capable of and eager to deal competently with these problems when the microbiologists who are their teachers make an effort to approach the subject in an intelligent manner. Hopefully, those who have not had access to such teachers will find it possible to correct some of their deficiencies on the basis of the content of these pages.

It may not be advisable for many veterinarians to carry out antimicrobial susceptibility testing in practice. A commitment to offer the service requires a rational basis and a thorough understanding of what the tests cost and of what they accomplish in the process of caring for patients.

If we consider only two alternatives, namely, susceptibility versus resistance of a microorganism to a given drug, it stands to reason that random choice alone should lead to a 50 percent chance of being correct. The likelihood of choosing correctly is enhanced to 80 percent and beyond if in lieu of a random process we substitute knowledge of the drug's pharmacology in the light of the disease process. Knowledge of the species identity of the disease-producing agent can further enhance the probability of choosing correctly. It thus follows that antimicrobial susceptibility testing is intended to narrow even further the chances of choosing incorrectly. It should not come as a surprise that there are increasing expenses as the gap is narrowed.

The effort invested in preparing antibiograms is almost certainly wasted if procedures fail to exceed 90 percent accuracy in predicting susceptibility or resistance for a given agent-drug combination. Moreover, failure to utilize methods that generate clinically useful data can compromise a practitioner's ability to select drugs for therapy to which agents are susceptible. Improper procedures generate test plates riddled with artifacts. Such plates are less reliable in predicting the usefulness of a drug than following its manufacturer's recommendations for use without doing a susceptibility test at all.

GENERAL CONSIDERATIONS

Antimicrobial chemotherapeutic agents are selectively toxic: they hamper the vital biochemical processes of infecting bacteria without seriously disturbing those of the host when used at recommended dosage levels. Agents in current use are of two general types, namely the antibiotics and the synthetic drugs. Antibiotics are substances originally discovered as products from living organisms, although today many are synthesized by artificial means. Synthetic drugs are laboratory-developed chemotherapeutic agents that have been shown to exhibit selective toxicity during extensive testing programs. Among them are the nitrofuran derivatives and the sulfonamides. Minute amounts of the antibiotics often suffice to hold back bacterial proliferation; larger amounts of the synthetic drugs are generally used to exert a comparable degree of inhibition.

The proper use of antimicrobial substances for therapeutic purposes demands considerable knowledge about the agents available and about their limitations. It is necessary to know, for example, the antimicrobial spectra of the commonly used drugs. Apart from the information provided on the subject in the package inserts and pharmacology texts, it is advisable to consult references where each drug class is exhaustively discussed. Early efforts by Lorian (1) along those lines should be supplemented by more recent texts such as the one by Pratt (2). Users simply must be aware that certain agents, such as the nitrofurans, methenamine mandelate, and nalidixic acid, are meant for use at specific body sites, and that others, such as ethionamide, isoniazid, pyrazinamide, and the antibiotics cycloserine and rifampin, are only useful against mycobacteria.

There is no point in demanding *in vitro* tests for susceptibility to methenamine mandelate because its *in vivo* activity is strongly dependent on a pH of 5 or less in the urine, a condition that is not met *in vitro*. Similarly, there is no point in demanding susceptibility tests for agents strictly intended for topical use at concentrations far in excess of those attainable with antimicrobial discs *in vitro*.

The pharmacology of drugs to be considered for therapy, including patterns of absorption and distribution in body tissues, is as important as their antimicrobial spectrum. For example, the oral administration of colistin and polymyxin B is of limited use because they are not absorbed from the intestine to a significant degree (1,2). Aminoglycosides are also poorly absorbed from the gastrointestinal tract, so they are as a rule administered parenterally except when neomycin or kanamycin are given orally to kill the bowel flora (2). Aminoglycosides are excreted mostly unchanged by the kidney, and very little of those administered parenterally reach the intestine. Again, susceptibility test results are applicable to the parenteral use of aminoglycosides; the highest concentrations attainable in the intestine may not be reflected by the *in vitro* tests. Aminoglycosides, novobiocin, penicillins, and tetracyclines are not favored in the treatment of central nervous system infections because they hardly penetrate cerebrospinal fluid and brain tissue in the absence of meningeal inflammation (1,2,3). The polymyxins fail to penetrate even when there is meningeal inflammation (2). Kanamycin does not penetrate muscle, and streptomycin fails to reach the testicles or the spleen (1). *In vitro* susceptibility to a drug, therefore, does not constitute an automatic endorsement of its use in therapy.

Drugs may also have dangerous side effects. All the aminoglycosides, namely amikacin, gentamicin, kanamycin, neomycin, streptomycin, and tobramycin, may seriously impair ear function (1,2,4). Along with the polymyxins, they are also nephrotoxic (1,2,4). Since aminoglycosides are excreted in active form by glomerular filtration, when renal function is impaired the parenteral levels may rise rapidly, with increased risk of toxicity and a need for readjusting dosage schedules (2).

In the acid urine of humans (2) as well as that of the dog and cat (4), certain sulfonamides may precipitate, actually contributing to renal failure. In humans, severe hematological disturbances can follow chloramphenicol treatment (1,2,4). In the cat, the slow dealkylation of chloramphenicol may lead to accumulation in and toxic effects to the bone marrow (3). Tetracyclines, novobiocin, lincomycin, erythromycin, and chloramphenicol are not uncommonly associated with gastrointestinal disorders, and hypersensitivity induced by lincomycin, streptomycin, and penicillin has been recorded (1,4). Death by anaphylactic shock following penicillin therapy is well documented for humans (1,2). Penicillin, lincomycin, streptomycin, erythromycin, and tetracyclines often precipitate fatal Gram-negative enterotoxemias in chinchillas, guinea pigs, and hamsters (5,6). Fatal streptomycin toxicity is seen in gerbils and in mice (5). Tetracyclines are known to be hepatotoxic in human patients with compromised renal function, and they are irritative enough to cause thrombophlebitis under certain circumstances (2).

The specific conditions of a disease process also have a bearing on the choice of antimicrobial therapy. In the case of anaerobic infections, the aminoglycosides are ineffective because their uptake by bacteria involves an energy-dependent active transport system that is not operative in the absence of oxygen (7). Quinones that function in aerobic metabolism thus play a central role in transporting these antibiotics to their intraribosomal site of action. Their inactivity in facultative organisms growing anaerobically and their absence from obligate anaerobes must be considered. Another circumstance that demands extreme caution is the patient suffering from a Gram-negative septicemia and kept marginally stable with a bacteriostatic agent like chloramphenicol. Indiscriminately switching these individuals to bactericidal drugs when susceptibility test results become available could conceivably be fatal if the change precipitates a massive bacterial kill with subsequent endotoxic shock.

Combination therapy—that is, the simultaneous administration of two or more antimicrobial agents—is commonly practiced in the initial therapy of severe infections as well as for coping with multiple pathogens in a single host and as a deterrent to the emergence of resistant microbes (8,9,10,11). Due to complex synergistic and antagonistic effects between antimicrobial agents, further pharmacological considerations must be made by the clinican using this approach (10,11). Broad-spectrum combinations may be suboptimal for a particular microorganism; they may lure clinicians into a false sense of security, they may be difficult to administer properly, and they may increase the rate of adverse drug reactions (10). Nonetheless, synergism between certain penicillins and cephalosporins has been established (10), and many compatible antibiotic combinations have been proposed (9,11). The most successful example of antimicrobial synergism is the sequential inhibition of a common biochemical pathway by the combination of trimethoprim and a sulfa (11).

In choosing antimicrobial agents for therapy, veterinarians must consider which agents are licensed for veterinary use and the different dosages, contraindications, and dangerous side-effects in the various animal groups. They may find that certain drugs are not licensed for use in the food animal species because they generate tissue levels unacceptable for human consumption, and that other drugs must be withdrawn at specified times before slaughter. In large animal therapy and in prophylactic regimes for herd or flock health programs, they may find that costs and availability in quantity must be weighed as much or more than other factors in selecting drugs for use. Antimicrobial agents should never be prescribed for trivial reasons, and those selected for use should have as narrow a spectrum as possible (8).

In view of the foregoing, it should be clear that a laboratory-generated antibiogram is a tool whose use requires considerable judgment. It is not meant for cursory reading by laymen or herdsmen, nor is it intended as a statement of dictated choices: The list of alternatives provided carries no implication of uncritical acceptance by clinicians. Obviously, it is not a substitute for the careful medical evaluation of possible courses of action. Its purpose is to help clinicians assess their choice of chemotherapy by providing for them a measure of the potential usefulness of various antimicrobial agents. That much the technologist at the bench can provide, but the ultimate choices and responsibility for their consequences lie strictly in the clinician's hand. Although treatment may have to be initiated before the antibiogram becomes available, clinicians generally abandon agents to which pathogens show resistance *in vitro*. In this sense, a laboratory report listing susceptibility or resistance to a particular antimicrobial agent becomes an endorsement of its use or of its withdrawal. Unfortunately, it is possible to err too much on the conservative side, suggesting resistance to drugs to which there is susceptibility, as well as to err by suggesting false susceptibility if correct procedures are not followed. Either error can be fatal if it deprives the patient of the drug of choice. In the face of resistance, switching to alternative choices rather than merely increasing the dosage of an administered drug is advisable because of the risk of toxicity to the host when accepted therapeutic levels are exceeded.

The only absolute criterion for evaluating the effectiveness of antimicrobial therapy is the clinical response of patients in the course of treatment (12). Host defense mechanisms may themselves act in synergism or antagonism with an antibiotic (13,14); its concentration at the infected site is a function of the site itself, the antibiotic's rate of diffusion from body fluids, and its inactivation or binding by body proteins. None of these factors play a role in the creation of an antibiogram.

Since susceptibility is defined as the quality or state of being susceptible, that is, the capability of or capacity for being acted upon, impressed, or moved, as by a drug, this term has been increasingly favored in usage. Sensitivity, on the other hand, is the quality or state of being sensitive, that is, the capacity of an organism or a sense organ to respond to stimulation, and it correlates well with irritability. In current literature, bacteria are generally spoken of as being susceptible to antimicrobial agents; "sensitivity to antibiotics" is still used, but professionals now often utilize such a term to denote hypersensitivity of the patient to drugs.

DETERMINATION OF SUSCEPTIBILITY

Dilution Methods

Dilution and diffusion are the two fundamental principles underlying techniques for the determination of susceptibility to antimicrobial agents (12,14,15,16,17,18,19). Dilution tests are of two types: tube dilution tests and plate dilution tests. The former entail exposing a measured number of microorganisms to increasing concentrations of antibiotics in liquid media; the latter involve a modification in which the agents are serially diluted in solid media. After incubation at 35°C, the lowest drug concentration that prevents visible growth is determined. This endpoint is defined as the minimal inhibitory concentration (MIC), which is expressed in micrograms or units per milliliter. This is a measure of the bacteriostatic effect of the agent on the microorganism (12). The original serial dilutions may be used as inocula for a second set in order to find out the minimal bactericidal, or lethal concentration (MBC or MLC) when necessary, which is rarely (19). Organisms are regarded as susceptible to a particular agent if its attainable blood levels are higher than the organism's *in vitro* MIC (17). Unfortunately, the current categorization of organisms as susceptible, intermediate, or resistant relates to dosage schedules used in humans, and it fails to consider the blood levels that may be attained with very high dosages of comparatively nontoxic penicillins and cephalosporins, the high urine levels achievable with certain antimicrobials, or the low tissue levels achieved with oral as opposed to parenteral administration in some cases (17).

Dilution tests yield accurate quantitative data (12,13,14,17,19,20) that allow direct comparisons between the different antimicrobial agents (13,20). They also provide information about synergistic and antagonistic effects by making it possible to test an organism with two or more antibiotics simultaneously (11,13). In addition, at the present time, considerable technical and interpretive difficulties in diffusion methodology for anaerobes make dilution tests preferable in that field (7). These facts notwithstanding, dilution procedures do have inherent limitations (10):

1. They require greater expertise than standard diffusion tests.
2. The handling of drug storage and dilution is subject to inconvenience and error.
3. Contamination is more difficult to detect.
4. The endpoint may be a less sensitive measure of resistant variants.
5. The endpoint may at times be more affected by variation in inoculum size.
6. Only discontinuous, arbitrary endpoints are possible, a fact that may be a problem with drugs with a narrow margin between toxic and therapeutic levels.

A major limitation of dilution tests in veterinary medicine is that the assay of antimicrobial agent levels in the body fluids of the different domestic species has not been routinely practiced, so that the attainable drug concentrations in tissues under routine dosage conditions are largely unknown. Reporting only a set of MIC values on a bacterial isolate is of limited value because the veterinary clinician would be forced to memorize for comparison drug levels extrapolated from the human species. To assume that dosage schedules employed in animal patients can reproduce levels attained in humans is logical only to the extent that both species are probably mammalian. Presuming that resistance to penicillin, as defined by an MIC too high for standard human dosages, necessarily applies to a mastitic cow given a massive dose of the drug could be misleading.

The traditional broth dilution tests are laborious, time-consuming, and costly in terms of equipment and personnel (13,16,17,19,20,21) so they are not suitable for routine clinical work, especially in small laboratories. For protocols and detailed information on them, the reader should refer to Finegold and Martin (12), Lennette (17), or Thrupp (19).

Accurate MIC determinations, however, are now within the reach of most clinical laboratories, thanks to advances in microdilution technology, automation, and inoculum replicators. An early example of efforts in this direction was the development of the Steers apparatus for plate dilutions (22). In the past decade various devices that automatically carry out serial dilutions for susceptibility testing have been marketed; the standardization and use of their microdilution techniques is extensively discussed in books edited by Balows (14), Hedén and Illéni (23), and Lorian (24). Quantitative procedures are the standard against which all others are measured (14,19). The major veterinary microbiology facilities, therefore, ought to consider committing at least one individual full-time to susceptibility testing and to charge such persons with the responsibility of developing at least semiautomated quantitative procedures for their laboratories.

Diffusion Methods

In diffusion techniques, a homogeneous film of the test organism is exposed to a concentration gradient of the antibiotic in an agar medium. An impregnated paper disc or strip, a tablet, or a ditch cut in the medium and filled with agar containing the antibiotic serve as reservoirs from which the antimicrobial agent diffuses out (12,13,14,16,17,18,21,25) after the drug is dissolved in water from the surrounding agar. This is a dynamic process: As time elapses, the concentration gradient changes, with the amount of drug per unit volume of agar dropping to increasingly lower levels, the outer limit of diffusion reaching farther and farther away, and the concentration peak near the disc becoming shallower. As diffusion progresses, microbial multiplication also proceeds. The speed of diffusion is a product of the interaction between the antimicrobial agent's molecules and the surrounding milieu, so the process is influenced by a tremendous number of variables. Among them are the amount of antibiotic in the reservoir, the density of the agar gel, the agent's diffusibility in aqueous solution, the ionic strength and composition of the medium, the geometry of disc placement, agar depth, and storage conditions for both agar and the drugs (12,13,14,16,17,18,21,25,26,27). The extent to which the antimicrobial agent has diffused out from its reservoir before a critical cellular density is reached governs the size of the growth inhibition zone (18). An often encountered misconception is the notion that the stated potency of antibiotic in the reservoir represents a "concentration" attainable *in vivo*. It cannot be emphasized enough that it represents only an amount, although it does create the concentration gradient that is the driving force behind diffusion (18).

Clinically relevant antibiograms appeared when it was realized that the growth inhibition zone size is directly proportional to a microorganism's susceptibility *provided* that all variables affecting the drug's diffusion are held constant (14,17,18,20,27). Two tests of an organism using different agar types will not generate inhibition zones of equal size around the same kind of antimicrobial disc. This crucial point escaped the attention of many early workers. A different point more widely understood is that agents to which an organism is equally susceptible yield different inhibition zone diameters if their diffusion rates differ enough (10,16,17,18,20,25,27,28). Polypeptides such as the polymyxins, for example, produce small inhibition zones due to their slow rate of diffusion, whereas the much smaller penicillin molecules diffuse rapidly and produce very large zones of inhibition.

Where inhibition zones develop, the number of organisms with which the plate was seeded is also a determinant of zone size. The size of the inoculum has actually been found to be the single most important variable influencing the size of the inhibition zone because the position of its border is determined when the critical cell mass is attained (18). Obviously, the more organisms in the inoculum, the greater the opportunity for generating visible growth before the antimicrobial agents reach inhibitory concentrations at a given distance from the disc. Conversely, the fewer organisms used to seed the plate, the fewer their chances for producing visible growth at that distance before inhibition takes over. No amount of the antimicrobial agent can prevent further growth if the bacteria reach a certain density (13,16,18), so heavy inocula will produce small inhibition zones.

Inoculum density is crucial when testing organisms resistant by virtue of drug-inactivating enzymes. Such organisms are falsely recorded as susceptible if the inoculum is light; if the inoculum is of appropriate density, enough preformed enzyme is transferred to the test system to allow microbial growth (18). Consequently, minor changes in inoculum density can lead to extreme changes in the result of susceptibility tests.

The size of the inhibition zone is determined at a point of time when a particular concentration of the drug reaches, for the first time, a density of growing cells too large for it to prevent growth (18). If the antimicrobial is added when incubation starts, the inhibition zone size is determined after this critical time. In most test systems, the critical time represents a period of about three to four hours (13,16,18). If the microorganisms are allowed to grow for some time before the drug reservoir is applied, the time elapsed before the threshold cell density is met is reduced, producing a smaller inhibition zone (18). Discs are therefore applied to plates as soon as possible and no later than fifteen minutes after inoculation of the organism (25).

The Bauer-Kirby Procedure

Without a standard, easily reproducible method that correlates well with dilution tests, the complex variables that influence diffusion techniques render them unreliable and difficult to interpret. Much research has been devoted to developing such standard procedures (13,15,16,18,21,25,29,30). The outcome of these efforts was the Bauer-Kirby (or Kirby-Bauer) single high-potency disc diffusion method (28), which is almost universally recommended for susceptibility testing (12,14,17,20,25,27). The method was introduced into veterinary practice (31,32) and has been officially sanctioned by the U.S. Food and Drug Administration for over a decade (33). Federal actions on behalf of the Bauer-Kirby procedure have been extensively discussed in scientific and technical circles (14,34,35,36). Completing the procedure was speeded up by the introduction of zone-interpretive devices or templates (37) and by reports that rapidly growing bacteria yield reliable zones in seven to eight hours of incubation (38,39,40). An alternative standardized procedure that is not used as widely is the agar overlay method of Barry, Garcia, and Thrupp (41).

The Bauer-Kirby disc method is based on the fact that for a given antibiotic, the size of the zone of inhibition is inversely related to the minimal inhibitory concentration (4,16,17,20,25,27). In other words, the size of the inhibition zone produced by the antibiotic increases as the MIC decreases, indicating that zone size is a measure of the MIC. In order to establish a correlation between zone size and MIC for a particular antimicrobial agent, the $\log_2$ of dilution-determined MICs is plotted against observed zone sizes produced by discs containing a fixed amount of antibiotic (14,17,20,25,27). At least 100 to 150 strains representing four or more bacterial species with comparable growth rates must be tested for any one drug disc type (17,25,27). By convention, zone diameters are plotted on the X axis and MIC values on the

Y axis. As illustrated in Figure 35-1, a "scattergram" of points is the result, for which a line of best fit or regression line must be calculated by the least squares method (25). A linear correlation between MICs and zone diameters produced by the disc is thus established, except in cases where the scattergram is bimodal.

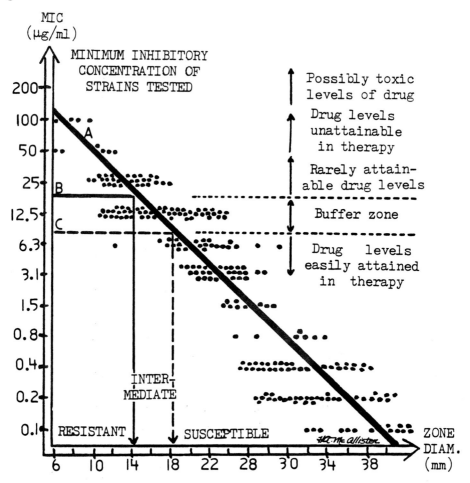

Figure 35-1. Scattergram determination of interpretive categories for agar growth inhibition zones produced by a particular high-potency disc of a given antimicrobial agent. A: line of best fit; B: upper breakpoint for susceptibility (resistance breakpoint); C: lower breakpoint for susceptibility.

Although regression analyses create graphs that could be used to determine the approximate MICs of an isolate from the inhibition zone sizes, such results would be inaccurate unless there is unreasonably stringent adherence to standard test conditions. A variation of a few millimeters in zone size could change graph-determined MICs by several orders of magnitude.

For this and other reasons, interpretive categories of susceptible, intermediate, and resistant have instead been established for each antimicrobial agent in common use (14,17,20,25,27). MIC breakpoints are selected (see Fig. 35-1) in order to create these categories. To do so, the attainable blood and tissue levels, the comparative response of many strains, and the clinical effectiveness of the agent as indicated by experience are all taken into account (14,20,25,27). The MIC breakpoints for susceptibility are lower than the drug levels attainable in blood or tissues (25). When scattergrams generate a bimodal distribution of susceptible and resistant strains, breakpoints are selected in the range of MICs that discriminate between the two populations as to their susceptibility.

The intermediate classification provided for in standard procedure categorizes strains that are neither clearly resistant nor fully susceptible, and it should be utilized. Strains falling into this category may be responsive to high dosages of the drug or to levels of it attainable in excretory organs. The most important function of the intermediate category, however, is to provide a buffer zone that minimizes the significance of minor variables that cannot be completely controlled in routine work (25).

In the future, the categorizations could be improved for veterinary use by taking into account blood and tissue drug levels seen routinely in the domestic animal species. Such refinements will require major research efforts at regional veterinary diagnostic laboratories and colleges of veterinary medicine.

Reliability of Diffusion Methods

Inability to understand the mechanics of diffusion rendered the early disc tests unreliable. Petersdorf and Plorde (13) examined the misconceptions built into multiple-potency disc methods popular until the early 1960s, and in 1960 some investigators (29) were compelled to propose a single low-potency disc method as a standard procedure. Although the proposal represented an improvement over the methods then prevailing, high-potency discs yield results that correlate better with the clinical response (36). Low-potency disc methods err heavily on the conservative side, because low-potency discs generate concentration levels in agar that are far under the levels attainable *in vivo* (30). They do, however, offer economy of space and media; up to twelve agents can be safely tested in a regular 10 cm plate. Research comparing multiple-disc, single-disc, serial dilution, and plate dilution procedures has led to the conclusion that single high-potency disc tests are superior to multiple-potency ones, since they are easier to perform and produce more accurate results (13,29).

The reliability of diffusion methods also improved with the realization that, on the average, about 4 mm difference in zone size define the limits between the two breakpoints for susceptibility and resistance. In laboratory practice, this means that ignoring even a single millimeter of the established zone interpretive categories can have a disproportionate impact on reported results. For example, if we select a single zone diameter, 17 mm, to define susceptibility for most or all antibiotics, strains at the limits for susceptibility and resistance may be inaccurately categorized with as many as a fourth of the antimicrobial agents that they are tested against. The reason for this becomes readily apparent when we consider that shifting the resistance breakpoint of cephalothin from 14 mm to 17 mm lowers the reference MIC from 32 µg/ml to about 13.4 µg/ml. A wide range of concentrations that may be attainable *in vivo* would consequently be ignored, and many strains of low or intermediate susceptibility would be listed as fully resistant. Moreover, shifting the lower level susceptibility breakpoint of the cephalothin from 18 mm to 17 mm raises the pertinent MIC from 10 µg/ml to 13.4 µg/ml, so that a number of other strains of intermediate susceptibility would end up listed as being fully susceptible. The largest number of errors would be introduced in reading zones for the penicillins, the polymyxins, carbenicillin, and ampicillin. Zones are therefore now almost invariably interpreted according to published tables subject to constant revision and that are promoted and sanctioned by the National Committee for Clinical Laboratory Standards (12,17,25,42).

Choice of Disc Potency

Since inhibition zones extending 3–16 mm beyond the disc's edge facilitate accurate measurements, a proper test aims at producing within those limits concentrations attainable *in vivo* (18,25). These concentrations must hold for the duration of the "critical time," that is, about four hours. Commercial low-potency discs reproduce the lower portion of this concentration range, but they yield a gradient that may not even begin to approximate the much higher levels attainable in urine. In the case of antibiotics that diffuse slowly, such as the polymyxins, the low-potency discs produce appropriate concentrations so close to the disc's edge that the inhibition zone of susceptible organisms may be overlooked. High-potency discs yield wider, easily measurable zones of inhibition with most antibiotics, and they produce a concentration gradient that includes levels attainable in urine.

Deterioration of antibiotics upon storage is to be expected (13); synthetic penicillins, for example, lose all their activity after exposure to high humidity at 37°C for three days (43). Therefore, in terms of reproducibility, low-potency discs are a disadvantage too. Significant deviations from the specified content are expected to occur in them because it is harder to deliver minute quantities accurately and it is easier to inactivate most or all of a small amount of an antibiotic through improper storage. The effect of these variables is minimized by the use of high-potency discs. The required potency of discs for Bauer-Kirby testing is always specified in zone interpretive tables.

SUSCEPTIBILITY TEST PROCEDURE

Recommended Approach

Individuals tempted to begin reading at this point are urged to turn back to the previous sections of this chapter; they may otherwise fall into the trap of introducing procedural modifications that render the test clinically meaningless.

In a recent textbook by Blood et al. (44) there is a statement on direct susceptibility testing that is correct but prone to misinterpretation. Direct susceptibility tests are those in which the inoculum is fresh clinical material. The textbook's authors do not advocate wholesale direct susceptibility testing. Direct tests are well outside the range of standard antibiograms and cannot be recommended (12,14,16,17,20,25,28). They are acceptable only when a pure culture of a pathogen is expected in the sample, and they represent an emergency procedure *invariably* subject to confirmation by standard techniques. The major flaw in the test is that the direct proportionality of inhibition zone size to susceptibility is lost. Since direct susceptibility tests yield only tentative results and can be a waste of time and media, they should not be demanded for routine cases. It is the clinician's responsibility to decide if a speedy but highly tentative answer may be beneficial to the patient, and there should be no surprise if the laboratory reports that the test failed to work.

Susceptibility tests should be limited to isolates widely regarded as pathogens or at least to those with a known potential for pathogenicity (16,17,21). This is necessary because the object of chemotherapy is to restrict the proliferation of such organisms while keeping interference with normal physiology at a minimum (8). An agent active against most or all of the observable intestinal flora, for example, can prove hazardous to herbivores, since they are adapted to deriving many nutrients through microbial action in their gastrointestinal tract. Similarly, the elimination of all commensals along with a pathogen in the oropharynx or the skin can open the door to a far more serious overgrowth by endogenous or exogenous yeasts. Since an antibiotic active against normal flora but not against an undetectable pathogen could actually enhance its proliferation, susceptibility tests should not be carried out on bacteria of very doubtful pathogenicity.

Pathogenicity is not a property inherent to the microorganism alone. Disease represents an interaction between the parasite and its host, and determinant factors are contributed by the latter (8). The same microorganism can be a commensal in one individual of a species and a pathogen in another, and its virulence can vary from one species to the next. Worse yet, an organism that may be safely regarded as a commensal in a particular area, as a *Proteus* in the intestinal tract, can be a primary pathogen in other tissues of the same host, such as the genitourinary tract. Debilitation of the host as a result of therapy that compromises its immune system or as a result of poor management also influences the pathogenic potential of colonizing commensals (26). For these reasons, the judgment as to the pathogenicity of any isolate should rest with those who have full access to the patients and their history.

Pitfalls of the Direct Test

The failings of direct susceptibility tests are not immediately apparent unless discussed. As a result, there is considerable temptation to perform them in veterinary practice. Practitioners tempted to use direct tests are also prone to carry them out in small facilities physically proximal to treatment and examining areas. The likelihood of contaminating test plates with airborne saprophytes under such conditions is very high, and this can lead to the development of inhibition zones that bear no relationship to the infectious agent.

Direct tests are justifiable in examining flora from samples collected under strict asepsis, like a spinal tap from a dog, or in the case of specimens with large numbers of a preponderant bacterial pathogen (17,20,25,28). Crucial problems are lack of control over the number of bacteria and over the composition of the inoculum. In veterinary practice, samples adequate for direct testing are rare. The reader is reminded that the single most important variable influencing the size of the inhibition zone is inoculum size, and in such tests there is no control over it. Most often, the bacteria are too few, leading to large zones of inhibition and to false reports of susceptibility. There may also be a mere scattering of colonies between the discs and no distinct inhibition zones.

A significant and less obvious problem is the presence of mixed floras in fresh clinical material. Quite often, the fastest-growing bacteria in these populations are incidental colonizers that play no role in the etiology of the disease. An appropriate example is a bovine lung yielding a preponderance of *Pasteurella haemolytica* along with a limited proportion of an *Enterobacter* sp. Standard procedures call for isolation and individual testing of the *Pasteurella*. A direct susceptibility test of the mixture would reflect the inhibition pattern of the fast-growing *Enterobacter* alone, hence misleading the clinician as to the drugs of choice. The test plate might not even have enough growth of the *Pasteurella* to outline faintly its antibiogram. Nutritionally fastidious pathogens such as beta-hemolytic streptococci, *Corynebacterium pyogenes, Listeria monocytogenes, Pasteurella haemolytica*, and many others can fail to grow when mixed with the faster-growing *Bacillus, Micrococcus*, or commensal Gram-negative enterics.

The behavior of mixed bacterial populations in routine diffusion tests is far more complex than that of single organisms. Each component has its own distinct and different antibiotic susceptibility pattern, and if all grow at the same rate, superimposing them creates an impression of complete or nearly complete resistance. This is because at least one member of the flora grows well around each disc. Moreover, if a commensal in the mixture produces an exoenzyme, such as a beta-lactamase, a penicillin-susceptible pathogen may grow near the penicillin disc, creating a false impression of resistance. Such an occurrence could lead to the selection of an alternate drug when in fact penicillin is the drug of choice.

Antibiograms of pure bacterial cultures often reveal resistant mutants growing within inhibition zones. Since they can rapidly replace the susceptible population *in vivo*, standard procedures call for a reading of resistance in such cases. In plates with mixed floras, these mutant colonies may be impossible to recognize as such due to the presence of numerous contaminants in the inhibition zones. This interpretive difficulty is common in direct susceptibility testing.

Material of Choice

The sample of choice for a susceptibility test is a pure culture of a potential pathogen. The inoculum should represent an adequate cross section of the population so that all susceptible as well as resistant cell types are included. To achieve this, four or more colonies of exactly the same type are suspended in 4 ml of trypticase soy or brain-heart infusion broth (12,14,17,25,28). These tubes are incubated at 37°C for several hours until a suspension of moderate cloudiness develops. The visual density of the suspension should be equivalent to that of a barium sulfate standard (12,14,17,18,25,28), so it may prove necessary to dilute the culture with sterile saline. The standard is prepared by adding 0.5 ml of 1% (11.75 g/liter) $BaCl_2 \cdot 2H_2O$ to 99.5 ml of 1% (0.36N) H_2SO_4. This standard, also known as a 0.5 McFarland nephelometer standard (17), must be placed in a tube of the same size as the culture tube and should be replaced monthly. Sealed standards stored in the dark at room temperature have a maximum shelf life of about six months (12). When high numbers of cases make it impractical to grow appropriately turbid broth subcultures, identical colonies may be selected directly and suspended in 4 ml of saline (17,25) so as to match the turbidity of a McFarland 1.0 standard (prepared by adding 1.0 ml of the 1% barium chloride solution to 99.0 ml of the 1% sulfuric acid).

A sterile cotton swab on a wooden applicator stick is used to transfer the diluted bacterial suspension to a plate; excess fluid must be squeezed out by rotating the swab against the sides of the tube. The plate is seeded uniformly by rubbing the swab against the entire agar surface in three different planes at roughly 60 degrees to each other (12,17,25).

Dilution rather than diffusion methods are currently preferred in clinical anaerobic bacteriology because adequate interpretive criteria from regression analyses have not yet been created for the latter (7,17). When diffusion methods are nonetheless used, the organisms are grown overnight in a modified thioglycollate medium without indicator enriched with hemin and menadione, and they are diluted in *Brucella* broth (7,14) prior to plating. Readers are urged to refer to the specialized literature on the subject if they must test anaerobes; simply attempting to use Bauer-Kirby procedures anaerobically has no rational basis and will not produce reliable results (7).

Media

The Bauer-Kirby technique calls for the use of Mueller-Hinton agar in susceptibility tests (12,14,17,18,20,25,28,33). Although other media could conceivably be used (14,18) and an inert, defined, and totally synthetic polyoxoethylene hydrogel-amino acid medium combination may minimize batch variation in the future (18), the standard zone size tables have been developed only for Mueller-Hinton agar and are not applicable to any other media (14,16,17,18,20,25,28).

Mueller-Hinton agar cooled to 50°C is poured into sterile Petri plates to a depth of 4 mm on a level surface (12,25). This is equivalent to about 60 ml in a 15 cm plate and about 25 ml in a 10 cm plate. The pH of each batch must be checked and should be 7.2–7.4 at room temperature (12,17). The test can be done by macerating a small amount of medium in distilled water or by using a surface electrode if available. Plates are stored at 4–8°C and those not to be used within a week must be kept in sealed plastic bags to prevent evaporation. Excess surface moisture is eliminated on the day of use by placing the plates in a non-CO_2 incubator with lids ajar for ten to twenty minutes. The total shelf life of Mueller-Hinton agar plates is two weeks (25).

There is sufficient variation in peptone contents, total solids, pH, glucose amount, and ionic concentration from one medium to the next to cause considerable differences in diffusion rates (14,20). Mueller-Hinton is a good compromise for routine susceptibility testing because it shows good batch-to-batch reproducibility, enables most pathogens to grow satisfactorily, and is low in tetracycline, trimethoprim, and sulfonamide inhibitors (14,25). Other agars are not as suitable for testing sulfonamides. These drugs act against bacteria by preventing the synthesis of para-aminobenzoic acid's (PABA) essential pteroyl derivatives through competitive inhibition. Most media contain an abundance of the main PABA derivative, folic acid, and observable inhibition around sulfonamide discs is therefore prevented by satisfying bacterial needs directly. Mueller-Hinton contains only traces of this antagonist (13,20,25).

Slowly growing, fastidious organisms usually require Mueller-Hinton enriched with 5% defibrinated whole blood. Alternatively, it may be supplemented with 1% hemoglobin plus either 1% Iso Vitale X® (BBL Microbiology Systems, Cockeysville, Md.) or Supplement VX (Difco Laboratories, Detroit, Mi.). Supplementation is required for organisms such as beta-hemolytic streptococci, *Corynebacterium pyogenes, Listeria monocytogenes,* and other comparable pathogens because the Bauer-Kirby procedure has been standardized for testing more rapidly growing bacteria like the Enterobacteriaceae (12,14,17,28). Supplementation helps them to attain a comparable rate of growth. If growth is too slow, results are unreliable (14,20,28), a point to keep in mind when testing organisms that grow optimally at 25–30°C. In general, if colonies are not macroscopically apparent within twenty-four hours of incubation, the diffusion test yields inaccurate results.

Selective or differential media such as MacConkey, Salmonella-Shigella, and eosin-methylene blue agars should never be used in susceptibility testing because they contain growth inhibitors and factors that can neutralize some antibiotics.

Mueller-Hinton agar is inadequate for use with anaerobes (7,17). If anaerobic diffusion susceptibility tests are done, its substitute is *Brucella* agar with 5% sheep blood supplemented with menadione (14,17).

Discs and Their Application

Bauer et al. (28) recommended the large (15 cm) Petri dishes for susceptibility tests, and they are widely used today. The discs should be no closer than 1.5 cm to the edge of the plate, and they should rest about 3.0 cm apart from each other (33). Crowding can cause overlapping zones and inaccurate readings. The large Petri dishes easily accommodate nine discs in an outer ring and three in the center, whereas no more than eight should be placed in small (10 cm) plates. Dispensers capable of dropping twelve antimicrobial discs on large Petri plates are commercially available. Multitipped sheets that make it unnecessary to employ disc dispensers have been marketed too. Although convenient to use, they can generate incomplete zones of inhibition that may overlap, so they are considered inferior to individual discs (21).

Susceptibility tests are usually carried out with commercially available paper discs impregnated with different antimicrobial agents. Since 1959, the Food and Drug Administration has enforced strict regulations controlling the manufacture of discs produced in the United States, and today, they are recognized as being of uniformly high quality (14). The disc potency selected for each antibiotic must match the level specified by the Bauer-Kirby method (12,14,17,20,25,28). To prevent loss of potency, discs are kept dehydrated and in a freezer at −14°C until ready to use. Cartridges that have been removed from their original container are stored under refrigeration and should not be used longer than a week (14,17,21). Discs must never be used beyond their expiration date (12,14,17). The discs are deposited on the plate mechanically or by hand after the inoculum has been allowed to dry for several minutes but not longer than fifteen (17,25,28). They then are pressed flat against the agar by applying firm but gentle pressure with an alcohol-flamed forceps (12,14,17,28). This insures full contact with the agar surface, without which the contents of the disc may not diffuse out in an even and stable manner. In addition, it prevents discs from falling off when the plate is inverted for incubation. Some commercial dispensers are designed to eliminate this hand tamping.

Drug Selection

In order to simplify routine susceptibility testing, it is necessary to limit sharply the number of drugs examined. Only one representative of any group of antimicrobial agents with similar chemical structure and action is therefore normally tested (14,16,17,20,21). Manufacturers are eventually expected to supply only one representative disc for each family of analogous antibiotics (14,34). At the present time, following this concept, the recommended representatives and their groups are as follows:

Representative	Group
1. *Ampicillin*	(amoxicillin, hetacillin)
2. *Cephalothin*	(cephaloridine, cephalexin, cefazolin, cephacetrile, cephradine, cefaclor, cephapirin)
3. *Clindamycin*	(lincomycin)
4. *Methicillin*	(cloxacillin, dicloxacillin, nafcillin, oxacillin)
5. *Penicillin G*	(phenethicillin, phenoxymethyl penicillin)
6. *Polymyxin B*	(colistin)
7. *Streptomycin*	(dihydrostreptomycin)
8. *Tetracycline*	(chlortetracycline, demeclocycline, doxycycline, methacycline, minocycline, oxytetracycline, rolitetracycline).

The class representative should be the least active drug in the group so that any differences that occur will be in the direction of "false" resistance rather than "false" susceptibility (17). Methicillin meets this criterion among the beta-lactamase resistant penicillins, but oxacillin or nafcillin discs are sometimes preferred because they are much more stable.

No one representative suffices for drugs from groups such as the aminoglycosides or the macrolides where resistance to one does not imply resistance to the others. Antibiotics that must be tested individually include amikacin, gentamicin, kanamycin, neomycin, streptomycin, and tobramycin (the aminoglycosides); erythromycin and oleandomycin (the macrolides); cefamandole and cefoxitin (the wide spectrum second generation cephalosporins); carbenicillin; chloramphenicol; and trimethoprim with sulfonamide.

Certain antimicrobial agents are ineffective under anaerobic conditions and should not be used in anaerobic susceptibility tests. These drugs are the aminoglycosides, the polymyxins, and the sulfonamides (17,26).

In veterinary practice, there are some differences in the selection of drugs for large versus small animals due to cost factors and tradition. As a result, some laboratories have found it useful to create separate selections for the two types of practice and a third selection for anaerobes. The actual selection used for a given isolate is not so much determined by the size of the animal patient as by the total biomass expected to undergo treatment. All domestic farm animals in flocks or herds, experimental laboratory animals in colonies, and large exotic species dictate a choice of the large animal set. Small domestic animals and large pets would dictate use of the small animal set. A given animal such as a rat would then be tested for with one or the other set, depending on whether it is a pet or an individual from a diseased colony. A fourth selection, for use in cases of mastitis, may also be desirable. Within an institution there should be agreement with the clinical staff as to the agents in use and therefore of interest. The choices should be periodically revised as new drugs become available or new information is acquired (25).

The widespread use of nitrofurans and sulfonamides in veterinary practice must be taken into account in creating a selection of drugs for routine use. Unfortunately, complete regression analyses on Mueller-Hinton agar have only been carried out with nitrofurantoin from the former and a few sulfas from the latter. Although accepted interpretive categories for all sulfonamides exist (12,14,17,28), it is not clear whether those for nitrofurantoin can or even should be extrapolated to other nitrofurans. No interpretive categories are currently provided in human microbiology sources for topically active antimicrobial agents nor for certain antibiotics of limited medical use, such as tylosin and the aminoglycoside spectinomycin. Considering the greater usefulness of the latter in the domestic animals, the veterinary profession ought to make certain that adequate Bauer-Kirby categories are created for them and that they are included in all standard interpretive tables. Commercially prepared discs that are locally unavailable may then be mass produced by manufacturers in the United States. McDonald and Bieberstein (45) have produced some very preliminary data pertaining to disc diffusion tests for tylosin.

An example of a selection of antimicrobial agents chosen for routine tests in a veterinary diagnostic laboratory during the mid-1970s is offered in Table 35-I. In addition to the sets of antibiotics routinely tested for, the laboratory should keep a number of additional agents on hand. These extra discs are very useful in examining antibiotic-resistant bacteria such as *Pseudomonas, Proteus*, and some enterococci. *Pseudomonas* in particular is often susceptible to sulfonamides other than those normally tested.

TABLE 35-I
ANTIMICROBIALS FOR ROUTINE TESTING

	Small Animal	Large Animal	Mastitis	Anaerobes
Ampicillin	test	test	test	test
Carbenicillin				test
Cephalothin	test	test	test	test
Clindamycin (Lincomycin)	test	test		test
Chloramphenicol	test	test	test	test
Erythromycin			test	test
Furaltadone			test	
Gentamicin	test	test	test	
Kanamycin			test	
Neomycin	test	test		
Nitrofurantoin	test			
Nitrofurazone		test		
Novobiocin			test	
Oxacillin			test	
Penicillin	test	test	test	test
Polymyxin B	test	test		
Streptomycin			test	test
Sulfachloropyridazine		test		
Sulfadimethoxine	test			
Trimethoprim + Sulfa	test			
Tetracycline	test	test	test	test

Incubation

An incubator without CO_2 is used in Bauer-Kirby testing because this gas will alter surface pH enough to affect the inhibition zone sizes significantly (12,17,25). Fastidious organisms that require increased CO_2 tension for growth do not lend themselves to susceptibility testing by standard disc methods; dilution test procedures are recommended for them.

An incubation temperature of 35°C is recommended because staphylococci with increased clinical resistance to the penicillins and cephalosporins are detectable at that temperature but not at 37°C (17). It is important to realize that a single plate placed on an incubator's metal shelf takes about one hour to warm to within 1°C of the incubator temperature (18). If plates are stacked in piles five deep and then placed in the incubator, however, the center plate takes up to four hours to reach that temperature (18). Such a delay can influence inhibition zone sizes, so for reference work test plates should not be stacked.

Interpretation

After overnight incubation, the diameter of each inhibition zone, including the 6 mm disc, is measured in millimeters using calipers or a ruler on the undersurface of the Petri dish (12,14,17,20,28). For measuring purposes, the end point is taken as complete inhibition of growth as determined by the naked eye except in the case of sulfonamides or when dealing with *Proteus* swarmers. Sulfonamides allow organisms to grow through several generations before inhibition takes effect, so a hazy zone representing up to 20 percent growth must be disregarded, and the margin of heavy growth is taken to determine zone size (12,14,17,28). *Proteus* swarming is not inhibited by all antibiotics, so a hazy veil of swarming into an inhibition zone must also be ignored (12,14,17,28). Properly inoculated plates are expected to feature homogeneous, nearly confluent colonial growth between the inhibition zones. The final results must be interpreted according to criteria listed in Table 35-II. Zone diameters can be read after incubation for six to eight hours if speedy completion of the test is important (38,39,40); these results are preliminary and subject to confirmation by the standard incubation time.

Individual measurement of each inhibition zone with sliding calipers is too time-consuming for busy laboratories or veterinary practices. Fortunately, Martin et al. (37) facilitated the reading procedure by developing a zone-interpretive device (ZID), versions of which are readily available. The device is constituted by an outline of discs on a 150 mm circle. A self-adhesive transparent overlay provides a series of zone-measuring tabs with two circles around the outline of each disc. The circles correspond to diameters representing susceptibility and resistance by Bauer-Kirby criteria. The tabs corresponding to different antibiotics can be placed on the ZID templates so that they coincide with the arrangement of antibiotics in the dispensers. Accurate and rapid categorization of the bacterial isolates is then carried out by simple inspection of plates against the template ("eyeballing"). Difco Laboratories (Detroit, Mi.) still supplies templates directly (Technical Information Bulletins 0345 and 0347: Antimicrobial Disc Zone Chart and Disc Zone Stickers, respectively); Baltimore Biological Laboratories (Cokeysville, Md.) provides them for a nominal fee through its suppliers (BBL 60639: Sensi-Disc® Zone Interpretation Kit).

Table 35-II

ZONE SIZE INTERPRETIVE CHART FOR ANTIMICROBIALS*

Antimicrobial Agent	Inhibition Zone Diameter (mm)			
	Disc Potency	Resis- tant	Inter- mediate	Suscep- tible
Amikacin	30 µg	≥ 14	15–16	≤ 17
Ampicillin – for Enterobacteriaceae and enterococci	10 µg	≥ 11	12–13	≤ 14
Ampicillin – for *Staph.* & penicillin G susceptible bacteria	10 µg	≥ 20	21–28	≤ 29
Ampicillin – for *Hemophilus*	10 µg	≥ 19	–	≤ 20
Bacitracin	10 U	≥ 8	9–12	≤ 13
Carbenicillin – for *Proteus* spp. and *Escherichia coli*	100 µg	≥ 17	18–22	≤ 23
Carbenicillin – for *Pseudomonas aeruginosa*	100 µg	≥ 13	14–16	≤ 17
Cefamandole	30 µg	≥ 14	15–17	≤ 18
Cefoxitin	30 µg	≥ 14	15–17	≤ 18
Cephalothin	30 µg	≥ 14	15–17	≤ 18
Chloramphenicol	30 µg	≥ 12	13–17	≤ 18
Clindamycin	2 µg	≥ 14	15–16	≤ 17
Colistin	10 µg	≥ 8	9–10	≤ 11
Erythromycin	15 µg	≥ 13	14–17	≤ 18
Gentamicin	10 µg	≥ 12	13–14	≤ 15
Kanamycin	30 µg	≥ 13	14–17	≤ 18
Methicillin	5 µg	≥ 9	10–13	≤ 14
Neomycin	30 µg	≥ 12	13–16	≤ 17
Nitrofurantoin – for urinary tract infections only	300 µg	≥ 14	15–16	≤ 17
Novobiocin – not applicable to media that contain blood	30 µg	≥ 17	18–21	≤ 22
Oxacillin or nafcillin	1 µg	≥ 10	11–12	≤ 13
Penicillin G – for Staphylococci	10 U	≥ 20	21–28	≤ 29
Penicillin G – for other microorganisms	10 U	≥ 11	12–21	≤ 22
Polymyxin B	300 U	≥ 8	9–11	≤ 12
Streptomycin	10 µg	≥ 11	12–14	≤ 15
Sulfonamides	250 or 300 µg	≥ 12	13–16	≤ 17
Tetracycline	30 µg	≥ 14	15–18	≤ 19
Trimethoprim + sulfadiazine	1.25 µg/23.75 µg	≥ 10	11–15	≤ 16
Tobramycin	10 µg	≥ 12	13–14	≤ 15
Vancomycin	30 µg	≥ 9	10–11	≤ 12

* An updated, expanded version of this table is available in the Second Supplement, *Performance Standards for Antimicrobic Disk Susceptibility* Tests, NCCLS Publications, 1982. Since this material is constantly updated the reader should obtain the latest version from the National Committee for Clinical Laboratory Standards, 771 East Lancaster Ave., Villanova, Pa. 19085.

When isolated colonies appear in an inhibition zone, either the culture is not pure or resistant mutants are present in the population. Mixed cultures with different proportions of unrelated bacteria usually stand out by developing a homogeneous field of the most common organism and scattered colonies of an obviously different character. Colonies of the contaminant appear in many of the inhibition zones. Such contaminated cultures require retesting unless the contaminant colonies are extremely few. Pure cultures with resistant mutants, on the other hand, do not usually have them in more than one or two inhibition zones. Resistant mutants represent significant growth (28) because they would replace the susceptible population if the drug were used for therapy.

Quality Control

Hopefully, the reader will now understand why incorrect techniques can lead to growth inhibition zones around antibiotic discs that are artifacts of procedure and that do not correlate with susceptibility *in vivo*. Disc diffusion tests that correlate well with *in vivo* results are entirely dependent on quality control measures. These include the use of properly prepared and stored supplemented and nonsupplemented Mueller-Hinton agar plates, fresh discs, adherence to the stated Bauer-Kirby protocol, and the capacity to absorb the cost of discarding outdated materials. Extra bacteriological equipment required for correct procedures includes a freezer for disc storage, refrigerator space for disc stocks in use, a non-CO_2 incubator kept at 35°C, reading templates, barium chloride turbidity standards, and, usually, disc dispensers. Veterinary practices distant from bacteriology reference laboratories should equip themselves accordingly.

Daily monitoring of accuracy in testing by means of the use of control strain cultures is expected today of all laboratories engaged in health care (12,14,17,25,46). High-quality veterinary practices that offer antibiogram service will want to do the same. The three strains recommended for the purpose are *Staphylococcus aureus* ATCC 25923, *Escherichia coli* ATCC 25922, and *Pseudomonas aeruginosa* ATCC 27853 (12,14,17,25,46).

To avoid variation and contamination, the control strains should be kept either lyophilized or frozen at −60°C in 15% glycerol broth or 50% inactivated calf serum in broth. These frozen stocks can be used to prepare fresh slants weekly or every two weeks. Alternatively, the laboratory could purchase control strains as single-use vials specifically designed for quality control.

The inhibition zones obtained with the monitor strains are measured to the nearest millimeter, and the results are compared with expected results (14,46). The maximum acceptable deviation is about 1.0–2.0 mm with most antibiotics and almost 4.0 mm with penicillin and ampicillin (14,46). Deviation trends as well as sudden procedural breakdowns can be detected this way. Laboratories should create their own records and subject them to standard statistical analysis (14,46). Additional information on quality control procedures is available from Balows and his colleagues (14,46).

The only risk inherent in failing to practice quality control in susceptibility testing is that of consuming vast sums of money and much time in clinically meaningless tests that do not enhance patient care.

REFERENCES

1. Lorian, V.: *Antibiotics and Chemotherapeutic Agents in Clinical and Laboratory Practice.* Springfield, Thomas, 1966.
2. Pratt, W. B.: *Chemotherapy of Infection.* New York, Oxford, 1977.
3. Keen, P.: *J Small Anim Pract, 16:*767, 1975.
4. Bryant, M. C.: *Antibiotics and their Laboratory Control.* Laboratory Aides Series. London, Appleton-Century-Crofts, 1968.
5. Williams, C. S. F.: *Practical Guide to Laboratory Animals.* St. Louis, Mosby, 1976.
6. Clark, J. D., Loew, F. M., and Olfert, E. D.: Rodents. In Fowler, M. E. (Ed.): *Zoo and Wild Animal Medicine.* Philadelphia, Saunders, 1978, pp. 457–478.
7. Rosenblatt, J. E.: Antimicrobial susceptibility testing of anaerobes. In Lorian, V. (Ed.): *Antibiotics in Laboratory Medicine.* Baltimore, Williams & Wilkins, 1980, pp. 114–134.
8. Hamilton-Miller, J. M. T.: *J Small Anim Pract, 16:*679, 1975.
9. Herrell, W. E.: *Clin Med, 76:*11, 1969.
10. Klastersky, J. (Ed.): *Clinical Use of Combinations of Antibiotics.* New York, Wiley, 1975.
11. Krogstad, D. J. and Moellering, R. C.: Combinations of antibiotics, mechanisms of interaction against bacteria. In Lorian, V. (Ed.): *Antibiotics in Laboratory Medicine.* Baltimore, Williams & Wilkins, 1980, pp. 298–341.
12. Finegold, S. M. and Martin, W. J.: *Bailey and Scott's Diagnostic Microbiology,* Part V, Antimicrobial Susceptibility Tests and Assays. St. Louis, Mosby, 1982, pp. 532–557.
13. Petersdorf, R. G., and Plorde, J. J.: *Ann Rev Med, 14:*41, 1963.
14. Balows, A. (Ed.): *Current Techniques for Antibiotic Susceptibility Testing.* Springfield, Thomas, 1974.
15. Anderson, T. G.: An evaluation of antimicrobial susceptibility testing. In Gray, P., Tabenkin, B. and Bradley, S. G. (Eds.): *Antimicrobial Agents Annual 1960.* New York, Plenum Press, 1961, pp. 472–477.
16. Petersdorf, R. G., and Sherris, J. C.: *Am J Med, 39:*766, 1965.
17. Lennette, E. H. (Ed.-in-Chief): *Manual of Clinical Microbiology,* 3rd. ed., Section V, Laboratory Tests in Chemotherapy. Washington, D.C., American Society for Microbiology, 1980, pp. 446–496.
18. Barry, A. L.: Procedure for testing antibiotics in agar media: Theoretical considerations. In Lorian, V. (Ed.): *Antibiotics in Laboratory Medicine.* Baltimore, Williams & Wilkins, 1980, pp. 1–23.
19. Thrupp, L. D.: Susceptibility testing of antibiotics in liquid media. In Lorian, V. (Ed.): *Antibiotics in Laboratory Medicine.* Baltimore, Williams & Wilkins, 1980, pp. 73–113.
20. Ryan, K. J., Schoenknecht, F. D., and Kirby, M. M.: *Hosp Prac, 5:*91, 1970.
21. Expert Committee on Antibiotics, Second Report. *WHO Tech Rep Ser 210,* 1961.
22. Steers, E., Foltz, E. L., and Graves, B. S.: *Antibiot Chemother, 9:*307, 1959.
23. Hedén, C. G. and Illéni, T. (Eds.): *Automation in Microbiology and Immunology.* New York, Wiley, 1975.
24. Thornsberry, C.: Automation in antibiotic susceptibility testing. In Lorian, V. (Ed.): *Antibiotics in Laboratory Medicine.* Baltimore, Williams & Wilkins, 1980, pp. 193–205.
25. Acar, J. F.: The disc susceptibility test. In Lorian, V. (Ed.): *Antibiotics in Laboratory Medicine.* Baltimore, Williams & Wilkins, 1980, pp. 24–54.
26. Prier, J. E. and Friedman, H. (Eds.): *Opportunistic Pathogens.* Baltimore, Univ. Park, 1974.
27. Wick, W. E., Preston, D. A., Hawley, L. C., and Griffith, R. S.: *Laboratory Testing of Bacterial Susceptibility to Antibiotics: Theory and Practice,* Rev. ed. Indianapolis, Eli Lilly & Co., 1974.

28. Bauer, A. W., Kirby, W. M. M., Sherris, J. C., and Turck, M.: *Am J Clin Pathol, 45*:493, 1966.

29. Marti-Ibáñez, F. (Ed.): *Antibiotics Annual 1959-1960*. New York, Antibiotica, Inc., 1960.

30. Hoette, I., and Struyk, A. P.: *J Lab Clin Med, 51*:638, 1958.

31. Owens, D. R., and Addison, J. B.: *Mo Vet, 22*:26, 1972.

32. Ladiges, W. C.: *J Am Anim Hosp Assoc, 10*:407, 1974.

33. Federal Register, Rules and Regulations: *Antibiotic Susceptibility Discs/Standardized Disc Susceptibility Test. Fed Regist 37*:20525, 1972; *Correction, 38*:2576, 1973.

34. Anonymous: *Clin Lab Forum, 3(1)*:2, 1971.

35. Sherris, J. C.: *Clin Lab Forum, 3(2)*:3, 1971.

36. Thornsberry, C., Smith, P. B., and Balows, A.: *Clin Lab Forum, 3(3)*:2, 1971.

37. Martin, W. T., Camp, H. M., and Bennett, J. V.: *Am J Clin Pathol, 51*:808, 1969.

38. Barry, A. L., Joyce, L. J., Adams, A. P., and Benner, E. J.: *Am J Clin Pathol, 59*:693, 1973.

39. Liberman, D., and Robertson, R. G.: *Antimicrob Agents Chemother, 7*:250, 1975.

40. Kluge, R. M.: *Antimicrob Agents Chemother, 8*:139, 1975.

41. Barry, A. L., Garcia, F., and Thrupp, L. D.: *Am J Clin Pathol, 53*:149, 1970.

42. National Committee for Clinical Laboratory Standards: *Second Supplement, Performance Standards for Antimicrobic Disk Susceptibility Tests*. Villanova, Pa., NCCLS Publications, 1982.

43. Griffith, L. J., and Mullins, C. G.: *Appl Microbiol, 16*:656, 1968.

44. Blood, D. C., Henderson, J. A., and Radostitis, O. M.: *Veterinary Medicine*, 5th ed. Philadelphia, Lea & Febiger, 1979, pp. 77–80.

45. McDonald, M. and Bieberstein, E.: *Am J Vet Res, 35*:1563, 1974.

46. Balows, A. and Gavan, T. L.: Quality Control Methods for in vitro Antibiotic Susceptibility Testing. In Lorian, V. (Ed.), *Antibiotics in Laboratory Medicine*. Baltimore, Williams & Wilkins, 1980, pp. 409–417.

SUPPLEMENTARY REFERENCES

Balows, A. (Ed.): *Current Techniques for Antibiotic Susceptibility Testing*. Springfield, Thomas, 1974.

Finegold, S. M. and Martin, W. J.: *Bailey and Scott's Diagnostic Microbiology*, Part V, Antimicrobial Susceptibility Tests and Assays. St. Louis, Mosby, 1982, pp. 532–557.

Hedén, C. G., and Illéni, T. (Eds.): *Automation in Microbiology and Immunology*. New York, Wiley, 1975.

Lennette, E. H. (Ed.-in-Chief): *Manual of Clinical Microbiology*, 3rd ed. Section V, Laboratory Tests in Chemotherapy. Washington, D.C., American Society for Microbiology, 1980, pp. 446–496.

Lorian, V. (Ed.): *Antibiotics in Laboratory Medicine*. Baltimore, Williams & Wilkins, 1980.

Chapter 36

IMMUNOFLUORESCENCE

THE TECHNIQUE OF IMMUNOFLUORESCENCE is usually referred to as the fluorescent antibody or FA procedure. Although the phenomenon has been known for several decades, it is only in the last decade that it has been applied to an appreciable extent for the identification of bacteria in clinical microbiology laboratories. Its use in public health laboratories has preceded its employment in veterinary laboratories, principally because of the greater availability of FA reagents for the identification of human pathogens.

Fluorescent antibody reagents are prepared by coupling a fluorescent dye to specific immune globulin. This conjugated antibody will unite with its corresponding bacterium, virus, etc. The union is detectable by the presence of a characteristic fluorescence when viewed by a fluorescence microscope.

The great advantage of the FA technique is its rapidity. It enables an identification to be made within an hour after the smear is made. Another advantageous feature is that dead organisms can be recognized as well as live ones. The method is particularly helpful in the identification of fastidious organisms and organisms that are difficult to identify by conventional means, e.g. leptospires, *Clostridium chauvoei, Cl. novyi*. It may obviate the need to work with cultures of organisms that are dangerous to man. The procedure may be used effectively for the identification of organisms in clinical materials as well as from cultures. This may result in a considerable saving of time and materials, e.g. in tularemia and brucellosis.

The major disadvantages of the FA technique are its costs, i.e. the cost of the reagents and the amount of time that is required to carry out the procedure. Isolation procedures can result in the recovery of various causal agents, but one FA reagent is specific for only one organism. Generally speaking, the FA procedures should be reserved for selected cases in which the suspicion is strong that a particular organism is involved, e.g. blackleg, malignant edema, tularemia, leptospirosis, or for the recognition of organisms that are difficult to propagate and identify.

Two procedures are used for the detection and identification of bacteria and fungi in smears. They are summarized briefly below:

1. Direct Procedure

   Labelled specific antibody + "homologous" organism = antigen-antibody union and fluorescence.
   Labelled antibody + unrelated organism = no antigen-antibody union and no fluorescence.
   The direct procedure is the one most commonly used for the identification of bacteria.

2. Indirect Procedure

   This is carried out in two steps:
   a. Unlabelled specific antibody (e.g. rabbit origin) + "homologous" organism = antigen-antibody union.
   b. + conjugated anti-rabbit globulin = fluorescence.

In the indirect test, the conjugated antiglobulin must have been prepared from the same animal species globulin as that from which the specific antiserum was obtained, e.g. if a specific antibacterial serum was prepared in a goat, the antiglobulin would be prepared by inoculating an animal with goat globulin.

The indirect test has the advantage that the same conjugated antiglobulin can be used for all antisera produced in the species for which the antiglobulin was prepared. This makes it practicable, if necessary, to prepare one's own antisera, e.g. in rabbits.

FLUORESCENCE

A substance is fluorescent if upon absorbing light of one wavelength it emits light of another wavelength.

The light required is in the short wavelength region of the spectrum, i.e. in the blue and ultraviolet. The source of this light is usually the mercury vapor high pressure arc. It is essential that the replacement bulb be of the same type supplied as the original equipment. If not, the power supply has to be suitably altered.

The requirements for fluorescence microscopy are schematically presented in Figure 36-1. The first filter employed is a heat filter with a good UV transmittance. It is used to protect the other filters and the specimen from too much heat. This filter effectively filters out the longer, heat-producing infrared wavelengths. This is usually followed by a filter with a pale, sky blue appearance that makes the background of the field of view in the fluorescence microscope even more completely black. It may also absorb some of the heat produced by the near-ultraviolet radiation. The next set of filters are designed to filter out all visible light or light of a wavelength longer than the exciting light. They are called exciter filters, and they should transmit completely all of the short wavelength exciting light. There are usually two sets of exciter filters. The one kind is very dark, practically black, and its transmission maximum is frequently in the 366 mμ region of the near-ultraviolet. This is used in conjunction with barrier filters (in the ocular) that are practically colorless. The others are a deep, dark blue and they transmit the blue light (420–450 mμ) as well as most of the UV. These blue light exciting filters are used with barrier filters of a yellow orange color.

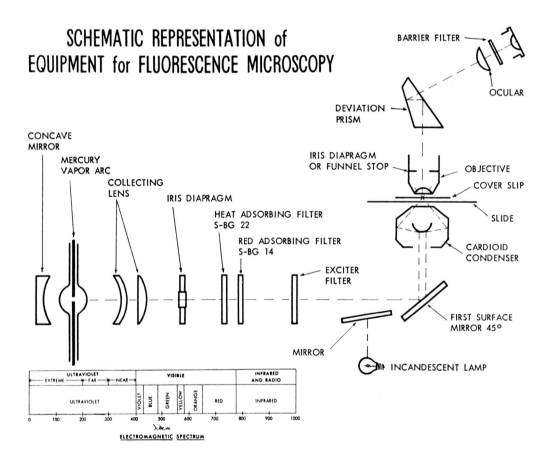

Figure 36-1. Schematic representation of equipment for fluorescence microscopy. From W. B. Cherry, in J. E. Blair, E. H. Lennette, and J. P. Truant (Eds.), *Manual of Clinical Microbiology*, 1970. Courtesy of American Society for Microbiology.

Darkfield condensers are usually employed because they render images of high contrast. Practically none of the exciting light enters the objective, and thus the darkfield condenser offers the best protection to the observer's eye. Of course, the UV-absorbing barrier filter must always be used.

The exciting light passes the condenser and is absorbed by the fluoro-chrome, usually fluorescein isothiocyanate (FITC), which then fluoresces or emits light of a longer wavelength in the visible spectrum. This light in the fluorescence microscope is yellowish green if the fluorochrome is FITC; if it is rhodamine, it is orange.

There have been two important, recent developments in immunofluores-cence (IF). These new developments have been incorporated into the many different pieces of equipment offered. There is now available a halogen lamp that is used with an interference filter for the excitation of such fluorochromes as FITC and rhodamine with maximal absorption peaks at the longer visible wavelengths. These halogen lamps are less expensive than either the mer-cury vapor or xenon lamps. They are suitable for routine IF procedures provided suitable interference and barrier filters are employed.

The other major innovation has been the introduction of the incident-light fluorescence. In this system, the exciting radiation of the desired wavelength is deflected downward by means of the dichroic, beam-splitting mirror through the objective to the specimen. In the case of blue light coming from the exciting filter, it is deflected downwards and is excited by FITC in the specimen to produce yellow green fluorescence. The latter longer wave radiation is transmitted upward through the dichroic mirror, which did not allow transmission of shorter wavelength blue light, and is seen through the microscope eyepiece.

Fluorescence microscopes are expensive instruments of varying complexity. It is strongly recommended that operators familiarize themselves with the instructional material supplied by the manufacturer. Information on the various combinations of filters employed for the earlier transmitted-light instruments is provided in Table 36-I. Cherry (2) has provided a more comprehensive table that includes information on combinations of filters employed with both transmitted and incident-light instruments.

PREPARATION AND FIXING OF FA SMEARS

The standard thickness of microscope slides is 1.2 mm, but some darkfield condensers require a slide of no more than 1 mm thickness. These "fluoro-slides" have two etched circles 15 mm in diameter that facilitate the location of the specimen and aid in its retention within the circle. No. 1 coverglasses (0.13 to 0.16 mm) are suitable for most microscope objectives. Slides and coverslips should be as clean as possible.

Various smears are employed depending upon the nature of the specimen. In order to conserve reagent, small areas of the slide are usually used. If the "fluoro-slides" are used, the smear is confined to the etched circles. Smears

TABLE 36-I
FILTER COMBINATIONS FOR FA WORK*

Type of Filter		
Exciter	Barrier	Comments
BG-12 Schott (S)	OG-1 (S)	Widely used in bacteriology
BG-25 (S)	OG-1 (S)	Equal to or better than BG-12
UG-1 (S) or		
5840 Corning (C)	GG-9 (S)	Many uses in bacteriology, especially for tissue sections, imprints, and clinical specimens containing tissue cells
UG-1 (S)	W2A or W2B†	Virus work, especially rabies; also excellent for bacteria in tissues
5840 (C)	W2A or W2B	Virus work, especially rabies; also excellent for bacteria in tissues
5113 (C)	W2A or W2B	Fungi
Heat-absorbing		
BG-22 (S)		Replaced by KG-1 or KG-2
KG-1 (S)		Similar to KG-2
KG-2 (S)		Higher T at 350 to 450 nm but lower heat absorption than KG-1
Red-absorbing		
BG-14 (S)		Replaced by BG-38
BG-38 (S)		Higher T at 350 to 450 nm with greater red absorption

*Modified slightly from W.B. Cherry, in J.E. Blair, E.H. Lennette, and J.P. Truant (Eds.): *Manual of Clinical Microbiology*, 1970, Chapter 76. Courtesy of the American Society for Microbiology, Washington, D.C.
†Kodak Wratten.

from tissue can be made by impression or by the smearing out of exudate, pus, fluid, etc., with a scalpel or an inoculating wire. If bacteria from a colony are to be examined, the smear is prepared in the same manner as for Gram's stain. Experience will indicate the concentration of organisms that is desirable.

The slide is air dried, and in the case of the smear from a colony of bacteria, it is usually lightly heat fixed; however, other recommendations for fixing may be called for with certain FA reagents. Smears from organs and tissue fluids are sometimes fixed in acetone for ten minutes, e.g. specimens from suspected cases of blackleg and malignant edema. The use of 10% formol buffered saline for fifteen to twenty minutes is recommended by some workers.

Slides fixed in formol saline or other chemical fixatives, excepting acetone and methyl alcohol, must be washed in buffered saline and dried before the FA stain is added.

STAINING AND MOUNTING OF SMEARS

The FA stain is added to the smear with a Pasteur pipette or with a large inoculating loop. Spreading of the stain over the whole smear may be facilitated with an inoculating loop or the end of a broken applicator stick. The slide is then placed in a moist chamber for fifteen to thirty minutes. Various chambers are employed, e.g. instrument trays or a large Petri dish. Moistened filter paper is placed on the bottom, and the slide is placed on two glass rods above the moist paper.

Various washing procedures are employed. The following has been found satisfactory: (a) shake off excess stain and rinse briefly in buffered saline, pH 7.5; (b) transfer to a second container of buffered saline and leave immersed for ten minutes; (c) remove and dip in a container of distilled water; (d) drain and air dry; (e) mount by adding to the smear a drop of buffered glycerol saline (9 parts glycerol to 1 part (v/v) of buffered saline pH 9.0) then coverslip. The coverslip can be fixed in position by adding a small amount of clear fingernail polish to each corner of the coverslip. If it is desired to keep FA-stained slides overnight, the mount is sealed all the way around with nail polish, then stored in the refrigerator. Some stained preparations can be stored at 4°C for several weeks or months.

In the indirect procedure, it is usually best not to dry the slide before adding the conjugated antiglobulin.

USING THE FLUORESCENCE MICROSCOPE

The instructions given below are reproduced through the courtesy of E. Leitz, Inc. (1).

1. Rack darkfield condenser down with substage rack and pinion control.
2. Apply drop of low viscosity, low fluorescence immersion oil to top of darkfield condenser. Avoid air bubbles.
3. Place specimen on the stage, turn on incandescent light.
4. Slowly rack condenser back up, and observe from the side.
5. When the oil contacts the lower surface of the slide a small flash of light will be seen. Observe the specimen now from above (not through the microscope yet!). A very small ring of light, caused by light scattering in the specimen, will be visible.
6. Carefully focus the darkfield condenser up further. The light ring will contract and become brighter. Finally a brightly illuminated spot will be visible. If the condenser is focused even higher up, the spot will expand again into a ring of light. Actually, what one sees is a cross section of the hollow cone of light coming from the darkfield condenser. The condenser should be left in the position where the bright spot is visible.

7. Swing in 10×, or 25× objective, look through the microscope and focus it onto the specimen. It does not matter at this moment if the specimen appears to be very poorly illuminated, or unevenly illuminated. All that should be achieved at this point is that the objective is focused properly, and as long as the specimen is visible at all, this is quite sufficient. If the field of view should at this point appear completely black, the condenser would be decentered to an extreme degree. Even a random adjustment of the condenser centering screws will remedy this situation.

8. Once the specimen appears in focus the fine adjustment of the microscope is left in that position. Now the condenser is centered and focused exactly. One should finally see the specimen in darkfield in the center of the field. Now the high power objective is brought into position, and focused. It is at this point often necessary to make a final correction in the condenser focus. This final focus correction may involve not more than an adjustment of about 50 µm up and down. Under high power, the full field should now appear in uniform dark-field illumination. The barrier filters may give an orange color to the incandescent light.

9. One can now be certain that the objective is properly focused, the specimen is in the field of view, the condenser is properly focused and centered.

10. Now one should switch to the exciter light source; one can then see the fluorescence, without uncertain adjustments.

Should the field be poorly illuminated, the first corrective measures must be taken at the alignment of the light source. It may especially be necessary to adjust the collector lens focus in the lamp housing, but if the mercury source had been aligned previously, such adjustments should not be necessary or should be minor.

When it is necessary to scan large areas of the slide, and when one can assume that one will find a reasonably bright fluorescence, one may with advantage work with a dry darkfield condenser. The same rule with respect to the relative apertures of objective and condenser applies here: a dark-field condenser of N.A. of 0.80 should, at the very most, be combined with an objective with an aperture of 0.75. A value of 0.65 for the objective aperture is, however, more practical.

APPLICATION OF THE DIRECT AND INDIRECT FA TECHNIQUES

The direct and indirect procedures are widely employed for the identification of microorganisms. Many reagents are available commercially. They are not all satisfactory, and they should be thoroughly tested on known positive and negative materials before being used routinely. A list of commercial sources of FA reagents and unlabelled sera is provided at the end of this chapter.

An attempt has been made below to list some of the microorganisms for which direct and indirect FA procedures have been used for identification. For additional information, readers should consult the discussions of the individual organisms in the text and the Supplementary References. The identification of fungi by FA has been summarized by Kaufman (3).

Agents	Remarks
Campylobacter fetus	Doesn't distinguish between subsp. *fetus* and subsp. *intestinalis*
E. coli	To identify common human enteropathogenic serotypes and K antigens
Klebsiella	Specific for some serotypes
Francisella tularensis	Specific
Brucella	Genus identification
Str. pyogenes (group A)	Specific
Listeria monocytogenes	Specific
Erysipelothrix rhusiopathiae	Specific
Bacillus anthracis	Not absolutely specific
Clostridium septicum	Specific
Cl. chauvoei	Specific
Cl. novyi	Specific
Cl. tetani	Specific
Cl. botulinum type A	Specific
Cl. botulinum type C	Specific
Cl. botulinum type E	Specific
Cl. sordellii	Specific
Leptospira spp.	Not species specific
Mycoplasma spp.	Available for some species; specific
Treponema hyodysenteriae	Specific
Francisella tularensis	Specific
Blastomyces dermatitidis	Specific
Coccidioides immitis	Specific
Sporothrix schenkii	Specific
Cryptococcus neoformans	Adequately specific
Histoplasma capsulatum	Specific

SOME SOURCES OF FA REAGENTS AND UNLABELLED ANTISERA

Antibodies Incorporated, P.O. Box 1560, Davis, California
BBL Microbiology Systems, Baltimore, Maryland
Burroughs Wellcome Co., Research Triangle Park, North Carolina
Cappel Laboratories, 237 Lacey Street, P.O. Box 37, West Chester, Pennsylvania
Case Laboratories, Inc., 515 North Halsted Street, Chicago, Illinois

Colorado Serum Company, 4950 New York Street, Denver, Colorado
Difco Laboratories, 920 Henry Street, Detroit, Michigan
Flow Laboratories, Inc., 828 West Hillcrest Boulevard, Inglewood, California
Melory, 6715 Electronic Drive, Springfield, Virginia
Microbiological Associated, Inc., 4733 Bethesda Avenue, Bethesda, Maryland
Miles Laboratories, Inc., Elkhart, Indiana

REFERENCES

1. Bartels, P.: *Fluorescence Microscopy: Principles, Applications and Instrumentation.* New York, E. Leitz, Inc.
2. Cherry, W. B.: In Lennette, E. H., Spaulding, E. H., Truant, J. P. (Eds.): *Manual of Clinical Microbiology,* 2nd ed. Washington, D.C., American Society for Microbiology, 1974, Chapter 4.
3. Kaufman, L.: In Lennette, E. H. (Ed.-in-chief): *Manual of Clinical Microbiology,* 3rd ed. Washington, D.C., American Society for Microbiology, 1980, Chapter 61.

SUPPLEMENTARY REFERENCES

Cherry, W. B., Goldman, M., and Carske, T. R.: *Fluorescent Antibody Techniques* (Public Health Service Pub. No. 729). Washington, D.C., U.S. Government Printing Office, 1961.

Cherry, W. B., and Moody, M. D.: *Bacteriol Rev,* 29:222, 1965.

Cherry, W. B.: In Lennette, E. H. (Ed-in-chief): *Manual of Clinical Microbiology,* 3rd ed. Washington, D.C., American Society for Microbiology, 1980, Chapter 49.

Goldman, M.: *Fluorescent Antibody Methods.* New York, Academic Press, 1968.

Hebert, G. A., Pittman, B., McKinney, R. M., and Cherry, W. B.: *The Preparation and Physiochemical Characterization of Fluorescent Antibody Reagents.* Atlanta, Public Health Service, Center for Disease Control, 1972.

Kawamura, A., Jr. (Ed.): *Fluorescent Antibody Techniques and Their Applications.* Baltimore, Univ Park, 1969.

Georgala, D. L., and Boothroyd, M.: In Gibbs, B. M., and Shapton, D. A. (Eds.): *Identification Methods for Microbiologists.* New York, Academic Press, 1966, Part B, p. 187.

Chapter 37

PREPARATION OF BACTERINS

IAGNOSTIC LABORATORIES are frequently called upon to prepare bacterins. Strictly speaking, an autogenous vaccine or bacterin is prepared from the organism causing disease in an individual animal. The use of such a product in that individual is called vaccine therapy and is now considered to be of little or no value (1). The bacterins prepared by Diagnostic Laboratories for groups of animals might be more correctly called "Custom Bacterins."

Some considerations that should be kept in mind when preparing bacterins are the following:

1. Bacterins should be prepared from fully antigenic bacteria recovered from diseased animals in the herd or flock in which the product is to be used. More than one strain should be used if at all possible.

2. Subcultures should be kept to a minimum as important antigens may be lost on transfer.

3. Cultures for bacterins should only be incubated until the end of the logarithmic phase of growth.

4. Bacterins are of most value in the preventions of infections in which the immune response is predominantly humoral.

5. Custom bacterins may not be effective in preventing a subsequent disease problem in the same herd.

The procedures outlined below, although simple, have been found adequate.

1. Media. Fluid media cultures require less processing than do solid media cultures, although heavier concentrations of organisms can be obtained from the latter. Tryptose phosphate, tryptose, trypticase, and infusion broth or agar yield satisfactory numbers of coliforms, salmonella, pasteurella, staphylococci, etc. If fastidious organisms such as streptococci are to be grown, sterile serum (2–5%) may be added. To prevent shock, the serum should, ideally, be from the same animal species for which the bacterin is being prepared.

2. The broth medium is dispensed in flasks, generally in 250 ml amounts. Smaller amounts may be used with a smaller inoculum. The requisite number of flasks are placed in the incubator so that their temperatures reach 37°C prior to inoculation. Tubes containing 5–10 ml of broth are inoculated with the strain being used. One tube is inoculated for each flask. A smaller

amount of inoculum is used for a flask or dish containing solid media. If a herd is involved, it may be desirable to prepare flasks from a number of different isolates. When turbidity is evident in the tubes, smears are stained to check for purity. If pure, each warm flask is inoculated with one tube of broth culture. Flasks are incubated overnight or longer if desired. Shaking and aeration may be used to increase yields.

3. Each flask is checked for purity by stained smear and culture. If pure, 0.5% formalin and sufficient merthiolate to yield a final concentration of 1/10,000 is added to each flask to suppress the growth of molds. The formaldehyde solution (formalin) used should be clear and show no evidence of polymerization. The amount of formalin required to kill depends upon the amount of organic material. The 0.5% concentration is more than adequate for most suspensions; 0.3% is less irritating and is usually sufficient for Gram-negative organisms. Incubate overnight at 37°C, after which the following sterility and safety tests are conducted:

A. 0.1 ml of bacterin (250 ml amount) is added to each of two tubes containing 10 ml of thioglycollate broth (previously heated to drive off oxygen) and to each of two tubes containing 10 ml of tryptose phosphate broth. Two tubes of each medium are incubated at 37°C and at room temperature for a week.

B. Each of three adult mice are inoculated subcutaneously with 0.2 ml of bacterin and observed for a week.

C. If the tubes of broth produce no growth and the mice remain healthy, the bacterin is considered safe for use.

D. If available, the species for which the bacterin has been prepared may be inoculated with the recommended dose in order to determine whether or not adverse reactions develop.

E. Bacterin should be thoroughly labeled with regard to use, dosage, storage, and precautions.

For individual animals, only small quantities of bacterin are required. In such cases, it is often convenient to grow the strain on blood agar. The growth is washed off with a small amount of 0.3% or 0.5% formalized saline, incubated, then diluted with 0.5% phenolized saline to the desired turbidity. The latter preservative is preferable for small animals because, unlike formalin, it produces no apparent irritation on injection. Similar purity and safety tests are required.

As mentioned above, "vaccine therapy" in an individual animal is widely considered to be of little or no value. The improvement sometimes seen is usually transient.

A convenient alternative to flasks is to dispense the media in the actual vaccine or bacterin bottles. These are available from veterinary or scientific supply houses in various sizes; 75 ml of medium is placed in a 100 cc bottle, 200 ml in a 250 cc bottle, and about 400 ml in a 500 cc bottle. The bottles are plugged with cotton, then sterilized. The rubber caps are stored in 70% ethanol until needed.

Although it is ordinarily not advisable to add incomplete Freund's adjuvant to a custom bacterin, one can without difficulty add an aluminum compound such as aluminum hydroxide (gel form). The addition of an adjuvant will usually extend the duration of protection. This would be most advisable when it is not feasible to give multiple injections of the bacterin. The following procedure is convenient and should be satisfactory for most bacterins.

At the end of the growth period, aluminum hydroxide gel is added in an amount equal to 15–25 percent of the volume of the culture (2,3). Formalin (0.3% to 0.5%) is added, and after thorough mixing, the bacterin is incubated at 37°C for eighteen to twenty-four hours. The aluminum hydroxide suspension Amphogel®* is a convenient preparation for addition to bacterins.

ADMINISTRATION OF BACTERINS

The dosage will vary with the size of the animal and the concentration and toxicity of the organisms. Gram-negative organisms contain a considerable amount of endotoxin, which can produce endotoxic shock. Anaphylactic shock may also be encountered. The label should state that if shock is encountered, epinephrine should be administered immediately. The general features of a satisfactory label are provided below. The doses given below cover the range commonly used. The smaller dose with revaccination once or twice is preferable to one large dose. Because some animals may have a hypersensitivity to an organism, e.g. in pyoderma caused by *Staph. aureus*, it may be advisable to give a series of small doses at weekly intervals. Suggested dosage range for bacterins and a sample label are given below.

Cats:	0.5–1 cc
Dogs (small):	0.5–2 cc
(large):	1.0–3 cc
Cattle (adults):	5.0–10 cc
Calves:	3.0–5 cc
Sheep and goats:	2.0–5 cc
Pigs:	2.0–5 cc
Horses:	5.0–10 cc
Chickens and turkeys:	0.5–2.0 cc

*Wyeth Laboratories, Inc., Philadelphia, Pennsylvania

Sample label:

CF-390-76 STAPHYLOCOCCUS AUREUS BACTERIN 5-10-76
Clinician: Dr. J. Doe, Kalamazoo, Michigan.
Contains 0.5% formalin and 1:10,000 merthiolate.
Store under refrigeration.
Shake bottle thoroughly prior to use.
Dose (dog): 1 cc; repeat at 2- to 3-week intervals to a maximum of 3 cc.
Administer subcutaneously or intramuscularly.
Caution: In case of anaphylactoid reaction, administer epinephrine.

REFERENCES

1. Carter, G. R.: *VM/SAC,* 77:741, 1982.
2. Matsumoto, M., and Yamamoto, R.: *Avian Dis,* 15:109, 1971.
3. Matsumoto M., and Yamamoto, R.: *Am J Vet Res,* 36:579, 1975.

Chapter 38

BACTERIOLOGY FOR
THE PRACTICING VETERINARIAN

V ETERINARY PRACTITIONERS are showing an increasing interest in carrying out a limited number of microbiological procedures in their clinics. The purpose of this chapter is to outline briefly some of the procedures that can be performed and to list media, reagents, and equipment that may be useful. For further details, reference should be made to the pertinent sections of the manual and to the Supplementary References.

Although a wide variety of prepared media are available commercially* at reasonable prices, it is usually not feasible for the practicing veterinarian to do much more than obtain primary cultures and conduct susceptibility tests. Except for the staphylococci and streptococci, he is usually not in a position to make other than tentative identifications of the organisms recovered. However, susceptibility tests can be conducted without the definitive identification of the organism.

SPECIMENS

Material consisting of discharges, exudates, and pus can be conveniently collected with sterile swabs. A portion of the plate is inoculated directly with a swab, then the whole plate is streaked over with an inoculating loop. Contaminants can be kept to a minimum by the careful collection of fresh material. Tissues and organs are first seared with a hot spatula, then incised aseptically with a scalpel. Material is transferred from the incision to plates on an inoculating loop or a swab. The inoculum should be generous and spread well over each plate, or sector of a plate, in order to obtain discrete colonies (Fig. 38-1).

DIRECT EXAMINATION

A direct examination in which material is examined in distilled water or in 10% sodium hydroxide under a coverslip is frequently helpful. A strong indication of a mycotic infection, such as blastomycosis or cryptococcosis, can often be obtained with a wet mount.

---

*Baltimore Biological Laboratories, Baltimore; Fisher Scientific Co. (various locations); Hyland Laboratories, Los Angeles; Difco Laboratories, Detroit; Colab Laboratories, Inc., Chicago Heights, Chicago.

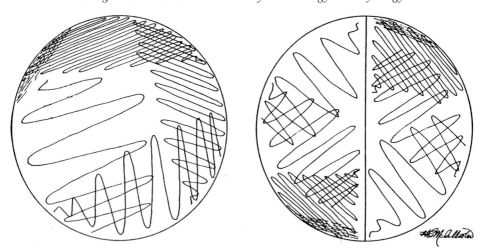

Figure 38-1. Procedures for inoculating plates in order to obtain individual colonies. The loop is flamed prior to changing the direction of the streaks (H. A. McAllister).

STAINING PROCEDURES

A basic and simple procedure that is frequently of value is Gram's stain. The solutions for Gram's procedure are available commercially.

The staining procedure and the preparation of the staining solutions are described in Appendix A. In doing Gram's stain, it is important that slides are clean and that a thin smear is made. If the slide is greasy or the smear is insufficiently heat fixed, the material may be washed off.

The lactophenol cotton blue stain described in Appendix D is useful for the examination of material from suspected fungous infections, especially ringworm.

Ready to use solutions for the Kinyoun and Ziel-Neelsen acid-fast staining procedures are available commercially.* These stains are used in the recognition of *Nocardia asteroides* and mycobacteria.

Koster's stain or the Modified Ziehl-Neelsen stain as described in Appendix A is useful for the demonstration of *Brucella* spp. in the chorionic epithelium of the placenta.

PRIMARY CULTURES

The most useful all-round medium is blood agar dispensed in plates. The commercial plates can be kept in their original package in the refrigerator for long periods without appreciable drying. Material from several tissues or swabs can be streaked on different sections of the same plate. Small incubators are available from several sources.

*Difco Laboratories, Detroit, Michigan; BBL Microbiology Systems, Baltimore, Maryland.

Table 38-I

SIMPLE PROCEDURES WHICH AID THE DIAGNOSIS OF VARIOUS DISEASES
(See Appropriate Chapters and Appendix A for Details)

Suspected Disease or Infection	Procedure
Vibriosis *(Campylobacter fetus)*	Gram or negative staining of fresh fetal stomach contents. Examination of the latter by phase contrast.
Actinobacillosis and Actinomycosis	Observe granules; crush granules and stain with Gram's stain.
Brucellosis	Plate agglutination test. Demonstration of organisms in cells of the chorion with Koster's stain or modified Ziehl-Neelsen stain.
Contagious footrot of sheep	Gram's stain of smear from lesion after paring away necrotic tissue.
Anthrax	Giemsa stain of blood, tissue, or organ smears.
Clostridia: Gas gangrene (blackleg and malignant edema)	Gram's stain of smears from suspected muscle.
Enterotoxemia	Gram's stain of smears from intestinal (small) mucosa or content
Mycobacteria	Kinyoun or Ziehl-Neelsen stain of smears from lesions.
Nocardiosis	Gram's and acid-fast stain (for partial acid fastness) of smears from lesions.
Streptothricosis or dermatophilosis	Giesma or gram stain of smears from moistened scabs and crusts.
Leptospirosis	Darkfield examination of urine centrifugate. Plate agglutination tests.
Fowl spirochetosis	Giemsa or Wright's stain of blood smears.
Dermatophytosis (ringworm)	Wood's lamp; NaOH wet mount examination for spores.
Cryptococcosis	NaOH wet mount of clinical specimens. India ink stain for capsules.
Blastomycosis	NaOH wet mount of clinical specimens.
Other fungous infections (direct examinations are often negative in sporotrichosis and histoplasmosis).	NaOH wet mount of clinical materials.
Pus, discharge, exudate etc.	Gram's stain. Tentative recognition of bacteria as streptococci, staphylococci, clostrida, *Bacillus*, diphtheroids, *Erysipelothrix* and *Listeria*, etc.
Treponema hyodysenteriae	Giemsa stain of smear from colonic mucosa or feces.

Other useful primary plate media are brilliant green agar and eosin-methylene blue agar (Appendix B). These are selective media favoring the growth of enteric pathogens. The appearance of the colonies of the enteric pathogens on these media is given in Appendix B and in Chapter 10.

The most prevalent potentially pathogenic bacteria recovered from animal tissues are those Gram-negative organisms giving rise to coliformlike colonies, e.g. *Pasteurella, Salmonella, E. coli, Pseudomonas*, and the Gram-positive organisms, e.g. staphylococci, streptococci, *Corynebacterium pyogenes*, and *Erysipelothrix rhusiopathiae*. With experience, some of these can be presumptively identified on the basis of colony appearance, Gram reaction, and morphology. The definitive identification of most of the aerobic bacteria, including those giving rise to coliformlike colonies, involves the use of differential media.

IMPORTANT CHARACTERISTICS OF
FREQUENTLY ENCOUNTERED BACTERIA

The characteristics listed below may be used to assist in the identification of important bacterial pathogens. Refer to Figure 38-2 for simplified procedures for the isolation and tentative identification of the most common Gram-negative bacteria.

Several commercial systems that obviate the need for the conventionally used "biochemical" differential media are available to facilitate and simplify the identification of enteric and other bacteria. They are listed in Chapter 4.

Gram-Negative Bacteria

Salmonella spp. Triple sugar iron agar (TSI): acid butt, alkaline slant, gas, and hydrogen sulfide variable. Agglutinates with polyvalent Salmonella O antiserum (poly A-1). Weak agglutination is obtained with some strains of *Proteus. Proteus* spp. frequently swarm and produce urease; *Salmonella* spp. do not. Slide agglutination tests using polyvalent Salmonella O antiserum (poly A-1) can be carried out using colonies from primary blood agar cultures.

E. coli. Colonies have a metallic sheen on eosin-methylene blue agar. TSI: acid slant and acid butt with gas bubbles in the butt. Some strains are beta-hemolytic.

Pasteurella multocida. Nonmotile. Characteristic odor. TSI: acid slant and acid butt, no gas; yellowing of agar not as marked as with *E. coli.* Does not grow as a rule on MacConkey's agar. Indole is formed (SIM or MIO medium).

Pasteurella haemolytica. Beta-hemolysis; nonmotile; grows on MacConkey's agar; indole is not formed.

Pseudomonas spp. Characteristic odor. Frequently beta-hemolytic with a greenish cast. Cytochrome oxidase production can be shown on a Patho-Tec® test paper (Appendix C).

SIMPLIFIED PROCEDURES FOR THE ISOLATION OF THE MOST COMMON GRAM NEGATIVE ORGANISMS

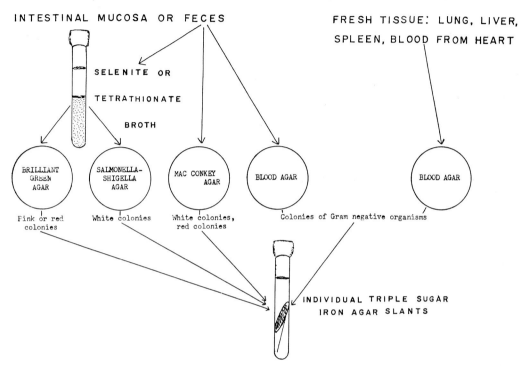

REACTIONS ON TRIPLE SUGAR IRON AGAR					
SLANT	BUTT			POSSIBLE IDENTITY	
COLOR	COLOR	GAS	H$_2$S	MOST LIKELY	LESS LIKELY
YELLOW	YELLOW	NEGATIVE	NEGATIVE	PASTEURELLA MULTOCIDA PASTEURELLA HEMOLYTICA ACTINOB. EQUULI ACTINOBACILLUS LIGNIERESII	AEROMONAS HYDROPHILA SHIGELLA SONNEI AEROMONAS SHIGELLOIDES SERRATIA SP.
YELLOW	YELLOW	BUBBLES	NEGATIVE	ESCHERICHIA COLI KLEBSIELLA SP. ENTEROBACTER (AEROBACTER) SP.	AEROMONAS HYDROPHILA PROVIDENCIA SP. PROTEUS RETTGERI
YELLOW	YELLOW	NEGATIVE	BLACK	SERRATIA SP.	
YELLOW	YELLOW	BUBBLES	BLACK	PROTEUS VULGARIS PROTEUS MIRABILIS	CITROBACTER
RED	YELLOW	NEGATIVE	NEGATIVE	SHIGELLA SP. SALMONELLA TYPHI	PROTEUS RETTGERI PROVIDENCIA SP. PASTEURELLA MULTOCIDA
RED	YELLOW	BUBBLES	NEGATIVE	PROTEUS MORGANII PROTEUS RETTGERI ENTEROBACTER HAFNIAE PROVIDENCIA SP. SHIGELLA FLEXNERI TYPE 6	SALMONELLA PARATYPHI A SALMONELLA SENDAI SALMONELLA CHOLERAESUIS S. ABORTIVOEQUINA SALMONELLA TYPHISUIS
RED	YELLOW	NEGATIVE	BLACK	SALMONELLA TYPHI SALMONELLA GALLINARUM	SALMONELLA PULLORUM
RED	YELLOW	BUBBLES	BLACK	SALMONELLA SP. PROTEUS MIRABILIS ARIZONA CITROBACTER	EDWARDSIELLA TARDA
RED	RED	NEGATIVE	NEGATIVE	PSEUDOMONAS SP. BORDETELLA SP. MORAXELLA SP. MIMA & HERELLEA SP.	ALCALIGENES SP. BRUCELLA SP.

Note:
Plate possible E. coli isolates on EMB agar to detect characteristic metallic sheen of the species.
Proteus sp. are urease positive; test possible Proteus isolates with urease Patho-Tec paper (Warner-Chilcott) or urea agar.
Test urease-negative Salmonella-like isolates with Salmonella polyvalent antiserum; if distinct agglutination occurs the isolate must be regarded as a presumptive Salmonella.

Figure 38-2.

Proteus. Some strains spread on the agar surface. Production of urease can be shown with a Patho-Tec test paper (Appendix C). TSI: same as salmonellas. Litmus milk: digested and peptonized.

Bacillus spp. Characteristic colonies; large Gram-positive, sporeforming rods; frequently beta-hemolytic; common contaminants. When they are recovered from fresh tissues or blood, one should suspect anthrax.

Gram-Positive Bacteria

Staphylococcus and *Streptococcus* spp. Characteristic colonies and morphology (see Chapters 16 and 17). In order to demonstrate the chain formation of the streptococci, it may be necessary to inoculate a fluid medium such as thioglycollate or BHI broth. Pathogenic strains of staphylococci are usually beta-hemolytic.

Corynebacterium spp. These widely occurring bacteria are small, pleomorphic, beaded rods that frequently show a palisade arrangement in smears. The most prevalent pathogen in this group is *C. pyogenes*, which is beta-hemolytic and produces small, streplike colonies after forty-eight hours incubation. Litmus milk is digested. Care must be taken to distinguish the corynebacteria from the streptococci.

Erysipelothrix rhusiopathiae. This slender Gram-positive rod produces a small colony with first a zone of greenish or alpha-hemolysis, which becomes a clear or beta zone on further incubation. TSI: After incubation for several days, hydrogen sulfide is usually produced along the stab; the slant and butt are yellow (acid). Morphologically, this species resembles listeria and diphtheroids.

Anaerobic and Microaerophilic Organisms

Cultivation of these organisms, including the pathogenic clostridia and the vibrios, is generally too involved for all but the very zealous. Gram-stained smears of fresh materials may yield useful information if anaerobic infections are suspected. *Campylobacter* can be demonstrated in fresh bovine fetal stomach contents with darkfield or phase microscopy and by means of the relief or negative stain described in Appendix A.

Bovine Mastitis

The procedures described in detail in Chapter 34 are recommended. If milk samples are taken with strict asepsis, blood agar alone may be used.

Selective media are of particular value to the veterinarian interested in the bacteriologic examination of milk. These selective media are described in Appendix B. Edward's medium is used for the recovery and presumptive identification of *Str. agalactiae* and other streptococci. Mannitol salt agar, Vogel-Johnson, and Baird-Parker are employed for the selective recovery of

Staph. aureus. The Camp test can also be used to presumptively identify *Str. agalactiae*.

Urine Cultures

The bacteriologic examination of urine is discussed in Chapter 4. A number of commercial screening tests are available to detect bacteriuria in human beings. There is no reason why they may not also be employed for animal urine. One of the more effective of these commercial tests is Testuria®.* In this test, a filter paper strip is dipped into an aseptically taken urine sample or a fresh midstream specimen, after which it is inoculated onto a small trypticase soy agar tray. The colonies are counted after overnight incubation. Zero to two colonies is considered negative, three to twenty-five colonies suspicious, and twenty-six or more positive. Another useful "system" that provides quantitative results is the Bactercult® tube.†

CULTIVATION OF FUNGI

Veterinarians are referred to Chapter 28 for the brief discussion of procedures. In subsequent chapters, each of the important pathogenic fungi is described briefly. The fungi as a general rule grow more slowly than do bacteria, and larger inocula are required to initiate growth. If a fungus is suspected, the plate should be sealed or placed in a sealed container to prevent dehydration and incubated from several days to several weeks at approximately 25°C. Fungi are occasionally encountered when one would ordinarily expect bacterial growth. The colonies of *Candida* spp. and *Cryptococcus* spp. can be mistaken for staphylococci and micrococci on blood agar.

Two commercially available media‡ are especially useful for the cultivation of fungi. They are Sabouraud dextrose agar with and without the inhibitors. Most of the dermatophytes causing ringworm can be readily cultivated on Sabouraud agar with inhibitors at room temperature, usually within one to four weeks. Although the species of dermatophyte recovered may require a specialist for definitive identification, the identification of the genera *Microsporum* and *Trichophyton* is not usually difficult (see Chapter 29).

The procedure followed by some veterinarians is to inoculate small bottles of Sabouraud agar with hairs pulled or scraped from the edge of the lesion. These are incubated at room temperature, and those that yield fungous

---

*Ayerst, 685 Third Avenue, New York, New York.

†Wampole Laboratories, Cranbury, New Jersey.

‡Trade names: Mycosel® (BBL) and Mycobiotic® (Difco); available in bottles ready to use from Derm Medical, P.O. Box 78595, West Adams Station, Los Angeles, California; Bacti-lab, P.O. Box 1179, Mountain View, California.

cultures after a week or more are dispatched to a laboratory for identification. Media are available in small bottles or special plates from several firms.

An improvement over the procedure just referred to is the use of Dermatophyte Test Medium (see Plate II). It makes possible the identification of dermatophytes with a reliability reported to be as high as 97 percent. More information on this medium including .commercial sources is given in Appendix D.

Table 28-I is provided as a guide for the selection of the appropriate media and incubation temperatures for the growth of important fungi. The usual incubation period required for adequate growth is also given.

A positive diagnosis of ringworm can sometimes be made on the basis of fluorescence of hair with a Wood's light and a microscopic examination of scrapings; however, a considerable number of cases that are negative by fluorescence and microscopic examination yield dermatophytes on culture.

SUSCEPTIBILITY TESTS

The performance of susceptibility tests is described in detail in Chapter 35. Mueller-Hinton medium is recommended for routine tests; blood agar is used for fastidious organisms. "Direct" susceptibility in which the presumed infectious material is plated on media prior to applying susceptibility discs is of limited value. The procedure is sometimes satisfactory with purulent material, but often specimens either do not yield sufficient organisms or mixed cultures are obtained.

If what appears to be a pure culture is obtained on primary culture, a Mueller-Hinton plate (or blood plate for fastidious organisms) is inoculated diffusely and evenly with several colonies by means of a swab.

Workers should refer to Table 35-I for assistance in the choice of susceptibility discs. The discs are best applied with the aid of dispensers. These are available for both large and regular size Petri plates. Details in regard to their use are given in Chapter 35.

A more reliable procedure than that just described is to suspend one or several similar colonies in 0.5 ml of sterile 0.85% sodium chloride solution or broth. A sterile swab is then dipped in the saline, or broth, and with this the plate is inoculated. The susceptibility discs are applied when the surface of the plate is thoroughly dry. Drying can be accomplished by inverting the plate and leaving it for a short period in the incubator or at room temperature. Susceptibility plates are incubated for not less than eighteen to twenty-four hours at 37°C. Some organisms may require a longer incubation period.

Readers are referred to Chapter 35 or to the literature accompanying the discs for the interpretation of the results.

SELECTIVE MEDIA SYSTEMS

A kit, Bactassay®,* consisting of several selective media has been designed especially for veterinary practice. It provides the following media in different compartments of a rectangular plate: brain-heart infusion agar, MacConkey's agar, mannitol salt agar, and streptosel agar. A section containing Mueller-Hinton agar is available for performing sensitivity tests. Detailed instructions are provided, along with detailed charts to aid in the interpretation of results. If carefully employed, this system can be used effectively.

If a clinic or hospital has the services of trained technologist, more involved procedures are feasible. However, regardless of the simplicity of the procedures, the results will usually not be satisfactory if the individual has not had some training in clinical bacteriology. Even the use of the so-called "quick systems" or kits requires a knowledge of basic bacteriological procedures and considerable attention to detail.

LIST OF EQUIPMENT AND MEDIA REQUIRED FOR BASIC PROCEDURES

Most of the items listed below are available from various scientific supply houses. Prepared media can be obtained from several sources.†

Small autoclave or pressure cooker
Incubator‡
Bacteriological inoculating loops
Two dissecting needles (for breaking up fungous colonies)
Sterile swabs (preferably in sterile tubes rather than envelopes)
Microscope slides
Coverslips (22 mm × 22 mm)
Bunsen burner, spirit lamp, or "compressed gas" torch
Microscope (with oil immersion objective)
Hand lens (for examining colonies)

Salmonella O antiserum (Poly A-1)
Disposable Petri dishes (for collection of materials)
Media:
Blood agar plates
Brilliant green agar plates
Triple sugar iron agar
Sabouraud dextrose agar
Sabouraud dextrose agar (with inhibitors; various trade names)
Brain-heart infusion broth
SIM, MIO medium
Litmus milk

*Pitman Moore, Inc., Washington Crossing, New Jersey.

†Gibco Diagnostics, Columbus, Indiana; BioQuest, P.O. Box 243, Cockeysville, Maryland (regional distributors available).

‡Jensen-Salsbery Laboratories, Kansas City, Missouri; Pitman Moore, Inc., Washington Crossing, New Jersey.

SUPPLEMENTARY REFERENCES

Pittawa, E. M., and Treen, V. L.: *Vet Rec, 97*:327, 1975.

Inverso, M.: In Catacott, E. J. (Ed.): *Animal Hospital Technology*. Wheaton, Illinois, American Veterinary Publications, 1971, p. 325.

Wilkins, R. J., and Vito, D.: *Vet Clin North Am, 6*:741, 1976.

Coles, E. H.: *Veterinary Clinical Pathology*, 2nd ed. Philadelphia, Saunders, 1974, Chapter 16.

Chapter 39

DIAGNOSTIC APPROACHES FOR FISH DISEASES

EMMETT B. SHOTTS, JR.

W HEN fish losses occur, the laboratory worker is often placed at a disadvantage by being presented an inadequate history along with a dead specimen. Diagnosis of maladies of fish is more involved than just the culture of a dead specimen. A knowledge of several associated disciplines is very important in determining the cause of death.

The most common problem associated with fish deaths is poor water quality, i.e. insufficient O_2 or presence of toxic compounds both natural (NO_2, NH_3) and introduced. The next most frequent problem associated with fish losses is parasites, the vast majority of which are external and are usually found either on the gills or finage. Complicating these two causes of fish disease, and a third cause in its own right, are infectious diseases including those caused by bacteria, viruses, and fungi. Two other factors that often compound the problems associated with a case are nutritional deficiencies and/or stress resulting from borderline environmental quality.

This discussion will be restricted to infectious diseases and primarily to bacteria associated with deaths of fish.

Specimen Submission

The ideal specimen is a moribund fish collected and transported to the laboratory in environmental water. The second choice is a pithed moribund fish frozen in tap water and submitted frozen. In either case, dead specimens are not suitable for culture unless submitted as referred to above. It has been documented that environmental organisms tend to invade and replicate significantly within thirty minutes following death. In the initial submission, the water quality should also be evaluated and the fish examined for parasites. In the latter situation, only bacterial cultures of internal organs should be done.

Organs for Culture

The most common site for bacterial culture in fish is the hind kidney. This organ is usually located along the ventral side of the "backbone" and is protected from cavity contamination by a protective membrane. There are two methods for dissecting the fish to obtain a kidney culture. In both cases the fish should be pithed before dissection. In the first, make a scissor incision starting at the cloaca and cut in a "half moon" fashion to a point just at the top of the gill. A second incision is made from the cloaca along the midline to the bottom of the gill. The tissue is pulled forward toward the gills and cut free from the fish. After reflection of the swim bladder(s) and visceral organs, the kidney should be apparent along the vertebrae as a dark reddish organ.

In the second method, external surface sterilization is important. The pithed fish is scaled, if necessary, along the central body area with subsequent disinfection using a paper towel soaked with either 70% alcohol or Roccal solution. The dorsal fin is clipped and using a hot searing spatula, a "saddlelike" sear is made using the center of the clipped dorsal fin for reference. A sterile incision in the seared area just through the backbone is made either using a scapel or large scissors. By downward pulling pressure, holding the head and tail portions, the fish kidney can be exposed somewhat like opening a shotgun for loading. If the fish is large and/or has been frozen this method is probably the more efficient. Very small fish often require special handling, and with some aquarium species two people are necessary to obtain suitable material for culture. Once the fish is open, a Gram stain of kidney tissue is made to determine if a bacterial cause should be considered.

Regardless of the method used, diagnosis should never be based on a single fish, but on data obtained from several moribund fish.

Diagnostic Approach

The results of the Gram stain will heavily influence how the case is handled beyond this point. The primary type of organism associated with bacterial diseases of fish is Gram-negative rods. This does not preclude the finding of Gram-positive cocci or rods on rare occasions.

The Gram-negative rods should be presumptively divided microscopically into rods 1–3 µm by 0.5–0.8 µm and thinner longer rods 2–12 µm by 0.4–0.8 µm. The former size is more characteristic of classic Gram-negative rods, while the latter would suggest the presence of the flexi-bacteria (formerly myxobacteria). Gram-positive organisms that may be observed include streptococci, mycobacteria, *Nocardia* sp., and a corynebacterialike organism belonging to the genus *Renibacterium*.

Media Selection and Bacterial Screening

Two options may be followed in media selection for the isolation of bacteria from diseased fish. The first is the use of a noninhibitory medium such as trypticase agar for transfer and identification. If this option is selected, it is strongly recommended that incubation be done at 20–25°C (out of the mesophilic range) to retard rapid overgrowth of potential contaminants. While most bacterial pathogens of fish grow at 37°C, two (*Aeromonas salmonicida* and *Pseudomonas fluorscens*) are retarded at temperatures over 30°C. After forty-eight hours incubation, noninhibitory isolation media may usually be examined and bacterial growth divided into two major groups on the basis of cytochrome oxidase for further identification procedures. In using non-inhibitory isolation media, the laboratory worker should pick and examine a statistical cross section of all the "Gram-negative-like colonies" on the plate, since the colonial morphology of this group is so similar. Following this approach the oxidase-positive colonies are most likely members of the *Pseudomonas, Aeromonas,* or *Vibrio* and may be further delineated using hanging drop motility, pigment, O/F glucose, and novobiocin or 0/19 sus-ceptibility into general groups pending any specific characterization by batteries of media inoculated at forty-eight hours. An effort should be made to establish a battery of media that does not include biochemical reactions prone to variation such as carbohydrate reactions. The oxidase-negative organisms associated with fish diseases belong primarily to the Enterobac-teriaceae. Although the genera *Edwardsiella* and *Yersinia* are of primary concern, other enteric organisms may cause the death of fish and should be identified using conventional enteric procedures (see Chapter 10).

Due to antigenic similarities among the enterics and associated Gram-negative rods, it is not reliable to depend too heavily on the use of conventional antisera to replace salient biochemical characteristics.

A second option available to the laboratory worker is the use of selective media for initial isolation with subsequent confirmatory tests. Media such as Pseudosel (BBL) for *Pseudomonas* isolation, Rimler-Shotts agar for *Aeromonas* (1), McConkey agar for Gram-negative rods, and Phenethyl alcohol agar (BBL) for Gram-positive organisms are but a few that may be developed into a useful protocol. One such schematic aid is shown in Figure 39-1.

While short Gram-negative rods are identified in some detail, it is rare that organisms suggestive of flexibacteria are grouped beyond a general term "columnaris" disease. Based on recent work, there appear to be at least three genera and perhaps six species capable of producing gill rot, fin rot, or so called "saddle back" in catfish. Diagnosis is based upon the presence of columns or "haystacks" of these organisms observed by direct microscopy of material from diseased fish.

Because of the infrequency of the association of Gram-positive organisms in bacterial diseases of fish, it is prudent that these cases be approached on an individual basis. The most frequent Gram-positive bacteria of fish is *Renibacterium salmonis*, which causes kidney disease (KD) in salmonids.

Fish disease attributed to primary fungous causes is very uncommon in

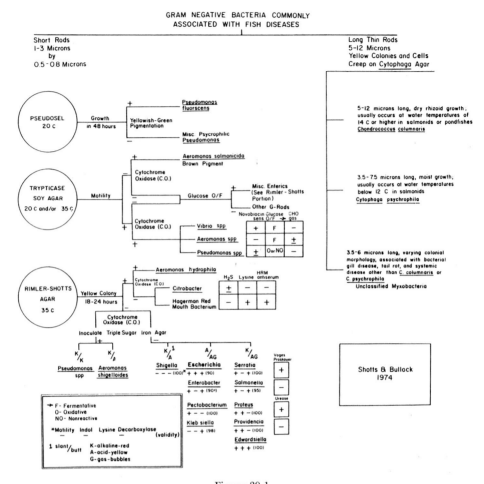

Figure 39-1.

the United States and infrequent in other parts of the world. Most of these may be grown on either Sabouraud's agar or blood agar incubated at 20°C. Microscopic observation of fruiting structures is criteria for identification.

In the absence of any of the above etiologic agents, viral disease may be considered. There is one primary viral disease of warm water cultured fish, viz. channel catfish virus disease (CCVD); another of some consequence in aquaria fish is lymphocystitis; as well, a large number of viruses are associated with salmonid culture.

There are some problems and points of confusion in the identification of some of the bacterial pathogens of fish. Some of the primary points of importance that should be given special note are discussed briefly below according to group. Consult the appropriate portions of this book for general identifying characteristics.

Aeromonas

Aeromonas hydrophila is perhaps the most common organism associated with the aquatic environment. It then becomes important that cultures be taken with care and from moribund fish to ensure the validity of findings. This organism is easily screened from isolates because it is oxidase positive, produces gas in most cases, and ferments glucose. A similar organism with similar characteristics, *Aeromonas soberia*, has been proposed, and while DNA homologies support the division, the biochemical differentiations are unreliable. Confusion often arises as to whether or not isolates are of significance. It is suggested that strains that ferment salicin with the production of gas and cause a rapid reversion (to alkalinity) (24–48 hours) of the TSI slant be considered of potential disease significance (Unpublished data, Hsu, T., Shotts, E. B., and Waltman, D., University of Georgia 1983).

Aeromonas salmonicida appears to be an obligate parasite of salmonids that produces disease; it also occurs in the intestine in carrier animals. There are atypical strains of this organism. Most notable are the strains causing "ulcer" disease in goldfish and other species. While the typical strain produces a melaninlike pigment, atypical strains as a rule do not. The atypical strain is very fastidious and appears as small pinpoint colonies after several days incubation at temperatures of 30°C or less. For this reason, it is necessary to take care in streaking plates to assure good separation. Atypical strains produce very compact colonies that fracture when pressure is applied and remain intact when teased with an inoculating needle. Fluorescent antibody staining has been used successfully in the identification of *A. salmonicida*.

Pseudomonas

Species of this genus are found frequently in the aquatic environment and from time to time are involved in deaths. Members of the *Ps. putida* group are often mistaken for *Ps. fluorscens* because growth temperature is not considered. *Ps. fluorscens* does not grow at 37°C, while *Ps. putida*-like organisms do. Although Pseudosel (BBL) is used as a selective medium for *Pseudomonas* spp., some isolates will not survive on it. There have been two reports of fish mortality associated with an encapsulated, motile *Pseudomonas* spp.

Vibrio

This genus is primarily associated with disease in brackish or salt water, although outbreaks in fresh water have been reported. The primary organisms involved are two serotypes of *Vibrio anguillarum*. In general, they are superficially indistinguishable from *Aeromonas*. Separation is quickly made as shown in Table 39-I.

Table 39-I

Reaction	Aeromonas	Vibrio
Arginine	+	−
Novobiocin suscept.	−	+
Gas from Carbohydrate	±	−
0/129* suscept.	−	+

* McDaniel (2)

Edwardsiella

Two members of this genus, *E. tarda* and *E. ictulurii*, have been associated with fish deaths. *Edwardsiella tarda* has been reported from both septicemic outbreaks and localized lesions in warm water fish in the U.S. This same organism is the primary cause of deaths in eels in the Far East. While the organism may be encountered in isolated instances, it is not of major importance in the U.S. Although the majority of isolates are H_2S positive, occasionally negative strains are isolated from fish. The second species, *E. ictulurii*, a very slow growing organism, was first described some five years ago in catfish fingerlings and has since become a major disease problem in all ages of catfish primarily in the Southeast. Primary differences between these two species are noted in Table 39-II.

Table 39-II

Reaction	E. tarda	E. ictulurii
H_2S	+	−
indole	+	−
42 C-growth	+	−

Biochemical characterization indicates close similarity among some thirty isolates of *E. ictulurii*. (Unpublished data, Waltman, W. D., Shotts, E. B., and Hsu, T. C., University of Georgia, 1983.)

Yersinia

Although a number of species of *Yersinia* have been associated with surface water, *Yersinia ruckeri* is the only member of the group associated regularly with fish mortality. This organism appears to be an obligate parasite and may be isolated from the distal part of the intestine in carrier fish. While the disease Hagerman red mouth was first associated with salmonids in the Hagerman Valley of Idaho, isolates have been recently identified in culture collections at the National Fish Disease Laboratory, Leetown, West Virginia, which precede the first cases in the West by some ten years. Further studies have shown the existence of at least three serotypes, the Hagerman, the O'Leary, and Australian types. Of these three, only the Hagerman has been shown to infect fish under experimental conditions via natural routes. Presently, regulatory problems are developing on the significance of isolations of the O'Leary type from fish being certified for movement. There are also diagnostic problems involved, and inexperienced personnel may confuse this organism with *Hafnia alvei*, a common enteric bacterium of aquatic environments. These two organisms may be separated as shown in Table 39-III.

Table 39-III

Reaction	Y. ruckeri	H. alvei
malonate	−	+
arabinose	−	+
rhamnose	−	+

Miscellaneous Enterics

On occasion other enteric organisms have been isolated as the primary cause of deaths in fish. These include *Proteus* spp., *Citrobacter freundi*, *Escherichia coli*, *Arizona hindshawii*, and *Salmonella* spp.

Streptococci

Streptococci, group B, have been recovered from some five extensive fish mortality problems in the past ten years. The most recent of these occurred in Taiwan in 1982. These isolates in retrospect appear to be very similar to isolates of streptococci isolated from diseased frogs in both Taiwan and the U.S.

Renibacterium

This genus for a number of years was thought to consist of a slow-growing, ill-defined species of *Corynebacterium*. In the past five years it has been accepted as a new species, *Renibacterium salmoninus*. Although studied extensively, *in vitro* cultivation is a slow and exacting process. Currently fluorescent antibody techniques are used for identification. The organism has only been found in salmonids.

Flexibacteria, Myxobacteria, Flavobacterium Complex

This very poorly defined group of bacteria have several things in common. First, they all have a yellowish orange pigment that turns red when 3% KOH is dropped on a colony. Second, they are very seldom cultured. Third, they are all interchangeably credited with causing "fin rot," "gill rot," and bacterial gill disease and are all diagnosed microscopically as "Columnaris" disease. Actually, there are at least three or more genera and an unknown number of species associated with this type of disease in fish. Presently this group constitutes an ill-defined complex of organisms with GC ratios ranging from about 30 to 70. It seems appropriate at present to refer to all these diseases as myxobacteriosis until more data is available.

Bacterial Reference Reagents

Direct FITC conjugates for *Aeromonas salmonicida, Flexibacter columnaris, Vibrio* spp., *Edwardsiella tarda,* and *Yersinia ruckeri* have been prepared, standardized, and lyophphilized for reference purposes. This material is available in limited quantities from the National Fish Health Research Laboratory, Route 3, Box 50, Kearneysville, West Virginia, 25430.

Viruses

When deaths persist in the absence of altered water quality, bacteria, or parasites there is the possibility that losses may be of viral origin.

Very few significant viruses are currently associated with aquarium fish (lymphocystis virus) or warm water fish (channel catfish virus); however, a large number of viruses are associated with salmonids. They include both RNA and DNA viruses; however, the majority are RNA. Some of the more commonly encountered viruses of salmonids are the viruses of infectious pancreatic necrosis (IPN), of viral hemorrhagic septicemia (VHS), of infectious hematopoietic necrosis (IHN), and the herpesvirus of trout. The number of viruses currently associated with fish disease reflects the interest and research in the disease problems associated with specific groups of fish.

If a viral disease is suspected, appropriate tissues should be taken from moribund fish and frozen at −70°C pending processing or forwarding to a speciality laboratory for virus isolation. One such laboratory is located at the National Fish Health Research Laboratories, Kearneysville, West Virginia.

Mycotic Organisms

A large number of fungi have been associated with disease in fish. Most of them belong to the *Oomycetes* or to the fungi imperfecti. The most common of these is *Saprolegnia* spp. This group of organisms appears as a mass of rapidly growing, nonseptate, multibranched cotton-wool-like mycelia. Reproduction of this organism is both sexual and asexual. The asexual form consists of diflagellated zoospores arising from a sporangium. This organism is usually associated with injured fish tissue and is most commonly noted at water temperatures of 15–20°C. If colonial growth is desired it may be obtained by inoculation of material onto either Sabouraud's agar or cornmeal agar, which is incubated at room temperature. Diagnosis is usually based on the microscopic examination of fresh material. Identity is based upon the presence of asexual sporangia containing zoospores and nonseptate mycelia. While a wide array of other mycotic organisms have been reported in fish, this organism is by far the most common.

A number of miscellaneous fungi have been isolated from systemic mycoses in fish. They include *Exophiala salmonis*, which causes cerebral mycetoma of trout; general infection caused by *Heterosporium* and *Scclecobasidium*; swim-bladder infection caused by *Phoma herbarum*; infection with *Ichthyophonus hoferi*; and dermal cysts caused by several species of *Dermocystidium*.

Although very infrequent in the United States, branchiomycosis has been reported and may produce extensive damage in fish. The *Branchiomyces* are unique because they grow intravascularly in the gills producing a massive infarctive necrosis. This disease is caused by either *B. sanguinus* or *B. demigrans*. It is usually found at temperatures of 20°C or higher and proliferates rapidly with mortality often in excess of 50 percent. Fungal infections have also been reported involving *Basidiobolus* spp. and *Fusarium* spp.

In general, fungi may be found on or in fish or fish eggs both as saprophytes and/or parasites. Their occurrence is usually associated with poor environmental quality such as high organic content, algal blooms, or highly fertilized water. Usually, the rapidity of spread is enhanced by an optimum temperature for the fungus involved.

Unless workers have had considerable experience in the identification of fungi, it is advisable to submit cultures to a research or reference laboratory for final identification. The procedures manual edited by Daniel (2) will be helpful in the identification of many of the fungi.

REFERENCES

1. Shotts, E. B., Jr., and Rimler, R.: *Appl Microbiol,* 26:550.
2. McDaniel, D. (Ed.): *Procedures for the Detection and Identification of Certain Fish Pathogens,* 2nd revision. Washington, D.C., American Fisheries Society, 1979.

SUPPLEMENTARY REFERENCES

Amlacher, E.: *Textbook of Fish Diseases* (translated by D. A. Conroy and R. L. Herman). Neptune City, New Jersey, TFH Publications, Inc., 1970.

Becker, C. D., and Fujihara, M. P.: *The Bacterial Pathogen Flexibacter Columnaris and Its Epizootiology Among Columbia River Fish,* Monograph 2. Washington, D.C., American Fisheries Society, 1978.

Boyd, C. E.: *Water Quality in Warmwater Fish Ponds.* Auburn, Alabama, Auburn University, Agricultural Experiment Station, 1979.

Brown, E. E., and Gratzek, J. B.: *Fish Farming Handbook.* Westport, CT, AVI Publishing Company, Inc., 1980.

Elkan, E., and Reichenbach-Klinke, H.: *Color Atlas of the Diseases of Fishes, Amphibians and Reptiles.* Neptune City, New Jersey, TFH Publications, 1974.

Hoffman, G. L.: *Parasites of North American Freshwater Fishes.* Los Angeles, CA, University of California Press, 1967.

McDaniel, D. (Ed.): *Procedures for the Detection and Identification of Certain Fish Pathogens,* 2nd revision. Washington, D.C., American Fisheries Society, 1979.

Roberts, R. J., and Shepherd, C. J.: *Handbook of Trout and Salmon Diseases.* Surrey, England, Fishing News (Books) Ltd., 1974.

Roberts, R. J.: *Fish Pathology*. London, Bailliere Tindall, 1978.

Shotts, E. B., and Bullock, G. L.: Bacterial diseases of fish: Diagnostic procedures for gram negative pathogens. *J Fish Res Board Can, 32*:1243, 1975.

Shotts, E. B., and Bullock, G. L.: Rapid diagnostic approaches in the identification of gram negative bacterial diseases of fish. *Fish Pathology, 10*:187, 1976.

Appendix A

STAINING PROCEDURES

STAINING solutions should be kept in airtight plastic or glass bottles. Because dyes are light-sensitive, they should be stored in a protective container, e.g. an amber glass bottle, or kept out of the direct sunlight. Ready-to-use solutions of a number of stains are available commercially.[*]

GRAM'S STAIN (HUCKER'S MODIFICATION)

1. Stock Crystal Violet
   Crystal violet   10 g
   Ethyl alcohol (95%)   100 ml
2. Stock Oxalate Solution
   Ammonium oxalate   1 g
   Distilled water   100 ml
   Crystal violet working solution: Mix 20 ml of solution no. 1 with 80 ml of solution no. 2. Additional dilution may be made if desired.
3. Gram's Iodine Solution
   Iodine crystals   1 g
   Potassium iodide   2 g
   Dissolve completely in 10 ml of distilled water, then add
   Distilled water to make   200 ml
   Store in amber bottle.
4. Decolorizer
   Ethyl alcohol (95%)   250 ml
   Acetone   250 ml
5. Counterstain
   Safranin   2.5 g
   Ethyl alcohol (95%)   100 ml
   Safranin working solution: The stock safranin is usually diluted 1:4 with distilled water.

---

[*]Difco Laboratories, Detroit, Michigan. BBL Microbiology Systems, Baltimore, Maryland.

Procedure for the Gram's Stain

Gram's stain is best performed on young cultures because older cultures often decolorize too readily.

1. Make a thin smear or film of clinical material. Bacteria from solid media are transferred to a drop of distilled water on a slide. A loopful of broth culture is placed directly on the slide. If the slides are not clean, material may wash off. New slides can be cleaned by soaking in 95% ethyl alcohol, then wiping dry with clean gauze. Allow the film to dry in the air. Drying is accelerated in the incubator. Fix the film by quickly passing the slide through the Bunsen flame several times. Proper fixing also prevents washing off.

2. The slide is flooded with crystal violet stain. Leave on thirty to sixty seconds.

3. Pour off the stain and wash the remaining stain off with the iodine solution, leaving the slide covered with iodine for one to two minutes.

4. Wash off the iodine and shake the excess water from the slide.

5. Decolorize with acetone alcohol until the decolorizer flows colorless from the slide. Some organisms are easily decolorized too much with acetone alcohol.

6. Counterstain with safranin for thirty to sixty seconds, and wash with water. Weak carbolfuchsin for sixty seconds can be substituted for safranin. To obtain weak carbolfuchsin, dilute Ziehl-Neelsen carbolfuchsin 1:10 or 1:20 with distilled water.

Gram's Stain of Smears from Lesions, Exudates, and Tissues

1. Alkaline Gentian Violet
   Solution A
Gentian or crystal violet	1.0 g
Distilled water	100.0 ml

   Solution B
Sodium hydrogen carbonate ($NaHCO_3$)	1.0 g
Distilled water	100.0 ml

2. Iodine Solution
Iodine	2.0 g
1 N Sodium hydroxide (NaOH, 40.01 g/liter)	10.0 ml
Distilled water	100.0 ml

3. Counterstain
   Safranin or carbolfuchsin as in Hucker's modification referred to previously

Staining Procedure

1. Air dry thinly spread films.
2. Flood slide with a freshly prepared mixture of gentian violet (solution A: 1.5 ml and solution B: 0.4 ml). Leave on for five to ten minutes.
3. Wash off with iodine solution, leaving the latter on for two to three minutes.
4. Wash with water, and blot water from surface of smear but do not dry.
5. Decolorize with acetone-ether (1:1). Ten seconds is usually sufficient.
6. Dry in air and counterstain for five to ten seconds with safranin or carbolfuchsin.
7. Wash with water, blot, dry, and examine.

ZIEHL-NEELSEN ACID-FAST STAIN

1. Carbolfuchsin Stain
   Basic fuchsin   0.3 g
   Ethyl alcohol (95%)   10.0 ml
   This solution is mixed with
   Phenol (melted crystals)   5.0 ml
   Distilled water   95.0 ml
2. Acid-Alcohol
   Hydrochloric acid (concentrated)   3.0 ml
   Ethyl alcohol (95%)   97.0 ml
3. Counterstain
   Methylene blue (certified)   0.3 g
   Distilled water   100.0 ml

Staining Procedure

1. Prepare a thin smear; dry and fix as described under the procedure for Gram's stain.
2. A strip of filter paper the size of the smear is placed on the slide covering the smear.
3. Flood the slide with carbolfuchsin stain and heat to steaming with a low Bunsen flame. Allow to stand for five minutes without further heating, after which the paper is removed and the slide washed in running water. The stain should not boil.
4. Decolorize to a faint pink color by several applications of acid-alcohol. The decolorizer is applied until no more color comes out. This usually takes approximately two minutes for films of average thickness. One-half percent aqueous sulfuric acid is recommended for decolorizing suspected *Nocardia asteroides*.

5. Wash with water and counterstain with methylene blue for twenty to thirty seconds.
6. Wash with water, dry, and examine.

Modified Ziehl-Neelsen Stain

This stain is useful for staining *Brucella* and chlamydia

1. Stock carbolfuchsin:

Basic fuchsin	1 g
Absolute methyl alcohol	10 ml
Phenol, 5%	90 ml

Staining Procedure

Smears are stained for ten minutes with a 1:10 solution of the stock carbolfuchsin, then washed with tap water. Decolorize with 0.5% acetic acid for twenty to thirty seconds, then wash and blot dry.

Brucella and chlamydia stain red against a blue background.

KINYOUN ACID-FAST STAIN

Basic fuchsin	4 g
Phenol (melted crystals)	8 ml
Ethyl alcohol (95%)	20 ml
Distilled water	100 ml

Dissolve the basic fuchsin in the alcohol and slowly add the water while stirring. The 8 ml of melted phenol is then added. Phenol crystals can be conveniently melted in a heated water bath.

Staining Procedure

1. Stain a fixed smear for three to five minutes with the staining solution described above. No heating is required.
2. The remaining procedures are the same as for the Ziehl-Neelsen method.

LOEFFLER'S METHYLENE BLUE STAIN

This stain, like safranin or carbolfuchsin, can be used as a general-purpose stain when morphology is the primary consideration. With this solution, the spores of spore-forming bacteria appear as unstained bodies within blue bacilli. The beading and granules of the corynebacteria may also be revealed with this stain.

After this stain ripens (12 months), it is particularly useful for staining the capsules of *Bacillus anthracis*. The capsule appears as an amorphous purplish material around the bacteria (McFadyean's reaction).

Methylene blue (1% in 95% ethanol)	30 ml
Potassium hydroxide (1% aqueous solution)	1 ml
Distilled water	100 ml

Staining Procedure

The fixed smear is stained with the above solution for one minute. Wash with water and dry.

MOTILITY EXAMINATION

Young organisms (six to eight hours) from media that do not contain fermentable substrates are examined. A method more convenient than the conventional hanging drop involves suspending a drop on the underside of a coverslip over a "perimeter" of petrolatum. The petrolatum is delivered to the slide from a 10.0 ml syringe through an 18 gauge needle. The coverslip is placed on the ridges of petrolatum in the same manner that one is placed over the concavity of a hollow-ground slide.

Because of sensitivity to oxygen, the hanging drop is not always satisfactory for clostridia. They can be examined in a capillary tube preparation, from cooked meat cultures, sealed at each end.

GIEMSA STAIN

Stock Solution

To 0.3 g of Giemsa powder (National Analine or other satisfactory source) is added 25.0 ml of glycerin and 25.0 ml of absolute, acetone-free methyl alcohol. The ready-to-use staining solution is available commercially.

If the stain does not go into complete solution, it should be filtered. One volume of the stock solution is diluted with ten volumes of distilled water. This diluted stain is ready for use.

Staining Procedure

1. Fix smear in methyl alcohol for three to five minutes.
2. Dry in air.
3. Immerse in diluted stain for twenty to thirty minutes. The staining period may be extended to an hour or longer as indicated by results.
4. Wash with distilled water.
5. Stand on end to dry, then examine.

WRIGHT'S STAIN

Staining Solution

Wright's stain powder (certified)	0.1 g
Absolute methyl alcohol (acetone-free)	60.0 ml

The stain is ground with alcohol in a mortar, then allowed to stand in a stoppered bottle for a week to ripen. The ready-to-use solution is available commercially.

Buffer Solution

Potassium phosphate (monobasic) KH_2PO_4	6.63 g
Sodium phosphate (dibasic) Na_2HPO_4	3.2 g
Distilled water	1000.0 ml

Staining Procedure

1. Thin smears are air dried.
2. Flood smear with stain, counting the drops.
3. Stain for one to five minutes. Experience will indicate the optimum time.
4. Add an equal amount of buffer solution and mix the stain by blowing an eddy in the fluid.
5. Leave the mixture on the slide for three to seven minutes.
6. Wash off by running water directly to the center of the slide to prevent a residue of precipitated stain.
7. Stand slide on end, and let dry in air.

KOSTER'S STAIN FOR BRUCELLA (SLIGHTLY MODIFIED) (1)

1. Films or smears are dried and fixed over a flame in the usual manner.
2. Wet smear under the tap. Add to the slide two drops of saturated safranin solution and five drops of 1 N potassium hydroxide solution. Mix and leave on slide for one to two minutes.
3. Wash under the tap.
4. Decolorize with 0.1% sulfuric acid solution and repeat this operation, with a total decolorization time of ten to twenty seconds.
5. Wash thoroughly.
6. Counterstain with ordinary carbol-methylene blue solution for two to three seconds. (Carbol-methylene blue solution: methylene blue 1 g, absolute ethyl alcohol 10 ml, and 100 ml of 5% aqueous solution of phenol.)
7. Wash with tap water, dry, and examine.

Brucella organisms, usually seen intracellularly, stain red, while other bacteria stain blue. Colonies of organisms from the cotyledons such as streptococci are stained clearly with this stain.

RELIEF OR NEGATIVE STAINING

Staining Solution

Nigrosin	10.0 g
Distilled water	100.0 ml

Boil this solution for thirty minutes, add 0.5 ml of formalin. Pass through filter paper and dispense in sterile tubes in 2 ml amounts.

Staining Procedure

1. Place a loopful of the bacterial suspension or clinical material on a slide.
2. Add an equal amount of nigrosin suspension and spread, making a moderately thin film.
3. Air dry and examine.

Bacteria appear white against a dark background. This method is especially convenient for organisms with poor staining properties such as spirochetes and vibrios.

INDIA INK WET MOUNT

Clinical material or organisms from cultures are mixed on a slide in a drop made up of loopfuls of distilled water and India ink. Experience will indicate the amount of India ink to use. A coverslip is added, and the preparation is examined. The large capsules of *Cryptococcus neoformans* show up strikingly by this method.

CAPSULE STAIN (HISS METHOD)

Mix a loopful of a suspension (physiological saline) of growth with a drop of normal serum on a microscope slide. Allow the slide to air dry, then heat fix. Flood the smear with 1% aqueous solution of crystal violet. Steam the slide gently for one minute, then rinse with 20% aqueous solution of copper sulfate. Capsules appear as faint blue haloes around purple to dark blue cells.

SCHAEFFER AND FULTON SPORE STAIN

Staining Procedure

1. Fix smear and flood with malachite green (5% aqueous solution).
2. Steam gently over a flame for thirty seconds.

3. Wash with water and stain with safranin (0.5% aqueous solution) for thirty seconds.
4. Wash with water, blot, dry, and examine.

FLAGELLA STAIN (2)

Flagella stain (powder) prepared according to the formulation of Leifson along with the staining procedure is available commercially.*

REFERENCES

1. Christoffersen, P. A.: Personal communication.
2. Leifson, E.: *J Bacteriol,* 62:377, 1951.

SUPPLEMENTARY REFERENCE

Gurr, E.: *A Practical Manual of Medical and Biological Staining Techniques,* 2nd ed. New York, Interscience, 1956.

---

*Difco Laboratories, Detroit, Michigan.

Appendix B

CULTURE MEDIA

A WIDE VARIETY of media are available for the cultivation of pathogenic bacteria and fungi. Only the more widely used ones are referred to in this appendix. All but several are available commercially in the dehydrated form. The name of the commercial source is given if the medium is one that is not available from most of the firms listed below. Commercial sources of the well-known media are not given, as they are widely available. Because the product of one firm is specified it does not necessarily indicate a preference. An equally satisfactory product may be available from another firm. The formulations and directions for the preparation of the commercially available dehydrated media are not given. These are available in the detailed and informative manuals and literature supplied by the manufacturers.

SOURCES OF COMMERCIALLY AVAILABLE MEDIA

The sources of the commercial media described in this section are indicated by the word or letters in the right-hand column below, corresponding with the full name and address of the firm given on the left.

Albimi Laboratories, Inc.   (Albimi)
   35–22 Linden Place
   Flushing, New York
BBL Microbiology Systems   (BBL)
   250 Schilling Circle
   P.O. Box 243
   Cockeysville, Maryland 21030
Case Laboratories, Inc.   (Case)
   1407 North Dayton Street
   Chicago, Illinois
Difco Laboratories, Inc.   (Difco)
   P.O. Box 1058A
   Detroit, Michigan 48232
Gibco Laboratories   (Gibco)
   P.O. Box 4385
   2801 Industrial Drive
   Madison, Wisconsin 53713

Oxoid U.S.A. Inc.   (Oxoid)
   9017 Red Branch Road
   Columbia, Maryland 21045
Pfizer Diagnostic Division   (Pfizer)
   199 Maywood Avenue
   Maywood, New Jersey

MEDIA FOR GENERAL PURPOSES

Collection of Blood for Media

In the smaller diagnostic laboratories, it is usually more convenient to collect blood in an anticoagulant solution. Blood collected in this manner has been found satisfactory for general purposes. Horse, bovine, and sheep blood is used.

The site over the jugular vein is clipped and generously disinfected with iodine. Blood is drawn into a 600 ml commercially prepared blood collection bottle containing 120 ml of "anticoagulant acid citrate dextrose solution" by means of a sterile, also commercially prepared, thirty-six-inch long "blood collection set" with siliconed needles. The needle on the one end of the collection set is inserted into the animal's vein, and when securely in place, the other needle on the other end is inserted into the blood collection bottle. Blood is drawn up to the 600 ml level. This fresh blood is distributed into sterile screw-cap tubes in 15 ml amounts and stored in the refrigerator until used.

If defibrinated blood is preferred, blood is taken sterilely into a bleeding flask containing glass beads for the collection of the fibrin. Ox, horse, and sheep defibrinated blood is available from a number of commercial sources.

Blood Agar

Various blood agar bases are available commercially. Trypticase soy agar (BBL) is widely used. Blood agar base is rehydrated, distributed in flasks in 250 ml amounts, and autoclaved. Flasks of the base medium are melted and placed in a water bath at 50°C. When the medium has reached 50°C, 15 ml of sterile blood is added to each flask. Plates are poured after mixing, then incubated before using.

BLOOD AGAR BASE IN TUBES. Blood agar base medium is dispensed in screw-cap tubes in 15 ml amounts, then autoclaved. When one or more inhibitory compounds are desired in blood agar, one or more tubes are melted, and when the temperature has reached 50°C, 1 ml of blood and the required amount of inhibiting compound is added (see Table B-I).

TABLE B-I

SOME INHIBITORY COMPOUNDS USED IN SELECTIVE MEDIA

Compound	Dilution in Media	To Aid the Isolation of	Convenient Solution	Amount Added to		
				10 ml	15 ml	100 ml
Crystal violet*	1:700,000	*Brucella*	1:14,000	0.2	0.3	2.0
Thallium or thallous acetate	1:4,000	Mycoplasmas	1:80	0.2	0.3	2.0
Sodium azide	1:2,000	Streptococci	1:20	0.1	0.15	1.0
Sodium azide	1:1,000	*Erysipelothrix*	1:20	0.2	0.3	2.0
Brilliant green*	1:40,000	*Vibrio fetus*	1:800	0.2	0.3	2.0
Brilliant green*	1:80,000	*Salmonella*	1:1,600	0.2	0.3	2.0
Potassium tellurite	1:10,000	*Listeria,* Streptococci, Staphylococci, Corynebacteria	1:200	0.2	0.3	2.0

*Prepare solution in 95% ethanol.

INHIBITORY COMPOUNDS USED IN MEDIA

The most commonly used inhibitory compounds and the groups of bacteria that they inhibit are listed in Table B-II.

TABLE B-II

DILUTION RANGE OF SOME INHIBITORY COMPOUNDS USED IN MEDIA

Compound	Principal Group Inhibited	Range of Dilutions Used
Crystal violet	Gram-positive organisms *Escherichia coli*	1:700,000–1:1,000,000
Brilliant green	*Shigella* to a lesser extent	1:25,000–1:80,000
Thallium or thallous acetate	Gram-negative organisms	1:1,000–1:8,000
Sodium azide	Gram-negative organisms	1:1,000–1:5,000
Potassium	Gram-negative organisms	1:10,000–1:35,000

Containers for Media

Screw-cap tubes or small bottles are widely used. A synthetic foam plug (diSPo® plug), ready for use, is available commercially* in various sizes. These plugs are not as pervious to air as are cotton plugs, and as a result, evaporation of media is slower.

---

*American Scientific Products, 1210 Waukegan Road, McGaw Park, Illinois 60085.

Culture Media

Tryptose Phosphate Broth (Difco) and
Trypticase Soy Broth (BBL)

These are useful general-purpose media. The growth of fastidious organisms is increased if 0.5–1.5 g of agar is added to each liter of broth.

Trypticase Soy Agar (BBL)

This medium may be used when a clear medium is required for general purposes.

Thioglycollate Media

Some of the thioglycollate media available and their corresponding uses are tabulated below.

Kind	Use
Fluid Thioglycollate Medium (Difco) (with dextrose; without indicator)	Isolation and growth of anaerobes; growth of various fastidious bacteria.
Thioglycollate Medium (BBL) (without indicator and dextrose)	Fermentation tests; anaerobes. Several drops of bromthymol blue (0.2% aq. sol.) added after incubation.
Thiol Medium (Difco) (with dextrose; without indicator)	*Campylobacter;* tissues containing antimicrobial agents.

Brain-Heart Infusion Agar (Difco)

This is a useful medium for the growth of fastidious bacteria and fungi. It is usually slanted in screw-cap tubes.

Brain-Heart Infusion Broth

A useful general-purpose broth.

Brain-Heart Infusion Broth with Agar

This medium is prepared by adding 1.5 g of agar to 1 liter of brain-heart infusion broth. Swabs are routinely placed in this medium, which supports the growth of many fastidious aerobes, microaerophilic and anaerobic bacteria.

OXIDATION AND FERMENTATION TESTS

The base media used for the conventional fermentation tests contain peptones that are broken down during bacterial growth to substances that are alkaline in reaction. Thus, acid must be produced in excess of the alkalinity derived from the breakdown of peptones if the indicator is to be altered.

The following indicators are commonly employed in carbohydrate base media.

Indicator Base	Medium	Acid When pH Falls Below
Phenol red	Broth base	6.0
Bromthymol blue	Broth base	6.0
Bromcresol purple	Peptone water	5.0
Andrade's indicator	Peptone water	5.5

The peptone water bases yield less alkalinity than do the broth bases such as the phenol red broth base described below, but they are not as nutritious.

Although the majority of the bacteria referred to in the manual ferment sugars, etc., some oxidize these substances, e.g. *Pseudomonas* spp. Those organisms that oxidize sugars may not produce sufficient acid in conventional liquid media to change the indicator. Oxidation may be demonstrated when these bacteria are grown on the surface of solid media. In the case of *Pseudomonas* spp., an ammonium salt as a nitrogen source may be substituted for the peptone. An ammonium salt carbohydrate broth is described by Cowan and Steel (1). The O–F test described below may be used to determine whether the breakdown of the sugar is by oxidation or fermentation.

Oxidation-Fermentation (O-F) Test

This test demonstrates whether the breakdown of sugars, etc., is by oxidation or fermentation. Workers are referred to Cowan and Steel (1) for a comprehensive description and discussion of the test. The bacterium in question is grown in two tubes of Hugh and Leifson's (2) broth, which is heated for five minutes in boiling water prior to inoculation. One of the tubes is covered after inoculation with Vaspar seal (one part petrolatum, one part paraffin). Those bacteria that oxidize show acid production in the open tube only, while those that ferment produce acid in both tubes.

Listed below are some important bacteria that break down sugars by oxidation rather than by fermentation.

Micrococci (some strains do not split sugars at all)

Neisseria (some strains do not split sugars at all)

Mycobacteria, Nocardia, *Pseudomonas* spp.

The results obtained with some bacteria in the O–F test are given in Table B-III.

Hugh and Leifson's O-F Medium (2)

This medium is used to determine whether an organism splits sugars by fermentation or oxidation (see above). It is available commercially as Bacto-OF Basal Medium (Difco).

TABLE B-III

RESULTS OBTAINED WITH THE OXIDATION-FERMENTATION (O-F) TEST*

Organism	Glucose		Lactose		Sucrose		Group
	Open	Covered	Open	Covered	Open	Covered	
Alcaligenes faecalis	—	—	—	—	—	—	I Nonoxidizers Nonfermenters
Pseudomonas aeruginosa	A	—	—	—	—	—	II
Bacterium anitratum	A	—	A	—	—	—	Oxidizers
Agrobacterium tumefaciens	A	—	—	—	A	—	Nonfermenters
Malleomyces pseudomallei	A	—	A	—	A	—	
Shigella dysenteriae	A	A	—	—	—	—	IIIa
Shigella sonnei	A	A	A	A	—	—	Fermenters
Vibrio comma	A	A	—	—	A	A	(Anaerogenic)
Salmonella enteritidis	AG	AG	—	—	—	—	IIIb
Escherichia coli	AG	AG	AG	AG	—	—	Fermenters
Aeromonas liquefaciens	AG	AG	—	—	AG	AG	(Aerogenic)
Aerobacter aerogenes	AG	AG	AG	AG	AG	AG	
Unclassified species	A	A	A	—?	Variable		IIIc
Some paracolon	AG	AG	A	—?			Fermenters Oxidizers

A = acid reaction;   AG = acid and gas formation;   — = no change or alkaline reaction
*From R.J. Hugh and E. Leifson, *J Bacteriol*, 66:24, 1953.

It is prepared as follows:

Peptone	2	g
Sodium chloride	5	g
Potassium phosphate (dibasic) K_2HPO_4	1.5	g
Agar	3	g
Distilled water	1000	ml
Bromthymol blue, 0.2% aqueous solution	15	ml

Dissolve ingredients by heating in water. Adjust the pH to 7.1; filter and add the bromthymol blue. Autoclave at 115°C for twenty minutes.

Add the desired sugars as sterile solutions to convenient amounts of the medium to give a final concentration of 1%, and dispense in 10 ml amounts in sterile tubes.

Some results obtained with this test are given in Table B-III.

Phenol Red Broth Base

This is a base medium for fermentation tests. Carbohydrates are added to give final concentrations of 0.5–1%. Some laboratories add to this base medium 0.15% agar. This aids the growth of fastidious organisms. If the carbohydrate broths are sterilized by Seitz or other filtration, they are dispensed aseptically in sterile tubes. Lactose, sucrose, xylose, arabinose, trehalose, and salacin are best sterilized by filtration. The other carbohydrate broths are tubed and, if desired, carefully sterilized in the autoclave at 116–118°C (10–12 lb.) for fifteen minutes.

Liquid media are usually dispensed in cotton-plugged tubes or in screw-cap tubes or vials. If the former are used, considerable dehydration takes place, even in the refrigerator. The small laboratory may find it convenient to dispense carbohydrate broths and other media in tubes (14 × 100 mm) closed with standard 00 rubber stoppers. The stoppers are pushed down to seal the tubes after autoclaving. The sterilized tubes of media can be stored indefinitely at room temperature if there is a shortage of refrigerator space.

Ten Percent Dextrose and Lactose Slants

These media are used in King's differential tables. Reactions readily differentiate *Acinetobacter calcoaceticus* from *A. lwoffi* (see Table 8-1).

The base medium consists of phenol red broth base to which has been added 1.5% agar. Dextrose and lactose are added to give a concentration of 10%. After careful sterilization (see above), the tubes are slanted. The culture to be examined is streaked on the slant surface of both media. A yellow color constitutes a positive reaction.

Trypticase Agar Base (BBL)

This is a useful medium for the study of fermentation reactions and demonstration of motility of the less fastidious anaerobes. The desired carbohydrate is added to give a final concentration of 0.5 or 1%. Thioglycollate medium (BBL) is preferred for fermentation tests of clostridia and other anaerobes.

CARBOHYDRATES DIFFERENTIATION DISCS (DIFCO)

These sterile discs contain a standardized amount of carbohydrate or other substrate for use in the differentiation and presumptive identification of bacteria based on fermentative or oxidative reactions. They may be used in a variety of carbohydrate-free nutrient broths, semisolid or solid media.

Perhaps their most important advantage, aside from affording a rapid test, is that of providing a wide range of carbohydrates for immediate use. It is difficult for the small laboratory to maintain a comprehensive stock of carbohydrate broths. The discs may also be added to the thioglycollate medium (without dextrose and indicator) for fermentation tests on the anaerobes. This use is especially convenient in view of the need for fresh media where the anaerobes are concerned.

The carbohydrate discs may be used with the following media:
1. Liquid media
    phenol red broth base
    purple broth base
2. Plating media
    phenol red agar base
    purple agar base
    Sanders agar
3. Semisolid media
    tryptic agar base
    cystine tryptic agar
    thioglycollate medium (without dextrose and indicator)

Liquid and Semisolid Media

Discs are transferred aseptically to either the liquid or semisolid media in tubes and inoculated heavily if a rapid test is required. If anaerobes are being tested, the thioglycollate broth should be heated in a boiling water bath for ten minutes prior to inoculation. After considerable growth has been obtained in this medium containing carbohydrate discs, several drops of 0.2% aqueous bromthymol blue are added to determine if fermentation has taken place. A positive reaction is indicated by a yellow color.

Solid Media

Discs are placed on the surface of the medium (2.5 cm apart at least) after it has been inoculated by streaking, smearing, or by the pour-plate procedure. Readings on solid media are made at four hours and at eighteen hours. Further details regarding the inoculation of media are supplied by the producer.

Indole Nitrate Medium (BBL)

This medium is employed to demonstrate indole production and the reduction of nitrate.

The trypticase media listed above are rehydrated by raising to boiling temperature, then tubed to a depth of several inches in screw-cap tubes. The media must be sterilized with care at 116–118°C (10–12 lb. pressure) for fifteen minutes. They are allowed to set in the upright position. For the growth of anaerobes, the media should be placed in a boiling water bath for ten minutes before inoculating.

Methyl Red and Voges-Proskauer Medium

Details of the MR–VP tests are given in Appendix C.

Peptone Water

Details of the procedure for the indole test using this medium are given in Appendix C.

S I M Medium (Difco)

This useful medium is used to test for the production of indole and hydrogen sulfide and to detect motility. The medium is inoculated with a straight stab to a depth of about two inches.

To test for indole, an oxalic acid test paper is suspended over the medium at the time of inoculation and held in place by the cotton plug (see Appendix C for preparation of the test paper). The formation of a pink color on the paper indicates indole production.

Motility is evidenced by diffuse growth producing a turbidity throughout the medium. Hydrogen sulfide production is indicated by darkening or blackening of the medium.

MIO Medium (Difco)

This useful medium is used in the identification of enterobacteria on the basis of tests for motility and for ornithine decarboxylase and indole production. The medium is tubed in 5 ml amounts and sterilized. The butt is inoculated with a straight wire to the bottom of the tube. After incubation for eighteen to twenty-four hours at 35–37°C, the motility and ornithine decarboxylase reactions are read. Motility is evidenced by a clouding of the medium extending from the stab line. A positive ornilhine decarboxylase test is indicated by a purple color throughout the medium. Three or four drops of Kovac's reagent are added to the top of the butt, and the tube is shaken gently. The appearance of a pink to red color in the reagent indicates the presence of indole.

Litmus Milk

Overheating will cause caramelization. Some laboratories add 0.1% peptone to litmus milk for the growth of anaerobes.

The important changes seen in this medium are the following:

Acidity. Lactose is fermented, and the indicator becomes red to pink.

Reduction. Indicator turns a pale pink or white.

Coagulation. This is produced by rennetlike enzymes or by acid production.

Peptonization and Digestion. This is produced by proteolytic enzymes. The medium clears as digestion proceeds.

Alkalinity. The indicator turns a darker blue.

Stormy Fermentation. The acid clot is broken up by gas production.

Urease Test Medium (BBL)

This medium, which is prepared according to the formula of Rustigan and Stuart, is suitable for the detection of urease activity of the brucellas.

Urea Agar Base (Difco)

Christensen's medium is used widely for the detection of urease production. A test paper (Patho-Tec*) is available for the rapid detection of urease.

Schaedler Broth (BBL)

This medium, which contains hemin, is useful for the cultivation of fastidious obligate and facultative anaerobes. If used instead of BHI semi-solid medium, 0.15% agar is added. Both of these media are tubed in 8–10 ml amounts prior to sterilization. Swabs can be placed in either medium after plate media have been inoculated.

Nutrient Gelatin (BBL)

The medium should gel at room temperature. The gelatin has been liquefied if the medium remains liquid after being kept in cold water or in the refrigerator until cold.

Stuart's Transport Medium (Oxoid)

Dispense in screw-cap tubes, leaving room for expansion during autoclaving. After cooling, tighten the caps. As a precaution against possible leakage, the caps may be taped.

Convenient vials of suitable transport media are available commercially (Colab) for the submission of clinical specimens, including urine, for the recovery of bacteria, fungi, and viruses.

Blood Culture Media

1. The diphasic medium is prepared by adding 50 ml of tryptose or trypticase broth (containing sodium citrate 5 g/1000 ml) to a five ounce or 150 ml square, clear glass screw-cap bottle containing as a slant 40 ml of tryptose or trypticase soy agar. The broth can be added from a flask or bottle prior to inoculation of the blood.

2. For the growth of anaerobes and various fastidious organisms, a thioglycollate medium is frequently used. This medium (without indicator but with dextrose) containing 0.5% sodium citrate is dispensed in four ounce bottles in 50 ml amounts. It is a good practice to inoculate one-half of the

---

*General Diagnostics Division, Warner-Lambert Co., Morris Plains, New Jersey.

blood specimen (heparinized, EDTA, or in sodium citrate) to No. 1 above and the other half to the 50 ml bottle of thioglycollate. The total blood specimen should be at least 10 ml. This amount of blood will not coagulate in 3 ml of 2% sodium citrate. Other satisfactory anticoagulants are sodium polyanethol sulfonate (Liquoid®*) and ethylenediaminetetraacetate (EDTA). Sterile Vacutainer tubes, containing sodium polyanethol sulfonate solution, are available* with a drawing capacity of 8.3 ml.

Bottles of prepared blood culture media are available commercially. The Liquoid blood culture bottle* containing the anticoagulant sodium poly-anethol sulfonate is reported to be more efficient than conventional media.

Bacteremia Analysis Kit†

The blood to be cultured is first lysed, then passed through a membrane filter to remove bacteria. The membrane is then placed on blood agar and incubated. Colonies develop on the membrane surface.

MEDIA FOR SPECIAL PURPOSES

Streptococcus

Edward's Medium (Modified) (Oxoid)

Beef extract	10 g
Peptone	10 g
Sodium chloride	5 g
Crystal violet (final conc.)	1:750,000
Thallous sulfate or acetate (final conc.)	1:3,000
Esculin	1 g
Agar	15 g
Distilled water	1000 ml

Final pH 7.4

After sterilization, cool to 50°C, add 5–7% sterile whole citrated blood, mix, and pour plates. A blood agar base may be substituted for the first three items and the agar listed in the formula above. The esculin and the following amounts of inhibitory compounds are then added:

Crystal violet	1:1,000 solution	1.3 ml
Thallous sulfate	1:10 solution	3.3 ml

---

*Roche Diagnostics, Nutley, New Jersey.

†Millipore Corporation, New York, New York.

Phenylethyl Alcohol Agar (BBL)

This medium is useful for the isolation of streptococci and micrococci from materials contaminated with Gram-negative organisms, particularly *Proteus*. The latter form visible colonies, but their size and number are much smaller than on ordinary media. *Proteus* does not spread on this medium.

Salt Broth

This medium is used to aid in the identification of streptococci (see Table 17-I). It is prepared by adding sufficient sodium chloride to BHI broth to give a final concentration of 6.5%. It is tubed in 3.0 ml amounts, then sterilized.

Methylene Blue Milk

This medium is used in the differentiation of the streptococci; enterococci and lactic streptococci grow in it. Dissolve in 900 ml distilled water and sufficient skim milk powder (Difco or BBL) to make 1 liter of solution, adjust to pH 7.4, and sterilize at 114°C for fifteen minutes. Add aseptically 100 ml of a 1% (w/v) aqueous solution of methylene blue (certified for bacteriological use), and tube in appropriate amounts.

Camp-Esculin Agar (See Camp Test, Appendix C)

This medium is used for the Camp-Esculin test (see Table 30-I and Fig. 34-1). It consists of blood agar with 0.1% esculin and 0.01% ferric citrate added prior to sterilization. A positive Camp test on this medium is identical to that seen on regular blood agar, viz. a semicircular or arrowhead-shaped zone of clear lysis in the beta or partial zone of hemolysis of the staphylococcus. A positive esculin test is indicated by a browning of the medium around the streptococcal colonies.

Streptosel Agar and Broth (BBL)

This medium is used for the selective recovery of streptococci and *Erysipelothrix insidiosa*. Coliforms, *Proteus, Pseudomonas,* and *Bacillus* spp. are markedly inhibited by sodium azide and crystal violet.

Staphylococcus

Mannitol Salt Agar (BBL)

This is a selective medium that inhibits the growth of almost all organisms but micrococci and staphylococci. Plates are inoculated and incubated for thirty-six hours. Colonies of nonpathogenic cocci are small and surrounded by red or purple zones, while the mannitol-fermenting pathogenic organisms have yellow zones. It should be noted that all acid-producing colonies are not *Staph. aureus*, as some strains of *Staph. epidermidis* and *Staph. saprophyticus* ferment mannitol under aerobic conditions.

Vogel-Johnson Medium (Difco)

This medium is used for the isolation and recognition of coagulase-positive mannitol-fermenting staphylococci. These organisms appear as black colonies surrounded by a yellow zone after twenty-four to forty-eight hours incubation at 37°C. Coagulase-negative, mannitol-negative staphylococci appear as black colonies in a red zone.

Baird-Parker Medium (Difco)

This is a selective medium for the isolation and recognition of coagulase-positive staphylococci. Prior to pouring plates, EY tellurite (tellurite egg yolk emulsion) enrichment is added to the agar base medium. Coagulase-positive staphylococci form black, shiny, convex colonies surrounded by a clear zone. Coagulase-negative staphylococci occasionally grow and produce black colonies with clear zones. They are readily distinguished by the irregular appearance of their colonies and by the wide opaque zones surrounding them.

DNase Test Agar (Difco)

This medium is used to demonstrate the deoxyribonuclease activity of microorganisms. It has been observed that DNase activity correlates closely with coagulase production in strains of *Staph. aureus*. For this reason, some laboratories have tested for DNase activity instead of coagulase production. It should be kept in mind that some strains of *Staph. epidermidis* also produce DNase.

Listeria

To suppress contaminants, sufficient potassium tellurite may be added to blood agar to give a final concentration of 1:5,000.

Erysipelothrix

Blood agar is satisfactory. Sodium azide yielding a final dilution of 1:1,000 may be added to suppress contaminants.

Streptosel Agar and Broth (BBL)

These media (see *Streptococcus* above) are of value in the recovery of *E. insidiosa* from contaminated specimens.

Clostridia

Cooked Meat Medium

Lean beef (or horse meat)	500 g
1 N Sodium hydroxide	25 ml
Distilled water	1000 ml

Remove as much fat and connective tissue as possible before grinding. After grinding, mix meat, water, and sodium hydroxide solution. Bring to boiling and stir. Cool to room temperature, then filter through several layers of gauze. Add water to restore to one liter. Then add

Peptone or trypticase	30 g
Yeast extract	5 g
Potassium phosphate (K_2HPO_4)	5 g

Adjust pH to 7.8 with 1 N NaOH. Dispense in 15 × 135 mm screw-cap tubes, using one part meat particles to three to four parts liquid (v/v) per tube. The total volume should be about 8 ml. Add a few iron filings to each tube, then autoclave at 121°C for fifteen minutes.

This medium is preferred to the dehydrated cooked meat preparations for the cultivation of the more fastidious anaerobes.

Cooked Meat Medium (Difco)

The most convenient way to prepare this medium is to add to each screw-cap tube 1.25 g of the dehydrated product. Ten milliliters of cold distilled water is then added to each tube. Sterilize at 121°C for fifteen minutes. Prior to use, the medium should be heated in a boiling water bath for ten minutes to remove oxygen.

Both of the meat media described above can be used to maintain stock cultures of anaerobes.

Egg Yolk Agar

Prepared plates are available from Gibco. This medium is used to demonstrate lecithinase, lipase, and proteolytic enzymes of clostridia. Lecithinase diffuses into the agar, producing a zone of opalescence around colonies. This can be specifically inhibited by the appropriate antiserum spread over half of the plate and dried before inoculation.

Egg yolk agar can be prepared by adding Oxoid egg yolk emulsion to Oxoid Columbia blood agar base (plus 0.5% NaCl) at the rate of 0.5–1.0 ml of emulsion to 10 ml of blood agar base (3). The egg emulsion is added to the base medium after the latter has been sterilized and cooled to 60°C.

Media for Fermentation Tests

The trypticase media described above may be used and also fluid thioglycollate medium (without dextrose and indicator). The latter medium is preferred for the more fastidious anaerobes. The filtered carbohydrate solutions are added to the latter medium prior to inoculation. If the carbohydrates are added before sterilization, autoclave carefully for fifteen minutes at 116–118°C. After incubation of the thioglycollate fermentative media, the presence or absence of acidity is determined by the addition of several drops of 0.2% aqueous solution of bromthymol blue.

Potassium nitrate (1%) may be added to the thioglycollate medium if nitrate broth is desired; lead acetate and oxalic acid test papers may be suspended over cultures for hydrogen sulfide and indole detection, respectively.

Obtaining an Atmosphere with Increased Carbon Dioxide

The simplest way of obtaining approximately 5% carbon dioxide is by the use of the candle jar. Media are placed in a screw-cap jar, after which a lighted candle is placed in the jar. The jar is sealed, and the candle, after burning to extinction, leaves about 5% carbon dioxide. A microaerophilic atmosphere suitable for the growth of many organisms can be obtained by simply replacing 10% of the volume of air with carbon dioxide.

The growth of many organisms is stimulated by 5–10% carbon dioxide. *Campylobacter fetus* subsp. *fetus* requires an atmosphere in which the oxygen is reduced by at least one-third. A mixture of 10% carbon dioxide, 5% oxygen, and 85% nitrogen is satisfactory. Replacement is accomplished by the water displacement method or by the use of a manometer.

Enteric Bacteria

MacConkey's Agar

This medium is usually put up in flasks in convenient amounts. Plates and slants are poured from the melted medium as required. Lactose-fermenting enteric bacteria produce red colonies, while non-lactose-fermenters do not. Included among those producing pale, colorless colonies are *Salmonella*, *Proteus* spp., and *Alcaligenes faecalis*.

Preparations without crystal violet (BBL) allow the growth of the enterococci, which appear as pink or colorless pinpoint colonies.

Selenite Broth, PLATE

This medium, which should not be autoclaved, is dispensed in screw-cap tubes to a depth of two inches. They are placed in flowing steam for thirty minutes. This is a selective, or enrichment, medium for the isolation of *Salmonella* and *Shigella. Proteus* and *Ps. aeruginosa* are not inhibited. Plate to solid media after not more than eighteen to twenty-four hours incubation.

Tetrathionate Broth

The base medium can be stored indefinitely, but after the addition of iodine, it should be used the same day. This is a selective and enrichment medium for the isolation of *Salmonella* and *Shigella.* Plate to solid media after not more than eighteen to twenty-four hours incubation.

SS Agar

This medium is used for direct plating and also for subculturing. It favors the growth of *Salmonella, Shigella,* and *Yersinia* while inhibiting Gram-positive organisms and many coliforms. Non-lactose-fermenters produce colorless colonies, while the lactose fermenters produce red or pink colonies.

Eosin-Methylene Blue Agar (without sucrose)

This medium is used for direct plating and also for subculturing from selenite and tetrathionate broths. Colonies have the following appearances:
E. coli: Colonies have a metallic sheen.
Enterobacter, Aerobacter, and *Klebsiella:* Brownish, frequently convex and mucoid with a tendency to coalescence.
Salmonella and *Shigella:* Transparent, amber to colorless.

Brilliant Green Agar

This is a highly selective medium recommended for the isolation of salmonellae directly from feces, less commonly from tissues, and also from previously inoculated enrichment broths. Large inocula may be used because of the strong selective capacity of this medium for salmonellae. *Salmonella* colonies appear as slightly pink white, opaque colonies surrounded by a brilliant red zone. Lactose– or sucrose-fermenting organisms produce colonies that are yellow green and surrounded by an intense yellow green zone.

XLD Agar (Gibco)

This xylose lysine desoxycholate agar is used for the isolation of enteric pathogens including *Salmonella* and *Shigella.*

Triple Sugar Iron Agar

This medium is dispensed in tubes that are one-third filled. After autoclaving, they are allowed to cool in the slanted position, the slant being such that a deep butt as well as a slant is obtained. Slants are inoculated from selected colonies. The slant is streaked and the butt stabbed. Observations are made after eighteen to forty-eight hours incubation.

Simmons Citrate Agar

Those organisms capable of utilizing citrate as a sole source of carbon will grow on this medium. Visible growth is usually accompanied by an alkaline (blue) change of the indicator.

Inhibition of Proteus Swarming

The separation of various bacteria from *Proteus* spp. is a recurring problem in the diagnostic laboratory. Contamination by *Proteus* is frequently due to the careless handling of specimens at necropsy. The procedures described below have been found effective.

GENERAL: Swarming is inhibited if the agar concentration in blood agar is increased to 4%.

GRAM-POSITIVE ORGANISMS: Plate on a medium consisting of one-half phenyl ethyl alcohol agar and one-half trypticase soy agar. This medium inhibits the growth of *Proteus* and other Gram-negative bacteria. Sodium azide (1:5000) will also prevent swarming.

GRAM-NEGATIVE ORGANISMS: MacConkey's agar and deoxycholate-citrate agar inhibit swarming. The inhibition by the former medium is increased if the agar concentration is raised to 4%.

Tergitol 7 Agar (BBL)

This is a selective medium for the recovery of coliforms. It has been employed in the diagnosis of colibacillosis. *Escherichia coli* produces yellow colonies with yellow zones, while the colonies of *Enterobacter* are greenish yellow. The spreading of *Proteus* is inhibited.

Rappaport Medium (4)

This is used as an enrichment medium for *Salmonella choleraesuis* and *S. pullorum*, both of which can be inhibited by tetrathionate broth.

It can be obtained commercially or prepared as follows:

SOLUTION A   Bacto tryptone (Difco or Oxoid equiv.)   5 g
                   Sodium chloride   8 g
                   Potassium dihydrogen phosphate   1.6g

	Distilled water	1000	ml
SOLUTION B	(Kept as Stock)		
	Magnesium chloride ($MgCl_2 \cdot 6H_2O$ A.R.)	40	g
	Distilled water — *make up to*	100	ml
SOLUTION C	(Stock)		
	Malachite green (Merck)	0.4g	
	Distilled water	100	ml
FOR USE:			
	SOLUTION A	100	ml
	SOLUTION B	10	ml
	SOLUTION C	3	ml

Distribute the medium in 10 ml volumes and sterilize by steaming for thirty minutes.

Media For Yersinia enterocolitica

M/15 Phosphate Buffered Saline (Oxoid) is used as an enrichment method. Specimens are added to the buffer and incubated for four to six weeks at 4°C.

SS Agar (Oxoid) is made selective for *Y. enterocolitica* by the addition of 2% sodium desoxycholate and alteration of the pH to 7.6 prior to sterilization by boiling. Food, feces, etc., can be directly plated on this medium. Details on the use of this medium and the appearance of colonies of *Y. enterocolitica* are given in *The Oxoid Manual.*

Pseudomonas

Seller's Differential Agar (Difco)

After rehydration of the medium by boiling, it is dispensed in tubes that are one-third filled. Autoclave for ten minutes at fifteen pounds pressure (121°C) and cool in the slanted position. The slant should provide a deep butt (one and one-half inches) as well as a slant (three inches). Immediately before inoculating, add 0.15 ml, or two drops, of a sterile 50% glucose solution to each tube by letting it run down the side of the tube opposite the slant. Inoculum is streaked on the slant, then stabbed into the butt. For differential reactions, see Table 5-I.

Cetrimide Agar (Pseudosel® Agar—BBL)

This is a selective medium for the isolation of *Pseudomonas aeruginosa*. The medium favors the production of pyocyanin and fluorescin. Most other organisms are completely inhibited by the quaternary ammonium compound cetyl trimethyl ammonium bromide (cetrimide).

Pasteurella and Yersinia

Three selective media for the isolation of *P. multocida, P. haemolytica,* and *Yersinia pseudotuberculosis,* respectively, have been described by Morris (5). See Chapter 11 for additional selective media.

Francisella

Blood Cystine Dextrose Agar (Difco)

This medium is used for the isolation of *Francisella tularensis.* After sterilization, the medium is cooled to 60–70°C, and 18 ml of whole rabbit blood is added for each 300 ml of medium. After mixing well, plates and slants are poured.

Bordetella

MacConkey's agar with 1% dextrose added is recommended as a selective medium for the isolation of *Bord. bronchiseptica* from the nasal passages of pigs (6). Small, grayish tan colonies with dark centers are produced in forty-eight hours. This medium was altered by the addition of 20 µg/ml of furaltadone (7) and later by adding 200 µg/ml of nitrofurantoin (8).

A more complex selective medium was used by Rutter (9).

Smith-Baskerville Selective Medium (10).

This medium makes possible the identification of *Bord. bronchiseptica* on primary culture. It is prepared as follows:

Bacto-peptone	20 g
NaCl	5
Agar	15
Distilled water	857 ml

The pH is adjusted to 7.2 and these ingredients are autoclaved at 121°C for fifteen minutes. The mixture is cooled and the following added:

A.	Gentamycin	0.5	µg/ml
	Penicillin	20	µg/ml (final
	Furaltadone	20	µg/ml conc.)
B.	Glucose (10% stock)	100	ml
	Lactose (10% stock)	100	ml
C.	Bromthymol Blue (0.2%)	40	ml

A stock solution of bromthymol blue is made as follows:

Bromthymol blue	1 g
0.1 N NaOH	25 ml
Distilled water	475 ml
Optional: Amphotericin B	10 µg/ml

The medium is dispensed in conventional Petri plates.

Brucella

Brucella spp. grow well on many media that support the growth of fastidious organisms, e.g. brucella agar (Albimi), tryptose agar (Difco), trypticase soy agar (BBL), and blood agar.

A Brucella Differentiation Kit, which includes media containing basic fuchsin and thionin, is available commercially (BBL).

"W" Medium (11)

To a liter of Albimi or Tryptose agar add:

Polymixin B sulfate	6,000 units
Actidione	100 mg
Bacitracin	25,000 units
Circulin	1,500 units
Crystal violet	1.4 mg

Add the crystal violet prior to sterilization by autoclaving at fifteen pounds for twenty minutes. The antibiotics are added after autoclaving.

This formula was modified by Weed (12) by leaving out the crystal violet (may inhibit the growth of some *Brucella* strains) and enriching with 5% blood.

Campylobacter

Blood Agar

The addition of mycostatin (300 units/ml) and brilliant green (1:40,000) may be helpful in the isolation of *Campylobacter fetus*. Some workers also recommended the addition to blood agar of 0.1% sodium thioglycollate. Plates should be fresh and preferably of greater than average thickness. Bacitracin, 2 units/ml, and novobiocin, 2µg/ml, have been incorporated to suppress contaminants.

Other media for the isolation of *C. fetus* have been described by Kuzdas and Morse (13) and Morgan (14).

Haemophilus

Chocolate Agar

Melt a flask of tryptose agar and cool to 50°C, then add 5–7% sterile citrated blood. Mix well and heat to 75–80°C in a water bath. Agitate the flask gently at short intervals without producing bubbles. When the blood is browned, the flask is cooled to 50°C and plates are poured.

The X-factor is supplied by the heated blood. The V-factor can be supplied by a growing culture of a staphylococcus, by the addition of Bacto-Supplement B® (Difco), or by the addition of 5–10% fresh yeast extract. The X– and V-factors can also be supplied through the use of impregnated discs.

Blood Agar

This medium, with the V-factor if required, supports the growth of the *Haemophilus* spp. recovered from animals.

X and V Factor Discs

The preparations described below are those of Marshall and Kelsey (15).

X-FACTOR. The red cells from 40 ml of blood are separated by centrifugation. To the red cells is added 100 ml of acetone containing 1.2 ml of concentrated hydrochloric acid. Filter, then add to the filtrate 100–120 ml of water to precipitate the hemin. The hemin is collected by filtration and washed with water, then dissolved in 25 ml of 0.1 M sodium phosphate (Na$_2$HPO$_4$) and sterilized at 115°C for ten minutes.

V-FACTOR. Fifty grams of yeast are suspended in 100ml 0.2M KH$_2$PO$_4$ and heated at 80° C for twenty minutes. Centrifuge to remove cells and sterilize by filtration. Store in freezer or refrigerator.

Discs 1 cm in diameter are cut from filter or other suitable absorbent paper. They are soaked in each of the above solutions or in a mixture of both. They are dried and stored in the refrigerator.

The discs are used on media that do not contain X– and V-factors. They are placed some distance from each other or placed about 2 cm apart if it is desired to see if both factors are required.

Acinetobacter

Herellea Agar (Difco)

This is a selective medium for the isolation of *Acinetobacter lwoffii* and *A. calcoaceticus*. Gram-positive organisms are inhibited, and Gram-negative fermenting bacteria can be distinguished from the nonfermenting *Acinetobacter calcoaceticus* by color differences.

Leptospira

Fletcher's Medium Base (Difco)

Used for the isolation of *Leptospira*.

Stuart's Medium Base (Difco)

Used for the isolation and maintenance of *Leptospira*.

Media for Susceptibility Tests

A number of media are available for use in conducting susceptibility tests. Mueller-Hinton agar is recommended for routine use with the disc method. For fastidious organisms, 5% blood may be added.

Mueller-Hinton Agar (BBL)

This is a beef infusion medium containing 0.15% starch. Mueller-Hinton broth is available for the broth dilution test tube method.

MEDIA FOR SUSCEPTIBILITY TESTING

Mueller-Hinton agar is currently the only agar recommended for antimicrobial susceptibility testing of aerobes by the disc diffusion method. It can be enriched by supplementation in order to enhance the growth rate of fastidious organisms. Supplements recommended (16) include 5% defibrinated whole animal blood (which may be chocolatized) or 1% hemoglobin plus 1% of a synthetic stimulant mixture (Iso Vitale X, BBL Microbiology Systems, Cockeysville, Md., and Supplement VX, Difco Laboratories, Detroit, Mi., are examples of the latter).

Although a few fastidious microorganisms that fail to grow well on supplemented Mueller-Hinton agar will grow adequately on trypticase soy and brain-heart infusion agars supplemented with blood and vitamins, such media should not be employed for routine disc susceptibility assays (16). Uncontrolled interactions with antimicrobial agents affect inhibition zone diameters in agars other than Mueller-Hinton, so that Bauer-Kirby interpretive criteria are inapplicable to them (17).

The recommended composition of Mueller-Hinton agar is as follows:

Dehydrated infusion from beef	300.0 g
Acid digest of casein	17.5 g
Corn starch	1.5 g
Agar	17.0 g
Distilled water to	1000 ml

Mix, heating if necessary. Autoclave at 116–121°C (12–15 pounds pressure) no longer than fifteen minutes. The pH subsequent to autoclaving should be

7.2–7.4. Due to the influence of magnesium and calcium ions on the susceptibility of *Pseudomonas aeruginosa*, the medium should contain 20–35 mg of magnesium and 50–200 mg of calcium per liter (16). Any attempts to adjust the pH must be done prior to autoclaving, using 1 N NaOH or 1 N HCl. Dehydrated Mueller-Hinton agar of satisfactory quality is available commercially from various sources. Commercial Mueller-Hinton broth for tube dilution methods is also available.

NON-SPOREFORMING ANAEROBIC BACTERIA

Modified Cary and Blair Transport Medium

Cary and Blair Medium (BBL)	2.5 g
CaCl$_2$, 1% solution	1.8 ml
Resazurin solution (0.1% w/v)	0.1 ml
L-Cysteine HCL	0.1 g
Distilled water	198.0 ml

Heat ingredients in a flask with glass beads until the agar has dissolved. Cool to 50°C in an atmosphere of O$_2$-free CO$_2$. Add cysteine. Adjust pH to 8.4. Continue to gas with O$_2$-free N$_2$, to maintain pH. Dispense 10 ml into tubes approximately 16 × 125 mm while gassing both tubes and flask with N$_2$ (or carry out this stage in an anaerobe chamber). Stopper with airtight butyl rubber stoppers and screw-caps. Steam in a press for fifteen minutes. Discard any tubes showing a pink color (i.e. not anaerobic).

Vitamin K$_1$—Hemin Supplement

1. Stock hemin solution: Dissolve 50 mg hemin in 1 ml 1 N NaOH. Add 100 ml distilled water, autoclave at 121°C for fifteen minutes.
2. Stock menadione (Vitamin K$_1$) solution: Dissolve 100 mg menadione in 20 ml 95% ethanol and filter sterilize.

Add 1 ml sterile menadione to 100 ml sterile hemin solution; add 1 ml Vitamin K-hemin solution per 100 ml sterile medium.

Cooked Meat Carbohydrate Medium (Anaerobes)

Chopped meat (BBL): Prepared as on manufacturer's directions, add 0.4% (w/v) glucose, 0.1% (w/v) cellobiose, 0.1% (w/v) maltose, and 0.1% (w/v) starch.

CHO Medium (Anaerobes)

CHO base (Difco)	26.0 g
Distilled water	900.0 ml

Mix and heat to dissolve. Autoclave at 121°C for fifteen minutes. Cool to 50°C; add 100 ml sterile distilled water for base control or 100 ml filter sterilized carbohydrate solution (6% w/v). Mix well, adjust pH to 7.0. Dispense 7 ml in 15 × 90 ml screw-cap tubes. Before use store twenty-four to forty-eight hours in anaerobic conditions with loosened caps, or boil for ten minutes, cool, and inoculate immediately. CHO fermentation medium can be prepared by making thioglycollate medium (BBL 135-C) without dextrose or indicator and adding 2 gm/l of yeast extract (Difco) and 1 ml of 1% aqueous bromthymol blue solution.

Esculin Broth (Anaerobes)

Dissolve ingredients by boiling, adjust pH to 7.2, dispense 7 ml into 15 × 90 mm screw-cap tubes. Autoclave for fifteen minutes at 121°C. Store for twenty-four to forty-eight hours anaerobically before use, or boil for ten minutes with loosened caps and inoculate immediately.

Indole-Nitrate Broth (Anaerobes)

Indole-nitrate medium (BBL)	25.0 g
Distilled water	1000.0 ml

Dissolve by boiling for one minute with frequent mixing, adjust pH to 7.2, dispense 7 ml into 15 × 90 mm screw-cap tubes. Autoclave for fifteen minutes at 121°C. Store at 4°C. Before use store twenty-four to forty-eight hours in anaerobic environment with loosened caps, or boil for ten minutes with loosened caps and inoculate immediately.

Thiogel medium (anaerobes)

Thiogel medium (BBL)	90.0 g
Distilled water	1000.0 ml

Heat water to 50°C, add Thiogel medium and let stand for five minutes. Boil for one minute with mixing and dispense in 7 ml aliquots into 15 × 90 mm screw-capped tubes. Autoclave for fifteen minutes at 118°C. Store at 4°C. Before use, store twenty-four to forty-eight hours in anaerobic environment with loosened caps, or boil for ten minutes, cool, and inoculate immediately.

Peptone-Yeast-Glucose Medium

A. Salts solution:

CaCl$_2$ (anhydrous)	0.2 g
MgSO$_4$	0.2 g
K$_2$HPO$_4$	1.0 g
KH$_2$PO$_4$	1.0 g
NaHCO$_3$	10.0 g
NaCl	2.0 g

Mix and dissolve $MgSO_4$ and $CaCl_2$ in 300 ml distilled water. Add 500 ml and slowly add remaining salts. Continue with swirling until salts are dissolved. Add 200 ml distilled water, mix, and store at 4°C.

B. Resazurin solution:

Dissolve one 11 mg resazurin tablet (Allied Chemical Co.) in 44 ml distilled water, or 25 mg powder (Baker, Difco) in 100 ml distilled water.

C. Peptone-yeast-glucose medium:

Peptone	20.0 g
Yeast extract (Difco)	10.0 g
Salts solution	40.0 ml
Cysteine HCl	0.5 g
Resazurin solution	4.0 ml
Distilled water	1000.0 ml

Adjust pH to 7.2, dispense 5 ml in 15 × 125 mm screw-cap tubes, autoclave at 121°C for fifteen minutes.

REFERENCES

1. Cowan, S. T.: *Manual for the Identification of Medical Bacteria*, 2nd ed., Cambridge, Cambridge U Pr, 1974.
2. Hugh, R., and Leifson, E.: *J Bacteriol*, 66:24, 1953.
3. Holdeman, L. V., Cato, E. P., and Moore, W. E. C. (eds.). *Anaerobe Laboratory Manual*, 4th ed. Blacksburg, Va. V.P.I. Anaerobe Laboratory, 1977, p. 145.
4. Iveson, J. B., and Kovacs, N.: *J Clin Path*, 20:290, 1967.
5. Morris, E. J.: *J Gen Microbiol*, 19:305, 1958.
6. Ross, R. F., Switzer, W. P., and Maré, C. J.: *Vet Med*, 58:562, 1963.
7. Farrington, D. O. and Switzer, W. P.: *J Am Vet Med Assoc*, 170:34, 1977.
8. Fuzi, M.: *Bakt Parasit Infect Hyg Abt 1 Orig*, 231:466, 1975.
9. Rutter, J. M.: *Vet Rec*, 108:451, 1981.
10. Smith, I. M., and Baskerville, A. J.: *Res Vet Sci*, 27:187, 1979.
11. Kuzdas, C. D., and Morse, E. V: *J Bact*, 66:502, 1953.
12. Weed L. A.,: *Am J Clin Path*, 72:482, 1957.
13. Kuzdas, C. D., and Morse, E. V.: *J Bacteriol*, 71:251, 1956.
14. Morgan, W. J. B.: *Vet Rec*, 69:32, 1957.
15. Marshall, J. H., and Kelsey, J. C.: *J Hyg (Camb)*, 58:367, 1960.
16. Lennette, E. H. (Ed.-in-chief): *Manual of Clinical Microbiology*, 3rd ed. Washington, D.C. American Society for Microbiology, 1980, Chapter 44.
17. Lorian, V. (Ed.): *Antibiotics in Laboratory Medicine*. Baltimore, Williams & Wilkins, 1980, Chapter 2.

SUPPLEMENTARY REFERENCES

BBL Manual of Products and Laboratory Procedures, 5th ed. Baltimore, Maryland, Bio-Quest, Division of Becton, Dickinson and Co., 1968.

Difco Supplementary Literature. Detroit, Michigan, Difco Laboratories, 1968.

Products for Microbiology — Technical Manual. Gibco Laboratories, Madison, Wisconsin.

The Oxoid Manual, 4th ed. Oxoid U.S.A., Inc., Columbia, Maryland, 1980.

Appendix C

REAGENTS AND TESTS

Nitrate Reduction Test

Reagents

Solution 1

    Alpha-naphthylamine   5 g

    5 N acetic acid (sp. gr. 1.041)   1000 ml

  Filter through clean absorbent cotton.

Solution 2

    Sulfanilic acid   8 g

    5 N acetic acid (sp. gr. 1.041)   1000 ml

Procedure

Add to 5 ml of trypticase nitrate broth culture 1 ml of solution 2, followed by 1 ml of solution 1 added drop by drop. If nitrite is present, a red, pink, or maroon color develops.

In the test for nitrate reduction, if the test is negative, either the nitrate has not been broken down or nitrite has been broken down further. Whether or not nitrate is present can be determined by adding a small amount of zinc dust. The amount is critical, so the same amount of zinc dust should be added to an uninoculated control tube. If nitrate is present, a red color is obtained as a result of the zinc reducing the nitrate to nitrite. If the zinc test does not indicate the presence of nitrate and the test for nitrite is negative, the nitrite has been broken down further and the organism is considered to give a positive nitrate reduction test.

Oxalic Acid Test Paper for Indole

Soak a piece of filter paper in saturated oxalic acid solution. Dry and cut into strips approximately 10 mm × 50 mm. These are suspended over the medium, e.g. SIM medium, trypticase, nitrate broth, and held in place by the cotton plug.

Lead Acetate Test Papers

Strips of filter paper (5 × 50 mm) are impregnated with saturated lead acetate solution, then dried in an oven at 70°C. These are suspended over a suitable medium and held in place by the cotton plug. Suitability of a medium can be determined by trial with a hydrogen sulfide producing strain. This test is many times more sensitive than the use of lead acetate agar or triple sugar iron agar. The production of hydrogen sulfide is only meaningful if the test procedure is given.

Solubility Test for Streptococcus pneumoniae

Prepare a 10% aqueous solution of sodium desoxycholate containing merthiolate 1:10,000 as preservative.

Procedure

Two drops of the sodium desoxycholate solution are added to 1 ml of a twenty-four-hour culture. If the organisms are pneumococci, clearing occurs rapidly, usually in less than a minute. Hold for ten to fifteen minutes before discarding as negative.

Coagulase Tests for Staphylococci

Production of coagulase by staphylococci is considered an important criterion of pathogenic potentiality. Fresh or commercially available lyophilized plasma may be used. Both rabbit and human plasma are satisfactory. Rabbit plasma may be obtained from blood taken sterilely by cardiac puncture. The blood is transferred quickly to a bottle or tube containing sodium citrate solution (10 ml blood to 1 ml of 5% citrate).

Procedure

To two drops of an overnight broth culture in a Kahn or similar tube, add 0.5 ml plasma diluted 1:5 with sterile physiological saline; mix. Two drops of a saline suspension of organisms from a solid medium are also suitable. Bring to 37°C in the incubator, or place in a 37°C water bath. If the strain is coagulase-positive, the plasma will clot, usually within three hours. Partial clotting is considered a positive test. Plasma should be checked before use with a known coagulase-producing staphylococcus.

A slide procedure is used as a presumptive test. Sufficient bacteria are emulsified in a drop of water on a microscope slide to yield a thick suspension. The suspension is stirred with a straight wire that has been dipped in suitable plasma. A positive reaction is indicated by clumping within five seconds. The factor tested for is sometimes referred to as the "clumping factor" rather than coagulase.

Methyl Red Test

Reagent

The methyl red solution is prepared by dissolving 0.1 g of methyl red in 300 ml of 95% ethyl alcohol and diluting to 500 ml with distilled water.

Procedure

To 5 ml of culture (incubated five days at 37°C in MR–VP broth) add five drops of methyl red solution. A positive reaction is indicated by a distinct red color, indicating acidity (pH 4.4–6.0). A yellow color constitutes a negative reaction.

Voges-Proskauer Test

This test is for acetylmethylcarbinol.

Reagents

Solution 1
    5% alpha-napthol in absolute ethyl alcohol.
Solution 2
    40% potassium hydroxide containing 0.3% creatine.

Procedure

Transfer 1.0 ml of a forty-eight-hour culture (37°C) grown in MR–VP broth to a Wasserman tube and add 0.6 ml of solution 1, then 0.2 ml of solution 2. Shake well and leave five to ten minutes. A bright orange red color develops and gradually extends throughout the broth if acetylmethylcarbinol has been produced.

An alternative procedure is to add 5 ml of 10% potassium hydroxide to 5 ml of culture. Mix well and allow to stand exposed to air. Observe at intervals of two, twelve, and twenty-four hours. A positive test is indicated by the development of an eosin pink color.

Indole Test

Reagent

p-dimethylaminobenzaldehyde	2 g
Ethyl alcohol, 95%	190 ml
Hydrochloric acid (conc.)	40 ml

Procedure

One milliliter of ether is added to a 5 ml portion of a forty-eight-hour culture of organisms (media: trypticase nitrate broth, peptone water, or beef-heart infusion broth). Shake well, then allow to stand until the ether rises to the top. Gently run 0.5 ml of the reagent down the side of the tube so that it forms a ring between the medium and the ether. If indole has been accumulated by the ether, a brilliant red ring will develop just below the ether layer. If there is no indole, no color will develop.

Oxidase Test

This is a test for production of cytochrome oxidase.

Reagent

Prepare a 1% solution (0.1 g in 10 ml distilled water) of *p*-aminodimethyl-aniline monohydrochloride or 0.5% of N,N,N′,N′-tetramethyl-*p*-phenylenedi-amine dihydrochloride.*

The dye is added to the distilled water and allowed to stand for fifteen minutes prior to using. The solution should be kept in a brown bottle and refrigerated. Solutions of the first compound are satisfactory for approximately a week, while solutions of the latter compound may be used for up to a month.

Procedure

The dye solution is added to portions of plate cultures containing suspicious colonies. Colonies producing cytochrome oxidase become pink, changing to red, then finally to black when solutions of the first compound are used. Solutions of the other compound stain cytochrome-oxidase-positive colonies a dark purple. Colonies treated with the dye can be used for Gram's stain.

Test papers are available commercially for the cytochrome oxidase test.[†] These are not considered as sensitive as the plate test.

Oxidase differentiation discs (Difco) containing *p*-aminodimethylaniline are available. They are employed by adding a suspension of the test organism to the disc or by placing the disc on the medium adjacent to isolated colonies. In the latter procedure, a drop of sterile distilled water is added to the disc. Diffusion of the reagent will turn oxidase-producing colonies a pink color that becomes black on standing. Oxidase-negative organisms are unchanged.

---

*N,N,N′,N′-tetramethyl-*p*-phenylenediamine dihydrochloride: Eastmen Organic Chemicals, Rochester, New York.

[†]Warner-Chilcott, Morris Plains, New Jersey.

Catalase Test

A slant culture is used. One milliliter of a 3% solution of hydrogen peroxide is poured over the growth. The slant is tilted so that the solution covers the growth. A rapid ebullition of gas indicates a positive reaction.

The test can also be performed transferring a small amount of growth from solid medium, preferably a single colony, to a microscope slide. A drop of fresh hydrogen peroxide (3%) is added, then a coverslip is applied. The production of gas bubbles constitutes a positive reaction.

The following procedure has been used for the demonstration of catalase production by vibrio. Organisms are grown for three days in semisolid medium, after which 5.0 ml of 3% peroxide is added to the culture. After mixing by inversion, the tube is observed for the presence of gas bubbles. A culture is considered positive if 2–3 mm of gas bubbles are produced.

Camp Test

This test is used for the presumptive identification of *Str. agalactiae*. This species is able to complete the partial lysis of red cells produced by the beta-hemolysin of a staphylococcus strain. Several techniques are employed. A beta-hemolytic staphylococcus can be streaked heavily across a blood plate, or beta-hemolysin can be supplied by a filtrate of a staphylococcus broth culture. If the latter procedure is used, a channel several millimeters wide is cut traversing the center of the plate. Into this channel is pipetted the staphylococcus filtrate. Strains of streptococci to be examined are streaked at right angles up the channel, or to the staphylococcus streak line.

Occasional strains of *Str. dysgalactiae* and *Str. uberis* will also give positive reactions. The latter splits esculin, while the former and *Str. agalactiae* do not. *Str. dysgalactiae* does not coagulate litmus milk, while the other two species do. Colonial appearances on Edward's medium are also differential.

The procedure outlined below has also been found satisfactory.

1. Draw a line across the center of a blood agar plate.
2. Streak the streptococcus culture at right angles and across the line.
3. Streak a beta-hemolytic *Staph. aureus* culture across the plate, directly over the line.
4. Incubate for eighteen to twenty-four hours and observe. If the streptococcus is a Group B, it produces a characteristically shaped clear zone of hemolysis in the beta zone of hemolysis produced by the staphylococcus (see Fig. 34-1).

Lancefield Grouping

The strain to be grouped is inoculated into 50 ml of broth (tryptose phosphate or other suitable broth) and incubated overnight. The bacteria are removed by centrifugation, and the deposit is suspended in 2 ml of N/20 hydrochloric acid in 0.85% sodium chloride solution (conc. HCl is approximately 12 N; a 1:200 dilution in physiological saline is satisfactory). The suspension should turn congo red paper blue (pH 3 or less); if it does not, add N/1 HCl until it does. Heat in a boiling waterbath for ten minutes. Remove the supernatant fluid and add several drops of 0.02% phenol red solution. Carefully add sufficient N/1 NaOH to turn the indicator red. Centrifuge, then remove the supernatant for use in the test.

Extracts and sera should be crystal clear. The latter can be clarified either by centrifugation or by filtration through a miniature paper-pulp filter. The filter is prepared by packing the neck of a Pasteur pipette with shreds of filter paper. Tests are carried out in capillary tubes or in small test tubes. In the latter, a small amount of extract is layered over a small amount of each serum. A positive reaction consists of the formation in ten to thirty minutes at room temperature of a white, flocculent precipitate at the junction of the extract and the homologous serum.

Test for Growth of Streptococci in 0.1% Methylene Blue Milk

Inoculate a tube of 0.1% methylene blue milk (see Appendix B) from a broth culture of the streptococcus. Incubate at 37°C for forty-eight hours. Examine for reduction of the methylene blue at twenty-four and forty-eight hours. All tests should be discarded at forty-eight hours, as continued incubation may produce irregular results.

Test for Ammonia Production

The strain to be tested is grown in peptone water for five days at 37°C. Nessler's reagent is then added. A brown color indicates ammonia production, while a faint yellow color is negative.

Nessler's Reagent

Dissolve 45.5 g of mercuric iodide (NF grade) and 34.9 g of potassium iodide in 500 ml of water in a 1 liter volumetric flask. Then add 200 ml of 10 M potassium hydroxide solution; mix. Dilute to 1 liter, and mix again. If a precipitate forms, let it settle and decant the clear solution.

Methylene Blue Reduction

Sufficient amount of a 1% aqueous solution of methylene blue is added to a twenty-four-hour broth culture to give a ratio of one part reagent to fifty parts culture. Incubate at 37°C and observe hourly for evidence of reduction (decolorization).

Phenylalanine Deaminase Test

Procedure

Apply a heavy loopful of the culture to one of the two reagent zones of the test paper* by rubbing with the loop for fifteen seconds to bring the culture into intimate contact with the reagents. The second reagent zone is used to run a negative control. Positive reaction: A darkened stain develops in the area within five to ten minutes. Color may vary from brownish to gray black.

Phenylalanine Agar (Difco)

Phenylalanine deaminase can be demonstrated in this medium. The test organism is grown on a slant of phenylalanine agar at 35–37°C for eighteen to twenty-four hours. Production of phenyl pyruvic acid from phenylalanine is determined by adding five drops of the test reagent to the growth on the slants in such a way as to cover the slant and loosen the growth. The production of a green color in one to five minutes indicates a positive test. The test reagent is prepared by dissolving 2.0 g of ammonium sulfate and 1.0 ml of 10% sulfuric acid in 5 ml of a half-saturated solution of ferric ammonium sulfate.

Decarboxylation of Arginine, Ornithine, and Lysine

Decarboxylase Test Medium, Moeller (Difco)

This is a base medium to which lysine, arginine, and ornithine or other amino acids may be added to demonstrate decarboxylase activity. Ten grams of the amino acid to be tested is added to 1 liter of the basal medium. Where DL-amino acids are employed, 2% concentration is used rather than 1%. No further adjustment of reaction will be required when lysine or arginine are used. Ornithine, being highly acidic, requires the readjustment of pH with sodium hydroxide before sterilization. Usually 1 liter of the medium in which 10 g L-ornithine has been dissolved requires the addition of 4.6 ml 10 N sodium hydroxide. Dispense the medium in 5 ml amounts into screw-cap tubes and sterilize in the autoclave for ten minutes at fifteen pounds pressure (121°C).

Inoculate the prepared media lightly from a twenty-four-hour agar slant culture. Aseptically overlay the inoculated broth and controls with 4–5 ml sterile mineral oil. Incubate at 35–37°C for four days. Positive decarboxylase reactions will be indicated by a change in the indicator to a reddish

---

*Warner-Chilcott Laboratories, Morris Plains, New Jersey.

violet color. A negative reaction is indicated by a yellow color in the tube.

The reaction of the medium containing the amino acids should be pH 6.0 after sterilization.

A test paper is available for the detection of lysine decarboxylase.*

Lysine Iron Agar (Difco)

This medium, which is tubed with a butt and a slant, is used in the differentiation of the enteric bacteria. Lysine decarboxylase production is evidenced by an alkaline reaction. Dextrose fermentation (yellow) and hydrogen sulfide production (black) are also shown in this medium.

The usual reactions with lysine iron agar after forty-eight hours are the following:

Salmonella	K/KG H_2S
Arizona	K/KG H_2S
Citrobacter	K/AG or K/AG H_2S
Providencia	K/AG or K/A
Proteus	K/AG, K/AG H_2S, or K/A
Shigella	K/A
Escherichia	K/A or N, K/AG, K/K, or K/KG
Enterobacter	K/AG or K/KG
Klebsiella	K/KG

K = alkaline, A = acid, N = neutral, G = gas, H_2S = hydrogen sulfide.

Malonate Broth (Difco)

This medium is used to determine if organisms, particularly of the enteric group, can utilize sodium malonate as a source of carbon. If malonate is utilized, an alkaline reaction is produced, resulting in the broth medium turning blue. Reactions of the various enteric species are given in Table 8-I.

Sodium Hippurate Hydrolysis

Reagent

Twelve grams of ferric chloride ($FeCl_3 \cdot 6H_2O$) are dissolved in 100 ml of a 2% aqueous solution of hydrochloric acid.

Medium

Sodium hippurate	10 g
Heart infusion broth	1000 ml

Dissolve the sodium hippurate in broth, tube in 1 ml amounts, and sterilize at 115°C for twenty minutes.

---

*Warner-Chilcott Laboratories, Morris Plains, New Jersey.

Procedure

Transfer 0.8 ml of the culture to a small tube, then add 0.2 ml of the reagent and mix immediately. After ten to fifteen minutes, observe for the presence of a permanent precipitate. The latter indicates the presence of benzoic acid resulting from hydrolysis of hippurate. If the culture is particularly turbid, the organisms can be separated by centrifugation and the test carried out using supernatant.

It is important that the prescribed amounts of culture and reagent be used because the benzoic acid can be dissolved by an excess of reagent. A control in which the sterile broth is used should be included.

Optochin or Ethylhydrocuprein (EHC) Inhibition Test

EHC specifically inhibits the growth of pneumococci. It is available impregnated in paper discs that are applied to inoculated plates in the same manner as a sensitivity disc. The zone of inhibition should be at least 5 mm. Small zones, as may be observed with some viridans streptococci, are ignored.

Tetrazolium Reduction

The reduction of tetrazolium salts in cultures is used as an indication of growth. This indicator is useful in the study of the mycoplasmas in that their growth in broth may be difficult to detect.

The compound 2,3,5-triphenyltetrazolium chloride is incorporated in PPLO broth at a concentration of 0.0005%. In an anaerobic atmosphere, it turns red upon being reduced to an insoluble formazan.

ONPG (*o*-nitrophenylß-D-galactopyranoside) Test

This test is used to detect potential lactose fermenters that in ordinary media either take several days to produce acid or do not produce it at all. This test detects the induced intracellular enzyme ß-galactosidase, one of two enzymes involved in lactose fermentation. The preparation of ONPG broth and the performance of the test is described by Cowan and Steel (1).

Turbidity Standards

McFarland Nephelometer Standards

These tubes of graded turbidity are useful for the rough quantitation of bacterial suspensions. They are available commercially (Difco).

Preservation of Bacteria

Lyophilization

Although somewhat laborious, lyophilization or freeze-drying is a very useful process for the preservation of almost all bacteria and mycoplasmas.

Several apparatuses are available commercially, including the Bellco* all-glass ampoule freeze-drying apparatus. Ampoules with firing-off constrictions are convenient when using a "crossfire" ampoule torch for sealing off. Although freeze-drying can be carried out using a good vacuum pump without a vacuum gauge, it is advantageous to have the latter. The lower the vacuum, at least 10 microns or less of mercury, the less residual moisture and consequently the greater survival.

Prior to being dispensed in ampoules, organisms are suspended in a protective medium such as the well-known Mist Dessicans®: one volume nutrient broth with 30% glucose, plus three volumes sterile inactivated serum. After lyophilization, the ampoules are sealed, then tested for vacuum using a high-frequency coil (Tesla type). Those that fail to "light up" are discarded.

Maintenance in Media

Many Gram-negative organisms and some Gram-positive ones can be maintained in a viable state in Stock Culture Agar® (Difco). The medium is dispensed in screw-cap or rubber-stoppered tubes. The unslanted medium is stabbed several times, incubated, then stored in the dark at room temperature. Some species such as *Pasteurella multocida* will die if stored at refrigerator temperature. Although many organisms can be stored for months in this medium, occasional cultures of a species will die.

Many aerobes and anaerobes, including clostridia, remain viable for many months in tubes of sealed cooked meat medium.

Deep Freeze

The procedure outlined below will maintain many fastidious organisms for long periods.
1. Grow on a blood plate or other suitable medium.
2. Place 0.5 ml of defibrinated blood in a small sterile tube.
3. Suspend loopfuls of bacteria in the blood.
4. Store in deep freeze, the colder the better.
5. To recover, remove the tube and allow the edge of the blood to thaw. Remove loopfuls of the melted blood from between the frozen plug of blood and the wall of the glass tube.
6. Plate on a suitable medium, then return the not completely thawed tube to the freezer.

---

*Bellco® Glass, Inc. Vineland, New Jersey

Liquid Nitrogen

Freezing in liquid nitrogen ($-196°C$) is the preferred method of long-term storage. The procedures involved in the storage of bacteria in liquid nitrogen and also in other ways have been described in detail (1).

REFERENCE

1. Gherma, R. L.: In Gerhardt, P. (Ed-in-chief): *Manual of Methods for General Bacteriology.* Washington, D.C. American Society for Microbiology, 1981, Chapter 12.

SUPPLEMENTARY REFERENCES

BBL Manual of Products and Laboratory Procedures, 5th ed. Baltimore, Maryland, Bio-Quest, Division of Becton, Dickinson and Co., 1968.

Cowan, S. T., and Steel, K. J.: *Manual for the Identification of Medical Bacteria.* Cambridge, Cambridge U Pr, 1966.

Cowan, S. T.: *Manual for the Identification of Medical Bacteria*, 2nd ed. Cambridge, Cambridge U Pr, 1974.

Difco Supplementary Literature. Detroit, Michigan, Difco Laboratories, 1968.

Appendix D

MYCOLOGICAL MEDIA AND STAINS

O nly a small number of useful media are described below. All are available commercially. Readers are referred to the Supplementary References of Chapters 28, 29, 30, 31, and 32 for additional media, reagents, and procedures.

Sabouraud Dextrose Agar

This medium without inhibitors is used routinely for the isolation of fungi associated with disease. It is available from several commercial sources.

Sabouraud Dextrose Agar (With Inhibitors)

This medium is similar to the regular Sabouraud agar except that it contains chloramphenicol for the reduction of bacterial growth and cycloheximide (actidione) for the suppression of saprophytic fungi.

It is available under various trade names, e.g. Mycosel® (BBL) and Mycobiotic® (Difco). Sabouraud media are usually dispensed in Petri dishes, large tubes as slants, or small prescription bottles. After inoculation, they are incubated at room temperature only. It should be remembered that the growth of the fungi listed and other saprophytic fungi is inhibited on this medium: *Allescheria boydii, Cryptococcus neoformans, Aspergillus fumigatus, Mucor, Absidia, Rhizopus.*

Sabouraud Broth

This broth is frequently used to study the character of growth of fungi in a liquid milieu.

Brain-Heart Infusion Agar

This is a good medium for the growth of the yeast phase of dimorphic fungi.

Chlamydospore Agar (Difco)

If a selective medium is desired onto which material can be inoculated directly, add per milliliter 40 units of penicillin, 40 µg streptomycin, and 20 µg of aureomycin. (See Candidiasis, Chapter 30, for appearance of chlamydospores in this medium.)

Birdseed Agar

Readers are referred to the *Manual of Clinical Microbiology* (1) for the preparation of this medium. *Cryptococcus neoformans*, unlike other *Cryptococcus* spp., produces brown colonies on this medium.

Levine Eosin-Methylene Blue (BBL)

Candida albicans produces characteristic germ tubes on this medium when incubated in a candle jar at 37°C.

Dermatophyte Test Medium (Pitman-Moore, Pfizer, Difco)

This is a recently developed selective and differential medium for dermatophytes. Bacterial and fungal contamination is greatly reduced. Dermatophytes change the color of the medium from yellow to red. Those bacteria and saprophytic fungi that grow on the medium do not change the yellow color of the medium, or they change it to red very slowly. Dermatophytes can be readily recovered on this medium and identified as such with a reliability as high as 97 percent. The medium should enable veterinarians to recover dermatophytes in practice, thus making possible a more rapid, positive diagnosis of ringworm. Cultures should be examined microscopically to confirm the presumptive identification of a dermatophyte.

Lactophenol Cotton Blue Stain

Staining Solution

Phenol crystals	20.0 g
Glycerin	40.0 ml
Lactic acid	20.0 ml
Distilled water	20.0 ml

The ingredients are dissolved by heating gently over a steam bath. When in solution, add 0.05 g of cotton blue dye.

Staining Procedure

A small portion of the mycelium is transferred to a drop of the staining solution on a slide. The mycelium is teased apart with needles, then covered with a coverslip. Permanent mounts are made by applying nail polish along the edges of the coverslip.

Clear Lactophenol

This preparation is prepared in the same manner as the lactophenol cotton blue, except that the cotton blue stain is not included. It is useful for mounts of clinical materials.

Ultraviolet Lamp or Wood's Lamp

Small, portable so-called black-light lamps* are available to provide the required long wave ultraviolet (wavelength: 3650 Angstrom units). These lamps are most effective at a close range (within twelve inches). Materials are examined in the dark.

Materials infected with *Microsporum canis, M. equinum*, and *M. audouini* fluoresce under a Wood's lamp. It should be kept in mind that a variety of substances fluoresce, including certain drugs, keratin, petrolatum, porphyrins, etc.

Preparation of Slide Cultures

The following materials are required:
1. Deep Petri dish.
2. A short piece of glass rod or tubing (approximately three-eighths inch in diameter) with a V bend. It must be short enough to sit in the bottom of a Petri dish. It serves as a support for a microscope slide.
3. One plate of cornmeal agar or other suitable medium for the promotion of sporulation.
4. Microscope slides and small coverslips (22 × 22 mm).

A microscope slide is placed in each Petri dish along with the V-shaped rod. The dish is wrapped in paper, then autoclaved.

The various steps are shown in Figure D-1. A square of cornmeal agar (approximately 15 × 15 mm) is aseptically cut and transferred to a sterile microscope slide within a sterile Petri dish. The slide should be sitting on the V-shaped rod.

By means of an inoculating loop with a short right-angle bend near its end, the agar block is seeded with the fungus to be cultured. This is done by making a short cut with the bent end of the needle into the dorsal central point of each side of the block. The amount of fungous colony applied to these cuts needs be just perceptible to the naked eye.

A coverslip is grasped with forceps and passed through the flame several times, then placed on the block of agar. The fungus will usually grow outwards under the coverslip and inward between the coverslip and the medium. The bottom of the Petri dish is covered with water to keep the agar block from drying out.

Incubate at room temperature.

With the more rapidly growing fungi, growth is apparent in several days, and the reproductive elements can be seen *in situ* with the low power objective of the microscope.

---

*Stroblite Co., Inc., 75 West 45th Street, New York, New York. Burton Medic-Quipment Co., El Segundo, California.

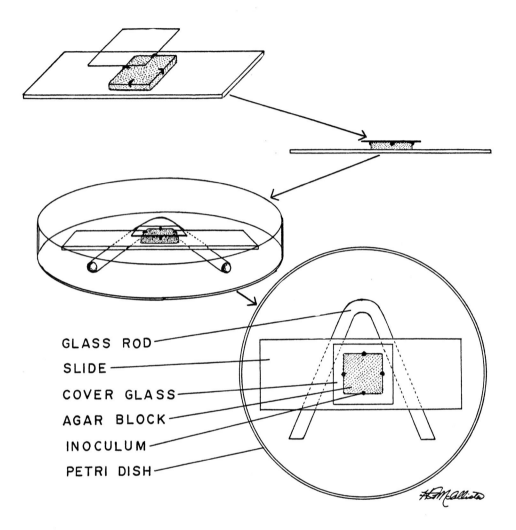

GLASS ROD

SLIDE

COVER GLASS

AGAR BLOCK

INOCULUM

PETRI DISH

Figure D-1. Steps in preparing a slide culture of a fungus (H. A. McAllister).

When development is considered optimal, the coverslip is removed with forceps and placed on a drop of lactophenol cotton blue stain previously added to the center of a microscope slide. The coverslip is then ringed with clear nail polish, thus furnishing a permanent mount.

Preservation of Fungi

Many fungi can be maintained at room temperature for several months provided the cultures are not allowed to dry out. They should be transfered to fresh media every two or three months.

Some fungi, but not all dermatophytes, may be maintained at refrigerator temperature with an interval between transfers of about four months.

Lyophilization is suitable for some fungi but not those that produce large spores. This rules out most of the pathogenic fungi.

Cultures on sealed Sabouraud agar slants can be maintained successfully in deep freeze at $-20°C$. Another procedure that is widely used is to cover slant cultures with a good grade of heavy, sterile mineral oil.

Preparation and Use of Gastric Mucin (Type 1701-W)*

The directions given by the producer are as follows: Measure 100 ml of distilled water. Place 5 g of the mucin in a mortar. Moisten well with a portion of the water. Let stand about thirty minutes. Rub the sticky mass until free from lumps, adding the rest of the water in small portions. There should be a uniform suspension. Place in suitable containers and autoclave at fifteen pounds pressure for fifteen minutes. When mixture is cool, adjust aseptically with sterile sodium hydroxide to pH 7.3.

If the mixture contains particles that will block the needle of the syringe, allow to settle a short time, remove the supernatant from the coarse sediment, place in tubes, and incubate for one day or more to insure sterility. It is then ready for use.

PRECAUTIONS. Do not adjust the pH before autoclaving. Do not attempt to get a solution free from suspended material. The material in the supernatant is essential for successful use of the mucin. Only the very coarse material should be discarded. Make certain that the mucin suspension is uniform before using.

Bacteria or fungi are suspended in the 5% gastric mucin prior to intraperitoneal inoculation. Mice are most frequently used. Gastric mucin enhances the pathogenicity for mice of *Histoplasma capsulatum*. Conversion of the mycelial to the yeast phase of some fungi may be accomplished by mouse inoculation.

REFERENCES

1. Vera, H. D., and Power, D. A.: In Lennette, E. H. (Ed.-in-chief): *Manual of Clinical Microbiology*, 3rd ed. Washington, D.C., American Society for Microbiology, 1980, Chapter 97.

---

*Wilson Laboratories Pharmaceutical Division, Wilson & Company, Inc., Chicago 9, Illinois

Appendix E

LABORATORY METHODS FOR RICKETTSIAE AND CHLAMYDIAE

STAINING METHODS

Preparation of Smears

The following clinical specimens are examined:
1. Fluids or discharges. Use one drop and smear over a small area of the slide.
2. Tissues. Section and make several impression smears from cut surfaces.
3. Yolk sac. Use the yolk sac membrane, previously washed with saline to remove excess yolk material and blotted on fibulous paper. Small pieces of yolk sac membrane are impressed to the slide in several places with forceps.

There are several stains that are satisfactory for demonstrating organisms of the Rickettsiales in smears from infected material. The stains used most often and deemed most reliable are those of Gimenez, Macchiavello, Castaneda, and Giemsa.

Gimenez Stain

A. Carbol basic fuchsin — stock solution

Basic fuchsin in 95% ETOH	10% w/v	100 ml
Aqueous phenol	4% v/v	250 ml
Distilled water		650 ml

Keep at 37°C for forty-eight hours before use. Basic fuchsin should be about 94%, not 84%. Harleco #677 99% and Allied Chemical #NA0434 92% work equally well.

B. Sodium phosphate buffer 0.1 M, pH 7.45

NaH_2PO_4	0.2 M	3.5 ml
Na_2HPO_4	0.2 M	15.5 ml
Distilled water		19.0 ml

C. Working solution basic fuchsin — mix fresh just before use.

Stock solution	4 ml
Na phosphate buffer	10 ml

Filter twice before use.

D. Staining procedure
   1. Make thin smear of tissue on a slide.
   2. Dry and lightly heat fix to inactive agent.
   3. Cover five minutes with basic fuchsin working solution.
   4. Rinse with warm distilled water.
   5. Counterstain twenty seconds with 0.8% malachite green oxalate.
   6. Wash and counterstain again for twenty seconds.
   7. Wash and blot the slide.
   8. Examine under oil. Intracytoplasmic and cell-free chlamydial elementary bodies will stain red, and background material will stain green.

Modified Macchiavello's Stain

Basic fuchsin*	0.50 g in 200 ml distilled water
Citric acid	0.5 g in 200 ml distilled water
Methylene blue*	2.0 g in 200 ml distilled water

Staining Procedure

   1. Smears are air dried, then lightly fixed with heat.
   2. Stain with basic fuchsin for five minutes.
   3. Wash with water and decolorize for fifteen seconds with citric acid.
   4. Wash with water and stain with methylene blue for twenty seconds.
   5. Wash with water and dry by gently blotting.
   6. Examine with the oil immersion objective. If the slide has been properly stained, the elementary bodies, which appear singly, in clusters, or in chains, will stain red. If overdecolorized, the elementary bodies will appear blue.

Castaneda's Stain (Bedson Modification)

1. Weiss mordant solution

Formalin	100.0 ml
Glacial acetic acid	7.5 ml

2. Formol blue

M/15 phosphate buffer, pH 7.0	180.0 ml
Azure II* or Unna's blue (1% in methyl alcohol)	20.0 ml
Formalin	10.0 ml

3. Safranin* 0.25% aqueous solution

---

*Filter through paper before using.

Staining Procedure

1. Smear is air dried and fixed in Weiss' mordant solution for two minutes.
2. Wash thoroughly with water.
3. Stain with formol blue for ten to twenty minutes.
4. Wash thoroughly with water.
5. Counterstain with safranin for five to eight seconds.
6. Wash thoroughly with water.
7. Examine with the oil immersion objective. Elementary bodies stain blue, while the tissue cells appear red.

Giemsa Stain

The preparation and staining procedure is given in Appendix A. Elementary bodies appear purple, but they may be difficult to differentiate from the similar staining of tissue cells and background.

Table E-1

Appearance of Rickettsiae and Chlamydial Developmental Forms
in Stained Cytological Preparations

Stain	Rickettsiae	Chlamydiae	Host cells or background debris
Giemsa	Purple to dark blue	EB purple; inclusions purple	Cytoplasm blue; nuclei dark purple
Gimenez	Bright red or greenish red	EB bright red; inclusions greenish red	Cytoplasm faint green-blue; nuclei intense green-blue
Macchiavello	Red	EB red; inclusions reddish-blue	Cytoplasm blue; nuclei intense blue

EB = elementary body

PREPARATION OF INOCULA

Clinical specimens received for the isolation of *Rickettsia* and *Chlamydia* are varied. Specimens include the following:

1. Blood. Grind clot in Bovarnick's buffer or tissue culture fluid using a Ten Broeck tissue grinder or a mortar and pestle.
2. Other fluids, discharges. Use undiluted unless the fluid is viscous. Otherwise, dilute and process as for blood.
3. Feces. Suspend feces in Bovarnick's or tissue culture fluid to give a 10% concentration. Grind and centrifuge (2000 G) to remove debris and fecal particles, and collect the supernatant and proceed as described below.
4. Tissues. Mince well with scissors, then grind thoroughly using a Ten Broeck tissue grinder or a mortar and pestle. Sufficient buffer is

employed to give approximately a 10% suspension. Centrifuge at 2000 G for twenty minutes to remove debris and fragments, then collect the supernatant.

If specimens are contaminated with bacteria, they have to be specially processed prior to being inoculated into chicken embryos. Several procedures are described below:

1. If the organisms being sought will propagate in laboratory animals, then the specimen usually does not have to be treated specially, as the animals can normally withstand bacterial contaminants unless they are excessive.
2. Antibiotics are employed as follows: 500 µg/ml of streptomycin, 75 µg/ml vancomycin, 50 µg/ml gentamycin, and 500 units of mycostatin.
3. Differential centrifugation (see Fig. 25-7)

   a. Centrifuge for thirty minutes at 2000 G.
   b. Withdraw 5 ml of the supernatant, being careful not to disturb the layers, and dilute 1:1 in buffer solution.
   c. The above procedure is repeated twice.
   d. Following the third cycle of centrifugation, withdraw several milliliters of supernatant and make two tenfold dilutions if chicken embryos are infected.
4. Inoculate the final supernatant and the two dilutions into different sets of chicken embryos.
5. The above procedure may be repeated using yolk sac material from the chicken embryo.

ISOLATION AND CULTIVATION

For practical purposes, all organisms in these groups must be cultivated in living host systems. Procedures for the inoculation of mice and the yolk sacs of embryonated chicken eggs are given below. See Table 25-I for the preferred host systems and route of inoculation for specific organisms.

Isolation of Chlamydia psittaci in Cell Culture under Enhancing Conditions through Centrifugation

1. Use mouse L cells (929) that have been grown on minimum essential medium (MEM) with 500 µg streptomycin and 75 µg vancomycin/ml for several passages—NO PENICILLIN!
2. Soak 3-dram shell vials in seven times the detergent, then wash thoroughly. Add one clean circular coverslip and foil cap. Dry sterilize in the oven.

3. Plant L cells to monolayer the coverslip, about 3×10^5 cells per vial. Use 1.2 ml of cell suspension in MEM plus 10% fetal calf serum (FCS). Prepare at least two vials per sample for isolation. Inoculate within twenty-four hours of planting.

4. To inoculate, remove cell culture medium and add 0.5–1.0 ml of inoculum. Place a UV-sterilized plastic cap on the vial and centrifuge twenty to thirty minutes at 2,000 rpm in the International centrifuge. Attempt to maintain cells at 37°C as much as possible. (Centrifuge shields may be prewarmed to 37°C to aid in maintaining the temperature, or centrifuge may be moved into a walk-in incubator.)

5. Remove the inoculum. If the inoculum contained debris, wash the cells with Dulbecco's solution plus streptomycin.

6. Add MEM plus additives (see below). For primary isolation or when bacterial contamination is possible, use gentamicin (50 μg/ml) and vancomycin (75 μg/ml).

> Additives to MEM for maintenance medium:
> 0.5% glucose
> 10% heat-inactivated FCS
> 1–2 μg cycloheximide/ml (use in one set of the vials if isolating an unknown strain and run parallel set without cycloheximide)
> 0.25 M HEPES
> 10 mM PO_4

7. Reading the results:

> Stain one set of coverslips with Giemsa for forty to sixty hours. Abortion strains require at least sixty hours.
> When isolating a strain, save a second vial for subpassage.
> If inclusions are small, allow further incubation of twelve to twenty-four hours before subpassage. Break cells with ultrasound treatment or freeze and thaw.

Enchancement of Chlamydial Uptake by Cultured Cells through DEAE–Dextran Treatment

1. Prepare a 1% stock solution of diethylamine-ethyl-dextran (Sigma Chemical Co., P.O. Box 14508, St. Louis, MO 63178) in Bulbecco's phosphate-buffered saline.

2. Add 0.2 ml of the 1% stock solution of DEAE–D to 100 ml of Dulbecco's solution for a final concentration of 20 mg/ml.

3. Wash the cell monolayer two to three times with DEAE–D Dulbecco's solution. Allow the last rinse to remain on the monolayer for about thirty minutes.

4. Decant the last rinse and add the inoculum for adsorption as usual.

5. Wash inoculated cell cultures with Dulbecco's solution, add suitable cell culture medium, and incubate.

Inoculation of Mice

Intraperitoneal Method

1. Young mice, three to five weeks of age, are preferred. A firm hold is obtained by gripping the tail between the little finger and the ring finger and by holding the loose skin at the back of the neck between the index finger and the thumb.
2. The abdomen is swabbed with a suitable disinfectant such as 70% ethanol or tincture of iodine. Using a tuberculin syringe with a 25 gauge needle, 0.2 ml of the inoculum is administered.
3. The mice are observed daily for evidence of infection.

HARVESTING TISSUES. Spleen, liver, and kidneys of all mice are harvested as follows:
1. The animal is swabbed with a disinfectant, then secured to a dissecting board by pinning all four legs. With sterile scissors and forceps, the walls of the abdomen and thorax are reflected to expose the viscera.
2. Using another pair of sterile scissors and forceps, the spleen, liver, and kidneys are collected (taking care not to cut the intestine) and placed in a sterile Petri dish.
3. Impression smears are made from sections of the organs for staining. Tissues are prepared for further passaging or frozen and stored for future reference ($-70°$C).

Intranasal Inoculation

1. Young mice, three to five weeks of age, are lightly anesthetized with ether. This facilitates the inhalation of the inoculum, as mice are not as likely to sneeze. It is easily accomplished by placing the mouse in a glass jar with a loose cap. The mouse is ready for inoculation when it shows signs of loss of equilibrium.
2. The mouse is held in the hand, abdomen up. Using a tuberculin syringe with a 22 gauge needle, several drops of the inoculum are placed at the openings of the nostrils. Each drop should be inhaled before the next is administered.
3. The inoculated mice are observed daily for such evidence of infection as hunched posture, labored respiration, and general depression.

HARVESTING LUNGS. The lungs of mice that have died and the lungs of those living mice with and without signs of infection are harvested. The procedure followed is essentially the same as that described previously for liver, spleen, and kidney.

The lungs are placed in a sterile Petri dish. Impression smears for staining are made from sections of the lungs. Tissue is prepared for further passaging if indicated or frozen and stored for future reference.

Yolk Sac Inoculation

Eggs should be from a flock receiving antibiotic-free feed. The temperature of incubation should be 36–37°C, and the humidity should be near the point of saturation.

1. Eggs, with five- to seven-day-old chicken embryos, are candled and the position of embryo and air space are marked.
2. The surface of the egg over the air sac is swabbed with tincture of iodine.
3. A hole is either punched (a needle pushed through a cork stopper works well) or drilled through the shell over the air space posterior to the embryo.
4. An inoculum of 0.3–0.5 ml is administered with a 22 gauge, one inch needle. The needle is inserted vertically to the hilt (Fig. E-1).
5. The needle is removed and the hole sealed with paraffin, collodion, or nail polish.
6. The inoculated chicken embryos are placed in the incubator and candled daily for evidence of loss of motility or death. Deaths occurring within twenty-four hours are normally due to contamination or trauma, and the eggs should be discarded.

Yolk Sac Harvest

1. The top of the eggs are swabbed with tincture of iodine.
2. The shell is cut, using sterile scissors, along the margin of the air space, thereby exposing the shell membrane.
3. The shell and chorioallantoic membranes are ruptured, and the chicken embryo with yolk sac is gently teased with sterile forceps into a sterile Petri dish.
4. The yolk sac is cut from the embryo, ruptured, and allowed to drain. The yolk sac membrane is stripped with dental forceps to remove remaining excess yolk.
5. Impression smears are made from small pieces of the yolk sac membrane, which are gently washed in saline and blotted on bibulous paper.
6. The rest of the yolk sac membrane is used for subpassage or frozen at −70°C for future reference after homogenization in Bovarnick's buffer.

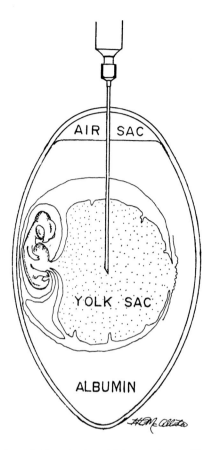

Figure E-1. Technique for inoculating six-day chicken embryos (H. A. McAllister).

Bovarnick's Buffer

Sucrose	0.218 M	74.62 g
KH$_2$PO$_4$	0.0038 M	0.52 g
K$_2$HPO$_4$	0.0072 M	1.64 g
Na glutamate	0.0049 M	0.82 g

Dissolve in 1000 ml distilled water and autoclave. Add 1% bovine serum albumin, adjust to pH 7.0, and add appropriate antibiotics.

Sulfadiazine Sensitivity

1. Grind one yolk sac with 2.5 ml Bovarnick's sucrose-phosphate buffer containing 500 µg streptomycin per ml (yields 1:10 dilution).
2. Transfer 0.2 ml of the 1:10 dilution to 1.8 ml Bovarnick's sucrose-phosphate buffer containing 500 µg streptomycin per ml, which gives

a 10^{-2} dilution. Continue in like manner to obtain dilutions of 10^{-3} through 10^{-8}. Use these dilutions as controls.

3. Transfer 0.2 ml of the 1:10 dilution to 1.8 ml Bovarnick's sucrose-phosphate buffer containing 500 µg streptomycin and 2 mg sulfadiazine per ml to prepare a 10^{-2} dilution. Continue in like manner to obtain dilutions of 10^{-3} through 10^{-8}.

4. Inoculate each of three or more seven-day-old developing chicken embryos with 0.5 ml of each dilution. Incubate until death occurs or for twelve days.

5. Compare the median chicken embryo lethal dose ($CELD_{50}$) for each of the two chlamydial preparations. A reduction of a hundredfold or more in the number of $CELD_{50}$ in suspension of 10^6 or more chlamydiae per ml indicates sulfadiazine susceptibility of the strain.

REFERENCES

1. Lennette, E. H., and Schmidt, (Eds.): *Diagnostic Procedures for Viral and Rickettsial Diseases*, 5th ed. New York, American Public Health Association, Inc., 1970, Chapter 22.
2. Storz, J.: *Proc 71st Ann Meeting US Livestock San Assoc*, 1967, pp. 585–595.

SUPPLEMENTARY REFERENCES

Lennette, E. H., and Schmidt, N. J. (Eds.): *Diagnostic Procedures for Viral and Rickettsial Infections*, 5th ed. New York, American Public Health Association, Inc., 1970, Chapter 22.
Lennette, E. H., Spaulding, E. H., and Truant, J. P. (Eds.): *Manual of Clinical Microbiology*, 2nd ed. Washington, D.C., American Society for Microbiology, 1974, Chapter 87.
Storz, J.: *Chlamydia and Chlamydia-Induced Diseases*. Springfield, Thomas, 1971.

Appendix F

LABORATORY METHODS FOR MYCOPLASMAS

LABORATORY METHODS FOR AVIAN MYCOPLASMAS

HARRY W. YODER, JR.

Mediums for Growth

Some typical formulations of mycoplasma mediums are as follows.

Mycoplasma Medium (Frey) (1)

Mycoplasma broth base*	22.5 g
Dextrose	10 g
Swine or horse serum	100–150 ml
Phenol red	25 mg
Penicillin G potassium	1000 U/ml
Thallous acetate (1:4000–1:2000)	2.5–5.0 ml 10% sol.
Distilled water	1,000 ml

Adjust pH to 7.8 and filter sterilize.

M. gallisepticum and *M. synoviae* ferment dextrose so readily that only 3 g/liter need be used for most routine purposes. Thallous acetate is used at the 1:2,000 level for potentially highly contaminated specimens like those collected on tracheal swabs. Reduced NAD must be added for the cultivation of *M. synoviae* (1.0 ml of a 1% solution of NAD plus 1.0 ml of a 1% solution of cysteine hydrochloride per 100 ml of medium).

In general, horse serum should be used in media for the cultivation of *Mycoplasma meleagridis*, swine serum for *M. synoviae*, and either swine or horse serum may be used in media for *M. gallisepticum*. Serum used in mycoplasma medium is usually heat inactivated (56°C for thirty minutes). Laboratory passaged cultures can be adapted to more variable ingredients.

Difco brain-heart infusion broth, beef-heart infusion broth, or PPLO broth may be used as the primary base medium supplemented with yeast autolysate, proteose peptone No. 3, dextrose, NAD, and various serums.

Avian meat infusion broth with 20% avian, swine, or horse serum plus

*Sources of Mycoplasma Broth Base (Frey): Gibco Diagnostics, 2801 Industrial Drive, Madison, Wisc. 53711; Scott Laboratories, Inc., 8 Westchester Plaza, Elmsford, N.Y. 10523.

yeast autolysate and 2,3,5-triphenyltetrazolium chloride is an excellent medium, especially for *M. gallisepticum*, but is more difficult to prepare.

AGAR PLATE MEDIUM. To 500 ml of any of the above described broth base media should be added 7.5 g (1.5%) agar, boiled to dissolve, adjusted to pH 7.8, and autoclave sterilized. The following filter-sterilized enrichments should be added after the agar base cools to 45°C:

	Per 500 ml base
Swine or horse serum	50–75 ml
Penicillin G potassium	1000 U/ml
Thallous acetate (1:4000)	1.25 ml 10% sol.
1% NAD solution	5.0 ml
1% Cysteine HCl solution	5.0 ml

Agar plates or slants should be poured and allowed to solidify.

LABORATORY METHODS FOR ANIMAL MYCOPLASMAS

OLE H. V. STALHEIM

Modified Hayflick's Medium

Bacto PPLO broth (Difco) w/o CV	2.1 g in 70 ml H_2O
Horse serum	20 ml
Fresh yeast extract	8 ml
Penicillin	1000 IU/ml (final conc.)
Deoxyribonucleic acid (Sigma)	0.002% (final conc.)
Thallium acetate	1:2000 (final conc.)

pH 7.6–7.8

Total volume 100 ml

Filter sterilize by passing through a 0.22 μ filter. Dispense in 1.8 ml quantities and store at 5°C. Optional additions are glucose (0.5% to 1.0%) and phenol red (1:5000).

Modified Hayflick's Medium with Agar

Bacto PPLO agar (Difco)	3.5 g in 70 ml H_2O
Horse serum	20 ml
Fresh yeast extract	8 ml
Penicillin G	1000 IU/ml (final conc.)
Thallium acetate	1:2000 (final conc.)

pH 7.2–7.6

Sterilize agar in an autoclave. Combine and filter sterilize balance of components. Heat fluid portion to 56°C in a water bath, combine with agar, and dispense in plates of individual choice. Store in a sealed bag at 5°C.

Modified Hayflick's Medium for Ureaplasmas

Mycoplasma broth base (BBL)	2.1 g
Distilled water	66 ml

Adjust pH to 5.4 with 2 N HCl. The final pH after sterilization and addition of all components should be 6.0±0.2. The exact pH required before sterilization depends upon the effect of the horse serum.

For semisolid medium, add 0.05 g Ionagar #2 or #2S; for solid medium, add 1.0 g.

Autoclave at fifteen pounds pressure for fifteen minutes. Cool (to 50°C for agar containing media), and add aseptically:

Horse serum	20.0 ml
Yeast extract, 25% (w/v), pH 6.0	10.0 ml
Urea, 25% (w/v)	1.0 ml
Phenol red, 0.5%	0.2 ml

Omit urea and phenol red for solid medium. Penicillin (500 IU/ml) may be added.

Gourlay and Leach Medium (2)

Hank's buffered salt solution	40 ml
Lactalbumin hydrolysate, 5% (w/v)	10 ml
Broth (Difco)	20 ml
Fetal calf serum (inactivated 30 min., 56°C)	20 ml
Glucose, 50% (w/v)	2 ml
Calf thymus DNA highly polymerized, 0.2 w/v	1 ml
Ampicillin (for injection)	0.15 mg/ml
Thallium acetate, 5% (w/v)	0.5 ml
Phenol red, 1% (w/v)	0.2 ml

Adjust pH value to 7.8; sterilize by filtration.

Friis Medium for Swine Mycoplasmas (3)

Hank's balanced salt solution	500 ml
Bacto brain-heart infusion (Difco)	8.2 g
Bacto PPLO broth without c/v (Difco)	8.7 g
Water750 ml	

Autoclave two to five minutes at 121°C

Yeast extract	60 ml
Phenol red, 0.5%	4.5 ml
Bacitracin	250 mg
Meticillin	250 mg

Thallium acetate	1:10,000
Porcine serum	250 ml

pH 7.4

Preparation of Yeast Extract

To 1800 ml of distilled deionized water, heated to 70°C, add 425 g yeast (bakers' cake form). Adjust pH to 4.6. Gently heat with continual agitation until boiling point is attained. Cool to refrigerator temperature, then freeze to −20°C. After a period in excess of three days, thaw at room temperature and adjust pH to 7.8. Centrifuge at 950 G for sixty minutes. Remove supernatant and filter first through No. 1 Whatman filter after addition of Celite followed by sterilization using Seitz sterilizing filter or Millipore (0.22 μ pores). Dispense sterile yeast extract in 100 ml volumes and store in freezer at −20°C.

M-96 Broth (Frey's Medium) (4)

Peptone CS (Albimi)	4 g
Peptone B (Albimi)	2 g
Peptone G (Albimi)	2 g
Yeast autolysate (Albimi)	2 g
Yeast extract	2 g
NaCl	5 g
KCl	0.4 g
$MgSO_4 \cdot 7H_2O$	0.2 g
Catalase	0.001 g
HEPES buffer	3.5 g
L-Arginine HCl	0.06 g
L-Glutamine	0.09 g
DNA	0.02 g
Distilled water	1 liter

Dissolve dry ingredients.

100X Eagle's MEM vitamin solution	10 ml
Glycerol	0.15 ml
Cholesterol (0.1% emulsion)	2 ml
Thallium acetate, 1%	25 ml
Penicillin	500,000 IU

Adjust pH to 7.4–7.5 with NaOH.

Porcine serum (clarified) and inactivated (30 min., 56°C)	170 ml

Filter sterilize and dispense in desired aliquots. Add 2 ml of DPN-cysteine solution to each 100 ml of broth on the day of use. Final concentration in broth, 0.01%.

M-96 Agar

Peptone CS (Alibimi)	4 g
Peptone B (Alibimi)	2 g
Peptone G (Alibimi)	2 g
Yeast autolysate (Alibimi)	2 g
NaCl	5 g
KCl	0.4 g
$MgSo_4 \cdot 7H_2O$	0.2 g
Catalase	0.001 g
HEPES buffer	3.5 g
L-Arginine HCl	0.06 g
L-Glutamine	0.09 g
Distilled water	500 ml

Dissolve dry ingredients.

100X Eagle's MEM vitamin solution	10 ml
Glycerol	0.15 ml
Cholesterol (0.1% emulsion)	2 ml
Thallium acetate, 1%	25 ml
Penicillin	500,000 IU

Adjust pH to 7.4–7.5 with NaOH.

Porcine serum (clarified) and inactivated (30 min., 56°C)	170 ml

Filter sterilize.

Ionagar #2	17 g
Distilled water	500 ml

Sterilize at fifteen pounds pressure for fifteen minutes.

Bring both broth and agar to 50°C and mix together. Add DPN-cysteine solution for a final concentration of 0.01%. Dispense into Petri dishes.

Preparation of Cholesterol Emulsion

1. Cholesterol (200 mg) is put in a sterile Petri dish and dissolved in diethyl ether. The ether is allowed to evaporate, and the process is repeated once more.

2. The sterile, recrystallized cholesterol is dissolved in 6–8 ml warm 95% ethyl alcohol.
3. The alcohol solution is drawn into a prewarmed glass Luer-Lok® syringe fitted with an 18 or 20 gauge needle.
4. Hold needle tip under the surface of 200 ml of distilled, deionized water and eject the cholesterol into it.

Clarification of Porcine Serum

1. Adjust pH of porcine serum to 4.4 with 1 N HCl (do not drop pH below 4.2).
2. Allow serum to stand at 4°C for eighteen to twenty hours.
3. Centrifuge at 9,000 G for thirty minutes; discard sediment.
4. Filter through No. 1 Whatman filter paper.
5. Adjust pH to 7.0 with 1 N NaOH.

Preparation of DPN-Cysteine Solution

Prepare a 1% solution (1 g/100 ml distilled water) of each of the following: Cozymase (coenzyme I) oxidized[*] and Cysteine HCl hydrate, A grade.[†] Mix the solutions together; the cysteine reduces the DPN. Filter sterilize, dispense in appropriate amounts, and store frozen at −20°C.

Dienes' Stain

Staining Solution

Methylene blue	2.40 g
Maltose	10.0 g
Azure II	1.25 g
Sodium chloride	0.25 g
Distilled water	100.0 ml

This solution is available commercially.[‡]

STAINING OF COVERSLIPS. Apply a thin film of the stain to clean coverslips by means of a cotton swab. The film should be uniform and light. When dry, they are ready for use and may be stored indefinitely.

Staining Procedure

A 1 cm square block is cut from an area containing suspected colonies and transferred to a microscope slide, colony side up. Place a treated coverslip,

---

[*]Nutritional Biochemical Corp., Cleveland, Ohio.

[†]Calbiochem, La Jolla, California.

[‡]Hyland Laboratories, Los Angeles, California.

stain side down, on the agar block. The staining reaction is complete within a few minutes. All colonies stain, but in about fifteen minutes, the bacterial colonies decolorize while the mycoplasma colonies retain their color. The preparation is examined under low or higher powers employing transmitted light.

It should be remembered that this stain does not distinguish L-type colonies from mycoplasma colonies.

REFERENCES

1. Frey, M. L., Hanson, R. P., and Anderson, D. P.: *Am J Vet Res, 29*:2163, 1968.
2. Gourlay, R. N., and Leach, R. H.: *J Med Microbiol, 3*:111, 1970.
3. Friis, N. F.: *Nord Vet Med, 27*:337, 1975.
4. Frey, L. M., Thomas, G. B., and Hale, P. A.: *Ann NY Acad Sci, 225*:334, 1973.

INDEX